Introduction to

TOPOLOGY AND MODERN ANALYSIS

INTERNATIONAL SERIES IN
PURE AND APPLIED MATHEMATICS

William Ted Martin and E. H. Spanier
CONSULTING EDITORS

Introduction to

TOPOLOGY AND MODERN ANALYSIS

GEORGE F. SIMMONS

Associate Professor of Mathematics
Colorado College

McGRAW-HILL BOOK COMPANY

New York   San Francisco   Toronto   London

INTRODUCTION TO TOPOLOGY AND MODERN ANALYSIS

14 15 16 – MAMM – 7 5

ISBN 07-057389-1

For Virgie May Hatcher
and Elizabeth B. Blossom

**TO EACH OF WHOM
I OWE MORE
THAN I CAN POSSIBLY EXPRESS**

Preface

For some time now, topology has been firmly established as one of the basic disciplines of pure mathematics. Its ideas and methods have transformed large parts of geometry and analysis almost beyond recognition. It has also greatly stimulated the growth of abstract algebra. As things stand today, much of modern pure mathematics must remain a closed book to the person who does not acquire a working knowledge of at least the elements of topology.

There are many domains in the broad field of topology, of which the following are only a few: the homology and cohomology theory of complexes, and of more general spaces as well; dimension theory; the theory of differentiable and Riemannian manifolds and of Lie groups; the theory of continuous curves; the theory of Banach and Hilbert spaces and their operators, and of Banach algebras; and abstract harmonic analysis on locally compact groups. Each of these subjects starts from roughly the same body of fundamental knowledge and develops its own methods of dealing with its own characteristic problems. The purpose of Part 1 of this book is to make available to the student this "hard core" of fundamental topology; specifically, to make it available in a form which is general enough to meet the needs of modern mathematics, and yet is unburdened by excess baggage best left in the research journals.

A topological space can be thought of as a set from which has been swept away all structure irrelevant to the continuity of functions defined on it. Part 1 therefore begins with an informal (but quite extensive) treatment of sets and functions. Some writers deal with the theory of metric spaces as if it were merely a fragment of the general theory of topological spaces. This practice is no doubt logically correct, but it seems to me to violate the natural relation between these topics, in which metric spaces motivate the more general theory. Metric spaces are therefore discussed rather fully in Chapter 2, and topological spaces are introduced in Chapter 3. The remaining four chapters in Part 1 are concerned with various kinds of topological spaces of special importance in applications and with the continuous functions carried by them.

It goes without saying that one aspect of this type of mathematics is its logical precision. Too many writers, however, are content with this, and make little effort to help the reader maintain his orientation in the midst of mazes of detail.   One of the main features of this book is the attention given to motivating the ideas under discussion.   On every possible occasion I have tried to make clear the intuitive meaning of what is taking place, and diagrams are provided, whenever it seems feasible, to help the reader develop skill in using his imagination to visualize abstract ideas.   Also, each chapter begins with a brief introduction which describes its main theme in general terms.   Courses in topology are being taught more and more widely on the undergraduate level in our colleges and universities, and I hope that these features, which tend to soften the austere framework of definitions, theorems, and proofs, will make this book readable and easy to use as a text.

Historically speaking, topology has followed two principal lines of development.   In homology theory, dimension theory, and the study of manifolds, the basic motivation appears to have come from geometry. In these fields, topological spaces are looked upon as generalized geometric configurations, and the emphasis is placed on the structure of the spaces themselves.   In the other direction, the main stimulus has been analysis. Continuous functions are the chief objects of interest here, and topological spaces are regarded primarily as carriers of such functions and as domains over which they can be integrated.   These ideas lead naturally into the theory of Banach and Hilbert spaces and Banach algebras, the modern theory of integration, and abstract harmonic analysis on locally compact groups.

In Part 1 of this book, I have attempted an even balance between these two points of view.   This part is suitable for a basic semester course, and most of the topics treated are indispensable for further study in almost any direction.   If the instructor wishes to devote a second semester to some of the extensions and applications of the theory, many possibilities are open.   If he prefers applications in modern analysis, he can continue with Part 2 of this book, supplemented, perhaps, with a brief treatment of measure and integration aimed at the general form of the Riesz representation theorem.   Or if his tastes incline him toward the geometric aspects of topology, he can switch over to one of the many excellent books which deal with these matters.

The instructor who intends to continue with Part 2 must face a question which only he can answer.   Do his students know enough about algebra?   This question is forced to the surface by the fact that Chapters 9 to 11 are as much about algebra as they are about topology and analysis.   If his students know little or nothing about modern algebra, then a careful and detailed treatment of Chapter 8 should make it possible to proceed without difficulty.   And if they know a good deal, then a quick survey of Chapter 8 should suffice.   It is my own opinion that education in abstract mathematics ought to begin on the junior level with a course in modern algebra, and that topology should be offered only to students who have acquired some familiarity, through such a course, with abstract methods.

Part 3 is intended for individual study by exceptionally well-qualified students with a reasonable knowledge of complex analysis.   Its principal purpose is to unify Parts 1 and 2 into a single body of thought, along the lines mapped out in the last section of Chapter 11.

Taken as a whole, the present work stands at the threshold of the more advanced books by Rickart [34], Loomis [27], and Naimark [32]; and much of its subject matter can be found (in one form or another and with innumerable applications to analysis) in the encyclopedic treatises of Dunford and Schwartz [8] and Hille and Phillips [20].[1]   This book is intended to be elementary, in the sense of being accessible to well-trained undergraduates, while those just mentioned are not.   Its prerequisites are almost negligible.   Several facts about determinants are used without proof in Chapter 11, and Chapter 12 leans heavily on Liouville's theorem and the Laurent expansion from complex analysis.   With these exceptions, the book is essentially self-contained.

It seems to me that a worthwhile distinction can be drawn between two types of pure mathematics.   The first—which unfortunately is somewhat out of style at present—centers attention on particular functions and theorems which are rich in meaning and history, like the gamma function and the prime number theorem, or on juicy individual facts, like Euler's wonderful formula

$$1 + \tfrac{1}{4} + \tfrac{1}{9} + \cdots = \pi^2/6.$$

The second is concerned primarily with form and structure.   The present book belongs to this camp; for its dominant theme can be expressed in just two words, *continuity* and *linearity*, and its purpose is to illuminate the meanings of these words and their relations to each other.   Mathematics of this kind hardly ever yields great and memorable results like the prime number theorem and Euler's formula.   On the contrary, its theorems are generally small parts of a much larger whole and derive their main significance from the place they occupy in that whole.   In my opinion, if a body of mathematics like this is to justify itself, it must possess aesthetic qualities akin to those of a good piece of architecture. It should have a solid foundation, its walls and beams should be firmly and truly placed, each part should bear a meaningful relation to every other part, and its towers and pinnacles should exalt the mind.   It is my hope that this book can contribute to a wider appreciation of these mathematical values.

George F. Simmons

[1] The numbers in brackets refer to works listed in the Bibliography.

A Note to the Reader

Two matters call for special comment: the problems and the proofs.

The majority of the problems are corollaries and extensions of theorems proved in the text, and are freely drawn upon at all later stages of the book. In general, they serve as a bridge between ideas just treated and developments yet to come, and the reader is strongly urged to master them as he goes along.

In the earlier chapters, proofs are given in considerable detail, in an effort to smooth the way for the beginner. As our subject unfolds through the successive chapters and the reader acquires experience in following abstract mathematical arguments, the proofs become briefer and minor details are more and more left for the reader to fill in for himself. The serious student will train himself to look for gaps in proofs, and should regard them as tacit invitations to do a little thinking on his own. Phrases like "it is easy to see," "one can easily show," "evidently," "clearly," and so on, are always to be taken as warning signals which indicate the presence of gaps, and they should put the reader on his guard.

It is a basic principle in the study of mathematics, and one too seldom emphasized, that a proof is not really understood until the stage is reached at which one can grasp it as a whole and see it as a single idea. In achieving this end, much more is necessary than merely following the individual steps in the reasoning. This is only the beginning. A proof should be chewed, swallowed, and digested, and this process of assimilation should not be abandoned until it yields a full comprehension of the overall pattern of thought.

Contents

PART TWO: OPERATORS

PART THREE: ALGEBRAS OF OPERATORS

PART ONE

CHAPTER ONE

Sets and Functions

It is sometimes said that mathematics *is* the study of sets and functions. Naturally, this oversimplifies matters; but it does come as close to the truth as an aphorism can.

The study of sets and functions leads two ways. One path goes down, into the abysses of logic, philosophy, and the foundations of mathematics. The other goes up, onto the highlands of mathematics itself, where these concepts are indispensable in almost all of pure mathematics as it is today. Needless to say, we follow the latter course. We regard sets and functions as tools of thought, and our purpose in this chapter is to develop these tools to the point where they are sufficiently powerful to serve our needs through the rest of this book.

As the reader proceeds, he will come to understand that the words *set* and *function* are not as simple as they may seem. In a sense, they are simple; but they are potent words, and the quality of simplicity they possess is that which lies on the far side of complexity. They are like seeds, which are primitive in appearance but have the capacity for vast and intricate development.

1. SETS AND SET INCLUSION

We adopt a naive point of view in our discussion of sets and assume that the concepts of an element and of a set of elements are intuitively clear. By an *element* we mean an object or entity of some sort, as, for example, a positive integer, a point on the real line (= a real number), or a point in the complex plane (= a complex number).   A *set* is a collection or aggregate of such elements, considered together or as a whole. Some examples are furnished by the set of all even positive integers, the set of all rational points on the real line, and the set of all points in the complex plane whose distance from the origin is 1 (= the unit circle in the plane).   We reserve the word *class* to refer to a set of sets.   We might speak, for instance, of the class of all circles in a plane (thinking of each circle as a set of points).   It will be useful in the work we do if we carry this hierarchy one step further and use the term *family* for a set of classes. One more remark: the words element, set, class, and family are not intended to be rigidly fixed in their usage; we use them fluidly, to express varying attitudes toward the mathematical objects and systems we study.   It is entirely reasonable, for instance, to think of a circle not as a set of points, but as a single entity in itself, in which case we might justifiably speak of the set of all circles in a plane.

There are two standard notations available for designating a particular set.   Whenever it is feasible to do so, we can list its elements between braces.   Thus $\{1, 2, 3\}$ signifies the set consisting of the first three positive integers, $\{1, i, -1, -i\}$ is the set of the four fourth roots of unity, and $\{\pm 1, \pm 3, \pm 5, \ldots\}$ is the set of all odd integers.   This manner of specifying a set, by listing its elements, is unworkable in many circumstances.   We are then obliged to fall back on the second method, which is to use a property or attribute that characterizes the elements of the set in question.   If P denotes a certain property of elements, then $\{x:P\}$ stands for the set of all elements x for which the property P is meaningful and true.   For example, the expression

$$\{x : x \text{ is real and irrational}\},$$

which we read *the set of all x such that x is real and irrational*, denotes the set of all real numbers which cannot be written as the quotient of two integers.   The set under discussion contains all those elements (and no others) which possess the stated property.   The three sets of numbers described at the beginning of this paragraph can be written either way:

$$\{1, 2, 3\} = \{n : n \text{ is an integer and } 0 < n < 4\},$$
$$\{1, i, -1, -i\} = \{z : z \text{ is a complex number and } z^4 = 1\},$$
and $\quad \{\pm 1, \pm 3, \pm 5, \ldots\} = \{n : n \text{ is an odd integer}\}.$

We often shorten our notation.   For instance, the last two sets mentioned might perfectly well be written $\{z : z^4 = 1\}$ and $\{n : n \text{ is odd}\}$.   Our purpose is to be clear and to avoid misunderstandings, and if this can be achieved with less notation, so much the better.   In the same vein we can write

$$\text{the } unit \; circle \; = \; \{z : |z| = 1\},$$
$$\text{the } closed \; unit \; disc \; = \; \{z : |z| \leq 1\},$$
and $\quad\quad\quad \text{the } open \; unit \; disc \; = \; \{z : |z| < 1\}.$

We use a special system of notation for designating intervals of various kinds on the real line. If a and b are real numbers such that $a < b$, then the following symbols on the left are defined to be the indicated sets on the right:

$$[a,b] = \{x : a \leq x \leq b\},$$
$$(a,b] = \{x : a < x \leq b\},$$
$$[a,b) = \{x : a \leq x < b\},$$
$$(a,b) = \{x : a < x < b\}.$$

We speak of these as the closed, the open-closed, the closed-open, and the open intervals from a to b. In particular, $[0,1]$ is the *closed unit interval*, and $(0,1)$ is the *open unit interval*.

There are certain logical difficulties which arise in the foundations of the theory of sets (see Problem 1). We avoid these difficulties by assuming that each discussion in which a number of sets are involved takes place in the context of a single fixed set. This set is called the *universal set*. It is denoted by U in this section and the next, and every set mentioned is assumed to consist of elements in U. In later chapters there will always be on hand a given space within which we work, and this will serve without further comment as our universal set.[1] It is often convenient to have available in U a set containing no elements whatever; we call this the *empty set* and denote it by the symbol $\emptyset$. A set is said to be *finite* if it is empty or consists of n elements for some positive integer n; otherwise, it is said to be *infinite*.

We usually denote elements by small letters and sets by large letters. If x is an element and A is a set, the statement that x is an element of A (or belongs to A, or is contained in A) is symbolized by $x \in A$. We denote the negation of this, namely, the statement that x is not an element of A, by $x \notin A$.

Two sets A and B are said to be *equal* if they consist of exactly the same elements; we denote this relation by $A = B$ and its negation by $A \neq B$. We say that A is a *subset* of B (or is contained in B) if each element of A is also an element of B. This relation is symbolized by $A \subseteq B$. We sometimes express this by saying that B is a *superset* of A (or con-

---

[1] The words *set* and *space* are often used in loose contrast to one another. A set is merely an amorphous collection of elements, without coherence or form. When some kind of algebraic or geometric structure is imposed on a set, so that its elements are organized into a systematic whole, then it becomes a space.

tains A). $A \subseteq B$ allows for the possibility that A and B might be equal. If A is a subset of B and is not equal to B, we say that A is a *proper subset of B* (or is properly contained in B). This relation is denoted by $A \subset B$. We can also express $A \subset B$ by saying that B is a *proper superset* of A (or properly contains A). The relation $\subseteq$ is usually called *set inclusion*.

We sometimes reverse the symbols introduced in the previous paragraph. Thus $A \subseteq B$ and $A \subset B$ are occasionally written in the equivalent forms $B \supseteq A$ and $B \supset A$.

It will often be convenient to have a symbol for logical implication, and $\Rightarrow$ is the symbol we use. If p and q are statements, then $p \Rightarrow q$ means that p *implies* q, or that if p is true, then q is also true. Similarly, $\Leftrightarrow$ is our symbol for two-way implication or logical equivalence. It means that the statement on each side implies the statement on the other, and is usually read *if and only if*, or *is equivalent to*.

The main properties of set inclusion are obvious. They are the following:

(1) $A \subseteq A$ for every A;
(2) $A \subseteq B$ and $B \subseteq A \Rightarrow A = B$;
(3) $A \subseteq B$ and $B \subseteq C \Rightarrow A \subseteq C$.

It is quite important to observe that (1) and (2) can be combined into the single statement that $A = B \Leftrightarrow A \subseteq B$ and $B \subseteq A$. This remark contains a useful principle of proof, namely, that the only way to show that two sets are equal, apart from merely inspecting them, is to show that each is a subset of the other.

Problems

1. Perhaps the most famous of the logical difficulties referred to in the text is *Russell's paradox*. To explain what this is, we begin by observing that a set can easily have elements which are themselves sets, e.g., $\{1, \{2,3\}, 4\}$. This raises the possibility that a set might well contain itself as one of its elements. We call such a set an *abnormal* set, and any set which does not contain itself as an element we call a *normal* set. Most sets are normal, and if we suspect that abnormal sets are in some way undesirable, we might try to confine our attention to the set N of all normal sets. Someone is now sure to ask, Is N itself normal or abnormal? It is evidently one or the other, and it cannot be both. Show that if N is normal, then it must be abnormal. Show also that if N is abnormal, then it must be normal. We see in this way that each of our two alternatives is self-contradictory, and it seems to be the assumption that N exists as a set which has brought us to this impasse. For further discussion of these matters, we refer the interested reader to Wilder [42, p. 55]

or Fraenkel and Bar-Hillel [10, p. 6]. Russell's own account of the discovery of his paradox can be found in Russell [36, p. 75].

2. The symbol we have used for set inclusion is similar to that used for the familiar order relation on the real line: if x and y are real numbers, $x \leq y$ means that $y - x$ is non-negative. The order relation on the real line has all the properties mentioned in the text:

(1') $x \leq x$ for every x;

(2') $x \leq y$ and $y \leq x \Rightarrow x = y$;

(3') $x \leq y$ and $y \leq z \Rightarrow x \leq z$.

It also has an important additional property:

(4') for any x and y, either $x \leq y$ or $y \leq x$.

Property (4') says that any two real numbers are comparable with respect to the relation in question, and it leads us to call the order relation on the real line a *total* (or *linear*) *order relation*. Show by an example that this property is not possessed by set inclusion. It is for this reason that set inclusion is called a *partial order relation*.

3.   (a) Let U be the single-element set $\{1\}$. There are two subsets, the empty set $\emptyset$ and $\{1\}$ itself. If A and B are arbitrary subsets of U, there are four possible relations of the form $A \subseteq B$. Count the number of true relations among these.

  (b) Let U be the set $\{1,2\}$. There are four subsets. List them. If A and B are arbitrary subsets of U, there are 16 possible relations of the form $A \subseteq B$. Count the number of true ones.

  (c) Let U be the set $\{1, 2, 3\}$. There are 8 subsets. What are they? There are 64 possible relations of the form $A \subseteq B$. Count the number of true ones.

  (d) Let U be the set $\{1, 2, \ldots, n\}$ for an arbitrary positive integer n. How many subsets are there? How many possible relations of the form $A \subseteq B$ are there? Can you make an informed guess as to how many of these are true?

2. THE ALGEBRA OF SETS

In this section we consider several useful ways in which sets can be combined with one another, and we develop the chief properties of these operations of combination.

As we emphasized above, all the sets we mention in this section are assumed to be subsets of our universal set U. U is the *frame of reference*, or the *universe*, for our present discussions. In our later work the frame of reference in a particular context will naturally depend on what ideas we happen to be considering. If we find ourselves studying sets of real numbers, then U is the set R of all real numbers.   If we wish to study sets of complex numbers, then we take U to be the set C of all complex numbers.   We sometimes want to narrow the frame of reference and to consider (for instance) only subsets of the closed unit interval [0,1], or of the closed unit disc $\{z:|z| \leq 1\}$, and in these cases we choose U accordingly.   Generally speaking, the universal set U is at our disposal, and we are free to select it to fit the needs of the moment.   For the present, however, U is to be regarded as a fixed but arbitrary set.   This generality allows us to apply the ideas we develop below to any situation which arises in our later work.

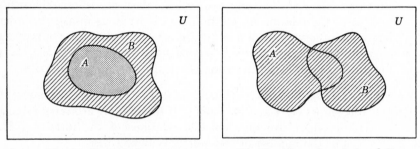

Fig. 1.   Set inclusion.   Fig. 2.   The union of A and B.

It is extremely helpful to the imagination to have a geometric picture available in terms of which we can visualize sets and operations on sets.   A convenient way to accomplish this is to represent U by a rectangular area in a plane, and the elements which make up U by the points of this area.   Sets can then be pictured by areas within this rectangle, and diagrams can be drawn which illustrate operations on sets and relations between them.   For instance, if A and B are sets, then Fig. 1 represents the circumstance that A is a subset of B (we think of each set as consisting of all points within the corresponding closed curve). Diagrammatic thought of this kind is admittedly loose and imprecise; nevertheless, the reader will find it invaluable.   No mathematics, however abstract it may appear, is ever carried on without the help of mental images of some kind, and these are often nebulous, personal, and difficult to describe.

The first operation we discuss in the algebra of sets is that of forming unions.   The *union* of two sets A and B, written $A \cup B$, is defined to be the set of all elements which are in either A or B (including those which are in both). $A \cup B$ is formed by lumping together the elements of A and those of B and regarding them as constituting a single set.   In Fig. 2, $A \cup B$ is indicated by the shaded area.   The above definition can also be expressed symbolically:

$$A \cup B = \{x : x \in A \text{ or } x \in B\}.$$

The operation of forming unions is commutative and associative:

$$A \cup B = B \cup A \qquad \text{and} \qquad A \cup (B \cup C) = (A \cup B) \cup C.$$

It has the following additional properties:

$$A \cup A = A, A \cup \emptyset = A, \text{ and } A \cup U = U.$$

We also note that

$$A \subseteq B \Leftrightarrow A \cup B = B,$$

so set inclusion can be expressed in terms of this operation.

Our next operation is that of forming intersections. The *intersection* of two sets A and B, written $A \cap B$, is the set of all elements which are in both A and B. In symbols,

$$A \cap B = \{x : x \in A \text{ and } x \in B\}.$$

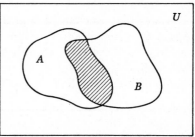

$A \cap B$ is the common part of the sets A and B. In Fig. 3, $A \cap B$ is represented by the shaded area. If $A \cap B$ is non-empty, we express this by saying that A *intersects* B. If, on the other hand, it happens that A and B have no common part, or equivalently that $A \cap B = \emptyset$, then we say that A *does not intersect* B, or

Fig. 3.   The intersection of A and B.

that A and B are *disjoint;* and a class of sets in which all pairs of distinct sets are disjoint is called a *disjoint class* of sets. The operation of forming intersections is also commutative and associative:

$$A \cap B = B \cap A \qquad \text{and} \qquad A \cap (B \cap C) = (A \cap B) \cap C.$$

It has the further properties that

$$A \cap A = A, A \cap \emptyset = \emptyset, \text{ and } A \cap U = A;$$

and since

$$A \subseteq B \Leftrightarrow A \cap B = A,$$

we see that set inclusion can also be expressed in terms of forming intersections.

We have now defined two of the fundamental operations on sets, and we have seen how each is related to set inclusion. The next obvious step is to see how they are related to one another. The facts here are given by the *distributive laws:*

$$A \cap (B \cup C) = (A \cap B) \cup (A \cap C)$$
and
$$A \cup (B \cap C) = (A \cup B) \cap (A \cup C).$$

These properties depend only on simple logic applied to the meanings of

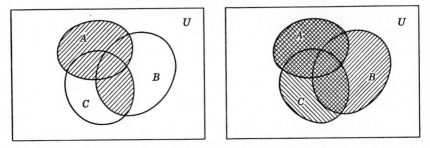

Fig. 4. $A \cup (B \cap C) = (A \cup B) \cap (A \cup C)$.

the symbols involved.   For instance, the first of the two distributive laws says that an element is in A and is in B or C precisely when it is in A and B or is in A and C.   We can convince ourselves intuitively of the validity of these laws by drawing pictures.   The second distributive law is illustrated in Fig. 4, where $A \cup (B \cap C)$ is formed on the left by shading and $(A \cup B) \cap (A \cup C)$ on the right by cross-shading.   A

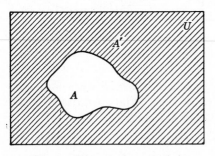

Fig. 5.   The complement of A.

moment's consideration of these diagrams ought to convince the reader that one obtains the same set in each case.

The last of our major operations on sets is the formation of complements.   The *complement* of a set A, denoted by A', is the set of all elements which are not in A.   Since the only elements we consider are those which make up U, it goes without saying—but it ought to be said —that A' consists of all those elements in U which are not in A. Symbolically,

$$A' = \{x : x \notin A\}.$$

Figure 5 (in which A' is shaded) illustrates this operation.   The operation of forming complements has the following obvious properties:

$$(A')' = A, \ \emptyset' = U, \ U' = \emptyset,$$
$$A \cup A' = U, \text{ and } A \cap A' = \emptyset.$$

Further, it is related to set inclusion by

$$A \subseteq B \Leftrightarrow B' \subseteq A'$$

and to the formation of unions and intersections by

$$(A \cup B)' = A' \cap B' \quad \text{and} \quad (A \cap B)' = A' \cup B'. \quad (1)$$

The first equation of (1) says that an element is not in either of two sets precisely when it is outside of both, and the second says that it is not in both precisely when it is outside of one or the other.

The operations of forming unions and intersections are primarily binary operations; that is, each is a process which applies to a pair of sets and yields a third. We have emphasized this by our use of parentheses to indicate the order in which the operations are to be performed, as in $(A_1 \cup A_2) \cup A_3$, where the parentheses direct us first to unite A_1 and A_2, and then to unite the result of this with A_3. Associativity makes it possible to dispense with parentheses in an expression like this and to write $A_1 \cup A_2 \cup A_3$, where we understand that these sets are to be united in any order and that the order in which the operations are performed is irrelevant. Similar remarks apply to $A_1 \cap A_2 \cap A_3$. Furthermore, if $\{A_1, A_2, \ldots, A_n\}$ is any finite class of sets, then we can form

$$A_1 \cup A_2 \cup \cdots \cup A_n \quad \text{and} \quad A_1 \cap A_2 \cap \cdots \cap A_n$$

in much the same way without any ambiguity of meaning whatever. In order to shorten the notation, we let $I = \{1, 2, \ldots, n\}$ be the set of subscripts which index the sets under consideration. I is called the *index set*. We then compress the symbols for the union and intersection just mentioned to $\cup_{i \in I} A_i$ and $\cap_{i \in I} A_i$. As long as it is quite clear what the index set is, we can write this union and intersection even more briefly, in the form $\cup_i A_i$ and $\cap_i A_i$. For the sake of both brevity and clarity, these sets are often written $\cup_{i=1}^{n} A_i$ and $\cap_{i=1}^{n} A_i$.

These extensions of our ideas and notations don't reach nearly far enough. It is often necessary to form unions and intersections of large (really large!) classes of sets. Let $\{A_i\}$ be an entirely arbitrary class of sets indexed by a set I of subscripts. Then

$$\cup_{i \in I} A_i = \{x : x \in A_i \text{ for at least one } i \in I\}$$

and $\quad \cap_{i \in I} A_i = \{x : x \in A_i \text{ for every } i \in I\}$

define their *union* and *intersection*. As above, we usually abbreviate these notations to $\cup_i A_i$ and $\cap_i A_i$; and if the class $\{A_i\}$ consists of a sequence of sets, that is, if $\{A_i\} = \{A_1, A_2, A_3, \ldots\}$, then their union and intersection are often written in the form $\cup_{i=1}^{\infty} A_i$ and $\cap_{i=1}^{\infty} A_i$. Observe that we did not require the class $\{A_i\}$ to be non-empty. If it does happen that this class is empty, then the above definitions give (remembering that all sets are subsets of U) $\cup_i A_i = \emptyset$ and $\cap_i A_i = U$. The second of these facts amounts to the following statement: if we require of an element that it belong to each set in a given class, and if there are no sets present in the class, then every element satisfies this requirement. If we had not made the agreement that the only elements under consideration are those in U, we would not have been able to assign a meaning to the intersection of an empty class of sets. A moment's consideration makes it clear that Eqs. (1) are valid for arbitrary unions and intersections:

$$(\cup_i A_i)' = \cap_i A_i' \quad \text{and} \quad (\cap_i A_i)' = \cup_i A_i'. \tag{2}$$

It is instructive to verify these equations for the case in which the class $\{A_i\}$ is empty.

We conclude our treatment of the general theory of sets with a brief discussion of certain special classes of sets which are of considerable importance in topology, logic, and measure theory. We usually denote classes of sets by capital letters in boldface.

First, some general remarks which will be useful both now and later, especially in connection with topological spaces. We shall often have occasion to speak of *finite unions* and *finite intersections*, by which we mean unions and intersections of finite classes of sets, and by a finite class of sets we always mean one which is empty or consists of n sets for some positive integer n. If we say that a class **A** of sets is closed under the formation of finite unions, we mean that **A** contains the union of each of its finite subclasses; and since the empty subclass qualifies as a finite subclass of **A**, we see that its union, the empty set, is necessarily an element of **A**. In the same way, a class of sets which is closed under the formation of finite intersections necessarily contains the universal set.

Now for the special classes of sets mentioned above. For the remainder of this section we specifically assume that the universal set U is non-empty. A *Boolean algebra of sets* is a non-empty class **A** of subsets of U which has the following properties:

(1) A and $B \in \mathbf{A} \Rightarrow A \cup B \in \mathbf{A}$;

(2) A and $B \in \mathbf{A} \Rightarrow A \cap B \in \mathbf{A}$;

(3) $A \in \mathbf{A} \Rightarrow A' \in \mathbf{A}$.

Since **A** is assumed to be non-empty, it must contain at least one set A. Property (3) shows that A' is in **A** along with A, and since $A \cap A' = \emptyset$ and $A \cup A' = U$, (1) and (2) guarantee that **A** contains the empty set and the universal set. Since the class consisting only of the empty set and the universal set is clearly a Boolean algebra of sets, these two distinct sets are the only ones which every Boolean algebra of sets must contain. It is equally clear that the class of all subsets of U is also a Boolean algebra of sets. There are many other less trivial kinds, and their applications are manifold in fields of study as diverse as statistics and electronics.

Let **A** be a Boolean algebra of sets. It is obvious that if $\{A_1, A_2, \ldots, A_n\}$ is a non-empty finite subclass of **A**, then

$$A_1 \cup A_2 \cup \cdots \cup A_n \quad \text{and} \quad A_1 \cap A_2 \cap \cdots \cap A_n$$

are both sets in **A**; and since **A** contains the empty set and the universal set, it is easy to see that **A** is a class of sets which is closed under the formation of finite unions, finite intersections, and complements. We now go in the other direction, and let **A** be a class of sets which is closed under the formation of finite unions, finite intersections, and complements. By these assumptions, **A** automatically contains the empty set and the universal set, so it is non-empty and is easily seen to be a Boolean algebra of sets. We conclude from these remarks that Boolean algebras of sets can be described alternatively as classes of sets which are closed under the formation of finite unions, finite intersections, and complements. It should be emphasized once again that when discussing Boolean algebras of sets we always assume that the universal set is non-empty.

One final comment. We speak of Boolean algebras *of sets* because there are other kinds of Boolean algebras than those which consist of sets, and we wish to preserve the distinction. We explore this topic further in our Appendix on Boolean algebras.

Problems

1. If $\{A_i\}$ and $\{B_j\}$ are two classes of sets such that $\{A_i\} \subseteq \{B_j\}$, show that $\cup_i A_i \subseteq \cup_j B_j$ and $\cap_j B_j \subseteq \cap_i A_i$.
2. The *difference* between two sets A and B, denoted by $A - B$, is the set of all elements in A and not in B; thus $A - B = A \cap B'$. Show the following:

$$A - B = A - (A \cap B) = (A \cup B) - B;$$
$$(A - B) - C = A - (B \cup C);$$
$$A - (B - C) = (A - B) \cup (A \cap C);$$
$$(A \cup B) - C = (A - C) \cup (B - C);$$
$$A - (B \cup C) = (A - B) \cap (A - C).$$

3. The *symmetric difference* of two sets A and B, denoted by $A \bigtriangleup B$, is defined by $A \bigtriangleup B = (A - B) \cup (B - A)$; it is thus the union of

their differences in opposite orders.   Show the following:

$$A \vartriangle (B \vartriangle C) = (A \vartriangle B) \vartriangle C;$$
$$A \vartriangle \emptyset = A; A \vartriangle A = \emptyset;$$
$$A \vartriangle B = B \vartriangle A;$$
$$A \cap (B \vartriangle C) = (A \cap B) \vartriangle (A \cap C).$$

4. A *ring of sets* is a non-empty class **A** of sets such that if A and B are in **A**, then $A \vartriangle B$ and $A \cap B$ are also in **A**.   Show that **A** must also contain the empty set, $A \cup B$, and $A - B$.   Show that if a non-empty class of sets contains the union and difference of any pair of its sets, then it is a ring of sets.   Show that a Boolean algebra of sets is a ring of sets.

5. Show that the class of all finite subsets (including the empty set) of an infinite set is a ring of sets but is not a Boolean algebra of sets.

6. Show that the class of all finite unions of closed-open intervals on the real line is a ring of sets but is not a Boolean algebra of sets.

7. Assuming that the universal set U is non-empty, show that Boolean algebras of sets can be described as rings of sets which contain U.

3. FUNCTIONS

Many kinds of functions occur in topology, in a great variety of situations.   In our work we shall need the full power of the general concept of a function, and since its modern meaning is much broader and deeper than its elementary meaning, we discuss this concept in considerable detail and develop its main abstract properties.

Let us begin with a brief inspection of some simple examples.   Consider the elementary function

$$y = x^2$$

of the real variable x.   What do we have in mind when we call this a function and say that y is a function of x?   In a nutshell, we are drawing attention to the fact that each real number x has linked to it a specific real number y, which can be calculated according to the rule (or law of correspondence) given by the formula.   We have here a process which, applied to any real number x, does something to it (squares it) to produce another real number y (the square of x).   Similarly,

$$y = x^3 - 3x \quad \text{and} \quad y = (x^2 + 1)^{-1}$$

are two other simple functions of the real variable x, and each is given by a rule in the form of an algebraic expression which specifies the exact manner in which the value of y depends on the value of x.

The rules for the functions we have just mentioned are expressed by formulas. In general, this is possible only for functions of a very simple kind or for those which are sufficiently important to deserve special symbols of their own. Consider, for instance, the function of the real variable x defined as follows: for each real number x, write x as an infinite decimal (using the scheme of decimal expansion in which infinite chains of 9's are avoided—in which, for example, $\frac{1}{4}$ is represented by .25000 . . . rather than by .24999 . . .); then let y be the fifty-ninth digit after the decimal point. There is of course no standard formula for this, but nevertheless it is a perfectly respectable function whose rule is given by a verbal description. On the other hand, the function $y = \sin x$ of the real variable x is so important that its rule, though fully as complicated as the one just defined, is assigned the special symbol *sin*. When discussing functions in general, we want to allow for all sorts of rules and to talk about them all at once, so we usually employ noncommittal notations like $y = f(x)$, $y = g(x)$, and so on.

Each of the functions mentioned above is defined for all real numbers x. The example $y = 1/x$ shows that this restriction is much too severe, for this function is defined only for non-zero values of x. Similarly, $y = \log x$ is defined only for positive values of x, and $y = \sin^{-1} x$ only for values of x which lie in the interval $[-1,1]$. Whatever our conception of a function may be, it should certainly be broad enough to include examples like these, which are defined only for some values of the real variable x.

In real analysis the notion of function is introduced in the following way. Let X be any non-empty set of real numbers. We say that a *function* $y = f(x)$ is defined on X if the rule f associates a definite real number y with each real number x in X. The specific nature of the rule f is totally irrelevant to the concept of a function. The set X is called the *domain* of the given function, and the set Y of all the values it assumes is called its *range*. If we speak of complex numbers here instead of real numbers, we have the notion of function as it is used in complex analysis.

This point of view toward functions is actually a bit more general than is needed for the aims of analysis, but it isn't nearly general enough for our purposes. The sets X and Y above were taken to be sets of numbers. If we now remove even this restriction and allow X and Y to be completely arbitrary non-empty sets, then we arrive at the most inclusive concept of a function. By way of illustration, suppose that X is the set of all squares in a plane and that Y is the set of all circles in the same plane. We can define a function $y = f(x)$ by requiring that the rule f associate with each square x that circle y which is inscribed in it. In general, there is no need at all for either X or Y to be a set of

numbers.   All that is really necessary for a function is two non-empty
sets X and Y and a rule f which is meaningful and unambiguous in
assigning to each element x in X a specific element y in Y.

With these preliminary descriptive remarks, we now turn to the
rather abstract but very precise ideas they are intended to motivate.

A *function* consists of three objects: two non-empty sets X and Y
(which may be equal, but need not be) and a rule f which assigns to each
element x in X a single fully determined element y in Y.   The y which
corresponds in this way to a given x is usually written $f(x)$, and is called
the *image* of x under the rule f, or the *value* of f at the element x.   This

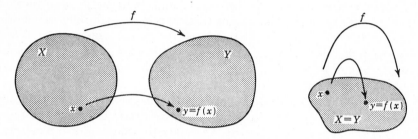

Fig. 6.   A way of visualizing mappings.

notation is supposed to be suggestive of the idea that the rule f takes the
element x and does something to it to produce the element $y = f(x)$.
The rule f is often called a *mapping*, or *transformation*, or *operator*, to
amplify this concept of it.   We then think of f as mapping x's to y's, or
transforming x's into y's, or operating on x's to produce y's.   The set X
is called the *domain* of the function, and the set of all $f(x)$'s for all x's
in X is called its *range*.   A function whose range consists of just one
element is called a *constant function*.

We often denote by $f:X \to Y$ the function with rule f, domain X,
and range contained in Y.   This notation is useful because the essential
parts of the function are displayed in a manner which emphasizes that it
is a composite object, the central thing being the rule or mapping f.
Figure 6 gives a convenient way of picturing this function.   On the
left, X and Y are different sets, and on the right, they are equal—in which
case we usually refer to f as a mapping of X into itself.   If it is clear
from the context what the sets X and Y are, or if there is no real need to
specify them explicitly, it is common practice to identify the function
$f:X \to Y$ with the rule f, and to speak of f alone as if it were the function
under consideration (without mentioning the sets X and Y).

It sometimes happens that two perfectly definite sets X and Y are
under discussion and that a mapping of X into Y arises which has no
natural symbol attached to it.   If there is no necessity to invent a

symbol for this mapping, and if it is quite clear what the mapping is, it is often convenient to designate it by $x \rightarrow y$. Accordingly, the function $y = x^2$ mentioned at the beginning of this section can be written as $x \rightarrow x^2$ or $x \rightarrow y$ (where y is understood to be the square of x).

A function f is called an *extension* of a function g (and g is called a *restriction* of f) if the domain of f contains the domain of g and $f(x) = g(x)$ for each x in the domain of g.

Most of mathematical analysis, both classical and modern, deals with functions whose values are real numbers or complex numbers.

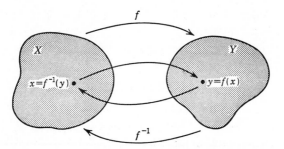

Fig. 7. The inverse of a mapping.

This is also true of those parts of topology which are concerned with the foundations of analysis. If the range of a function consists of real numbers, we call it a *real function;* similarly, a *complex function* is one whose range consists of complex numbers. Obviously, every real function is also complex. We lay very heavy emphasis on real and complex functions throughout our work.

As a matter of usage, we generally prefer to reserve the term *function* for real or complex functions and to speak of *mappings* when dealing with functions whose values are not necessarily numbers.

Consider a mapping $f : X \rightarrow Y$. When we call f a mapping of X *into* Y, we mean to suggest by this that the elements $f(x)$—as x varies over all the elements of X—need not fill up Y; but if it definitely does happen that the range of f equals Y, or if we specifically want to assume this, then we call f a mapping of X *onto* Y. If two different elements in X always have different images under f, then we call f a *one-to-one* mapping of X into Y. If $f : X \rightarrow Y$ is both onto and one-to-one, then we can define its *inverse mapping* $f^{-1} : Y \rightarrow X$ as follows: for each y in Y, we find that unique element x in X such that $f(x) = y$ (x exists and is unique since f is onto and one-to-one); we then define x to be $f^{-1}(y)$. The equation $x = f^{-1}(y)$ is the result of solving $y = f(x)$ for x in just the same way as $x = \log y$ is the result of solving $y = e^x$ for x. Figure 7 illustrates the concept of the inverse of a mapping.

If f is a one-to-one mapping of X onto Y, it will sometimes be convenient to subordinate the conception of f as a mapping sending x's over to y's and to emphasize its role as a link between x's and y's. Each x has linked to it (or has corresponding to it) precisely one $y = f(x)$; and, turning the situation around, each y has linked to it (or has corresponding to it) exactly one $x = f^{-1}(y)$. When we focus our attention on this aspect of a mapping which is one-to-one onto, we usually call it a *one-to-one correspondence*. Thus f is a one-to-one correspondence between X and Y, and f^{-1} is a one-to-one correspondence between Y and X.

Now consider an arbitrary mapping $f:X \to Y$. The mapping f, which sends each element of X over to an element of Y, induces the following two important *set mappings*. If A is a subset of X, then its *image $f(A)$* is the subset of Y defined by

$$f(A) = \{f(x): x \in A\},$$

and our first set mapping is that which sends each A over to its corresponding $f(A)$. Similarly, if B is a subset of Y, then its *inverse image $f^{-1}(B)$* is the subset of X defined by

$$f^{-1}(B) = \{x: f(x) \in B\},$$

and the second set mapping pulls each B back to its corresponding $f^{-1}(B)$. It is often essential for us to know how these set mappings behave with respect to set inclusion and operations on sets. We develop most of their significant features in the following two paragraphs.

The main properties of the first set mapping are:

$$\begin{aligned}
&f(\emptyset) = \emptyset; \quad f(X) \subseteq Y; \\
&A_1 \subseteq A_2 \Rightarrow f(A_1) \subseteq f(A_2); \\
&f(\cup_i A_i) = \cup_i f(A_i); \\
&f(\cap_i A_i) \subseteq \cap_i f(A_i).
\end{aligned} \tag{1}$$

The reader should convince himself of the truth of these statements. For instance, to prove (1) we would have to prove first that $f(\cup_i A_i)$ is a subset of $\cup_i f(A_i)$, and second that $\cup_i f(A_i)$ is a subset of $f(\cup_i A_i)$. A proof of the first of these set inclusions might run as follows: an element in $f(\cup_i A_i)$ is the image of some element in $\cup_i A_i$, therefore it is the image of an element in some A_i, therefore it is in some $f(A_i)$, and so finally it is in $\cup_i f(A_i)$. The irregularities and gaps which the reader will notice in the above statements are essential features of this set mapping. For example, the image of an intersection need not equal the intersection of the images, because two disjoint sets can easily have images which are not disjoint. Furthermore, without special assumptions (see Problem 6) nothing can be said about the relation between $f(A)'$ and $f(A')$.

The second set mapping is much better behaved. Its properties are satisfyingly complete, and can be stated as follows:

$$f^{-1}(\emptyset) = \emptyset; \quad f^{-1}(Y) = X;$$
$$B_1 \subseteq B_2 \Rightarrow f^{-1}(B_1) \subseteq f^{-1}(B_2);$$
$$f^{-1}(\cup_i B_i) = \cup_i f^{-1}(B_i); \quad (2)$$
$$f^{-1}(\cap_i B_i) = \cap_i f^{-1}(B_i); \quad (3)$$
$$f^{-1}(B') = f^{-1}(B)'. \quad (4)$$

Again, the reader should verify each of these statements for himself.

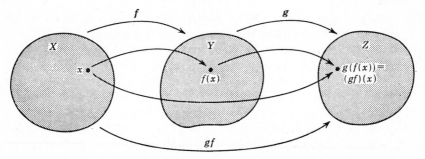

Fig. 8.   Multiplication of mappings.

We discuss one more concept in this section, that of the *multiplication* (or *composition*) of mappings.   If $y = f(x) = x^2 + 1$ and

$$z = g(y) = \sin y,$$

then these two functions can be put together to form a single function defined by $z = (gf)(x) = g(f(x)) = g(x^2 + 1) = \sin (x^2 + 1)$.   One of the most important tools of calculus (the chain rule) explains how to differentiate functions of this kind.   This manner of multiplying functions together is of basic importance for us as well, and we formulate it in general as follows.   Suppose that $f : X \to Y$ and $g : Y \to Z$ are any two mappings.   We define the *product* of these mappings, denoted by $gf : X \to Z$, by $(gf)(x) = g(f(x))$.   In words: an element x in X is taken by f to the element $f(x)$ in Y, and then g maps $f(x)$ to $g(f(x))$ in Z.   Figure 8 is a picture of this process.   We observe that the two mappings involved here are not entirely arbitrary, for the set Y which contains the range of the first equals the domain of the second.   More generally, the product of two mappings is meaningful whenever the range of the first is contained in the domain of the second.   We have regarded f as the first mapping and g as the second, and in forming their product gf, their symbols have gotten turned around.   This is a rather unpleasant phenomenon, for which we blame the occasional perversity of mathematical symbols.   Perhaps it will help the reader to keep this straight in his mind if he will remember to read the product gf from right to left: first apply f, then g.

Problems

1. Two mappings $f:X \to Y$ and $g:X \to Y$ are said to be *equal* (and we write this $f = g$) if $f(x) = g(x)$ for every x in X. Let f, g, and h be any three mappings of a non-empty set X into itself, and show that multiplication of mappings is associative in the sense that $f(gh) = (fg)h$.

2. Let X be a non-empty set. The *identity mapping* i_X on X is the mapping of X onto itself defined by $i_X(x) = x$ for every x. Thus i_X sends each element of X to itself; that is, it leaves fixed each element of X. Show that $fi_X = i_Xf = f$ for any mapping f of X into itself. If f is one-to-one onto, so that its inverse f^{-1} exists, show that $ff^{-1} = f^{-1}f = i_X$. Show further that f^{-1} is the only mapping of X into itself which has this property; that is, show that if g is a mapping of X into itself such that $fg = gf = i_X$, then $g = f^{-1}$ (*hint:* $g = gi_X = g(ff^{-1}) = (gf)f^{-1} = i_Xf^{-1} = f^{-1}$, or

$$g = i_Xg = (f^{-1}f)g = f^{-1}(fg) = f^{-1}i_X = f^{-1}).$$

3. Let X and Y be non-empty sets and f a mapping of X into Y. Show the following:
    (a) f is one-to-one $\Leftrightarrow$ there exists a mapping g of Y into X such that $gf = i_X$;
    (b) f is onto $\Leftrightarrow$ there exists a mapping h of Y into X such that $fh = i_Y$.

4. Let X be a non-empty set and f a mapping of X into itself. Show that f is one-to-one onto $\Leftrightarrow$ there exists a mapping g of X into itself such that $fg = gf = i_X$. If there exists a mapping g with this property, then there is only one such mapping. Why?

5. Let X be a non-empty set, and let f and g be one-to-one mappings of X onto itself. Show that fg is also a one-to-one mapping of X onto itself and that $(fg)^{-1} = g^{-1}f^{-1}$.

6. Let X and Y be non-empty sets and f a mapping of X into Y. If A and B are, respectively, subsets of X and Y, show the following:
    (a) $ff^{-1}(B) \subseteq B$, and $ff^{-1}(B) = B$ is true for all $B \Leftrightarrow f$ is onto;
    (b) $A \subseteq f^{-1}f(A)$, and $A = f^{-1}f(A)$ is true for all $A \Leftrightarrow f$ is one-to-one;
    (c) $f(A_1 \cap A_2) = f(A_1) \cap f(A_2)$ is true for all A_1 and $A_2 \Leftrightarrow f$ is one-to-one;
    (d) $f(A)' \subseteq f(A')$ is true for all $A \Leftrightarrow f$ is onto;
    (e) if f is onto—so that $f(A)' \subseteq f(A')$ is true for all A—then $f(A)' = f(A')$ is true for all $A \Leftrightarrow f$ is also one-to-one.

4. PRODUCTS OF SETS

We shall often have occasion to weld together the sets of a given class into a single new set called their *product* (or their *Cartesian product*). The ancestor of this concept is the coordinate plane of analytic geometry, that is, a plane equipped with the usual rectangular coordinate system. We give a brief description of this fundamental idea with a view to paving the way for our discussion of products of sets in general.

First, a few preliminary comments about the *real line*. We have already used this term several times without any explanation, and of course what we mean by it is an ordinary geometric straight line (see Fig. 9) whose points have been identified with—or coordinatized by—the

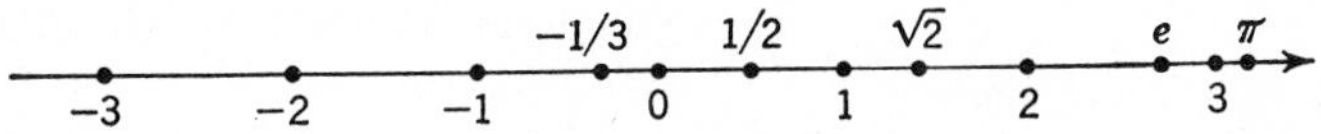

Fig. 9. The real line.

set R of all real numbers. We use the letter R to denote the real line as well as the set of all real numbers, and we often speak of real numbers as if they were points on the real line, and of points on the real line as if they were real numbers. Let no one be deceived into thinking that the real line is a simple thing, for its structure is exceedingly intricate. Our present view of it, however, is as naive and uncomplicated as the picture of it given in Fig. 9. Generally speaking, we assume that the reader is familiar with the simpler properties of the real line—those relating to inequalities (see Problem 1-2) and the basic algebraic operations of addition, subtraction, multiplication, and division. One of the most significant facts about the real number system is perhaps less well known. This is the so-called *least upper bound property*, which asserts that every non-empty set of real numbers which has an upper bound has a least upper bound. It is an easy consequence of this that every non-empty set of real numbers which has a lower bound has a greatest lower bound. All these matters can be developed rigorously on the basis of a small number of axioms, and detailed treatments can often be found in books on elementary abstract algebra.

To construct the coordinate plane, we now proceed as follows. We take two identical replicas of the real line, which we call the x *axis* and the y *axis*, and paste them on a plane at right angles to one another in such a way that they cross at the zero point on each. The usual picture is given in Fig. 10. Now let P be a point in the plane. We project P perpendicularly onto points P_x and P_y on the axes. If x and y are the coordinates of P_x and P_y on their respective axes, this process leads us from the point P to the uniquely determined ordered pair (x,y) of real numbers, where x and y are called the x *coordinate* and y *coordinate* of P. We can reverse the process, and, starting with the ordered pair of real numbers, we can recapture the point. This is the manner in which we establish the familiar one-to-one correspondence between points P in the plane and ordered pairs (x,y) of real numbers. In fact, we think of a point in the plane (which is a geometric object) and its corresponding ordered pair of real numbers (which is an algebraic object) as being—to all intents and purposes—*identical with one another.*

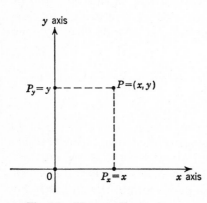

Fig. 10.   The coordinate plane.

The essence of analytic geometry lies in the possibility of exploiting this identification by using algebraic tools in geometric arguments and giving geometric interpretations to algebraic calculations.

The conventional attitude toward the coordinate plane in analytic geometry is that the geometry is the focus of interest and the algebra of ordered pairs is only a convenient tool. Here we reverse this point of view. For us, the *coordinate plane* is defined to be the set of all ordered pairs (x,y) of real numbers. We can satisfy our desire for visual images by using Fig. 10 as a picture of this set and by calling such an ordered pair a point, but this geometric language is more a convenience than a necessity.

Our notation for the coordinate plane is $R \times R$, or R^2. This symbolism reflects the idea that the coordinate plane is the result of "multiplying together" two replicas of the real line R.

It is perhaps necessary to comment on one possible source of misunderstanding. When we speak of R^2 as a plane, we do so only to establish an intuitive bond with the reader's previous experience in analytic geometry. Our present attitude is that R^2 is a pure set and has no structure whatever, because no structure has yet been assigned to it. We remarked earlier (with deliberate vagueness) that a space is a set to which has been added some kind of algebraic or geometric structure. In Sec. 15 we shall convert the *set* R^2 into the *space* of analytic geometry by defining the distance between any two points (x_1,y_1) and (x_2,y_2) to be

$$\sqrt{(x_1 - x_2)^2 + (y_1 - y_2)^2}.$$

This notion of distance endows the set R^2 with a certain "spatial" character, which we shall recognize by calling the resulting space the *Euclidean plane* instead of the coordinate plane.

We assume that the reader is fully acquainted with the way in which the set C of all complex numbers can be identified (as a set) with the coordinate plane R^2. If z is a complex number, and if z has the standard form $x + iy$ where x and y are real numbers, then we identify z with the ordered pair (x,y), and thus with an element of R^2. The complex numbers, however, are much more than merely a set. They constitute a number system, with operations of addition, multiplication, conjugation, etc. When the coordinate plane R^2 is thought of as consisting of complex numbers and is enriched by the algebraic structure it acquires in this way, it is called the *complex plane*. The letter C is used to denote either the set of all complex numbers or the complex plane. We shall make a space out of the complex plane in Sec. 9.

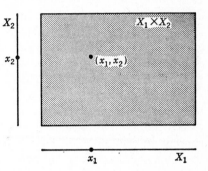

Fig. 11.   A way of visualizing $X_1 \times X_2$.

Suppose now that X_1 and X_2 are any two non-empty sets. By analogy with our above discussion, their *product* $X_1 \times X_2$ is defined to be the set of all ordered pairs (x_1,x_2), where x_1 is in X_1 and x_2 is in X_2. In spite of the arbitrary nature of X_1 and X_2, their product can be represented by a picture (see Fig. 11) which is loosely similar to the usual picture of the coordinate plane. The term *product* is applied to this set, and it is thought of as the result of "multiplying together" X_1 and X_2, for the following reason: if X_1 and X_2 are finite sets with m and n elements, then (clearly) $X_1 \times X_2$ has mn elements. If $f:X_1 \rightarrow X_2$ is a mapping with domain X_1 and range in X_2, its *graph* is that subset of $X_1 \times X_2$ which consists of all ordered pairs of the form $(x_1,f(x_1))$. We observe that this is an appropriate generalization of the concept of the graph of a function as it occurs in elementary mathematics.

This definition of the product of two sets extends easily to the case of n sets for any positive integer n. If $X_1, X_2, \ldots, X_n$ are non-empty sets, then their *product* $X_1 \times X_2 \times \cdots \times X_n$ is the set of all ordered n-tuples $(x_1, x_2, \ldots, x_n)$, where x_i is in X_i for each subscript i. If the X_i's are all replicas of a single set X, that is, if

$$X_1 = X_2 = \cdots = X_n = X,$$

then their product is usually denoted by the symbol X^n.

These ideas specialize directly to yield the important sets R^n and C^n. R^1 is just R, the real line, and R^2 is the coordinate plane. R^3—the set of all ordered triples of real numbers—is the set which underlies solid analytic geometry, and we assume that the reader is familiar with the manner in which this set arises, through the introduction of a rectangular coordinate system into ordinary three-dimensional space.   We can draw pictures here just as in the case of the coordinate plane, and we can use geometric language as much as we please, but it must be understood that the mathematics of this set is the mathematics of ordered triples of real numbers and that the pictures are merely an aid to the intuition.   Once we fully grasp this point of view, there is no difficulty whatever in advancing at once to the study of the set R^n of all ordered n-tuples $(x_1, x_2, \ldots, x_n)$ of real numbers for any positive integer n. It is quite true that when n is greater than 3 it is no longer possible to draw the same kinds of intuitively rich pictures, but at worst this is merely an inconvenience.   We can (and do) continue to use suggestive geometric language, so all is not lost.   The set C^n is defined similarly: it is the set of all ordered n-tuples $(z_1, z_2, \ldots, z_n)$ of complex numbers. Each of the sets R^n and C^n plays a prominent part in our later work.

We emphasized above that for the present the coordinate plane is to be considered as merely a set, and not a space.   Similar remarks apply to R^n and C^n.   In due course (in Sec. 15) we shall impart form and content to each of these sets by suitable definitions.   We shall convert them into the *Euclidean* and *unitary n-spaces* which underlie and motivate so many developments in modern pure mathematics, and we shall explore some aspects of their algebraic and topological structure to the very last pages of this book.   But as of now—and this is the point we insist on—neither one of these sets has any structure at all.

As the reader doubtless suspects, it is not enough that we consider only products of finite classes of sets.   The needs of topology compel us to extend these ideas to arbitrary classes of sets.

We defined the product $X_1 \times X_2 \times \cdots \times X_n$ to be the set of all ordered n-tuples $(x_1, x_2, \ldots, x_n)$ such that x_i is in X_i for each subscript i.   To see how to extend this definition, we reformulate it as follows.   We have an index set I, consisting of the integers from 1 to n, and corresponding to each index (or subscript) i we have a non-empty set X_i.   The n-tuple $(x_1, x_2, \ldots, x_n)$ is simply a function (call it x) defined on the index set I, with the restriction that its value $x(i) = x_i$ is an element of the set X_i for each i in I.   Our point of view here is that the function x is completely determined by, and is essentially equivalent to, the array $(x_1, x_2, \ldots, x_n)$ of its values.

The way is now open for the definition of products in their full generality.   Let $\{X_i\}$ be a non-empty class of non-empty sets, indexed by the elements i of an index set I.   The sets X_i need not be different from one another; indeed, it may happen that they are all identical replicas of a single set, distinguished only by different indices.   The *product* of the sets X_i, written $P_{i \in I} X_i$, is defined to be the set of all functions x defined on I such that $x(i)$ is an element of the set X_i for each index i. We call X_i the *ith coordinate set*. When there can be no misunderstanding about the index set, the symbol $P_{i \in I} X_i$ is often abbreviated to $P_i X_i$. The definition we have just given requires that each coordinate set be non-empty before the product can be formed. It will be useful if we extend this definition slightly by agreeing that if any of the X_i's are empty, then $P_i X_i$ is also empty.

This approach to the idea of the product of a class of sets, by means of functions defined on the index set, is useful mainly in giving the definition. In practice, it is much more convenient to use the subscript notation x_i instead of the function notation $x(i)$. We then interpret the product $P_i X_i$ as made up of elements x, each of which is specified by the exhibited array $\{x_i\}$ of its values in the respective coordinate sets X_i. We call x_i the *ith coordinate* of the element $x = \{x_i\}$.

The mapping p_i of the product $P_i X_i$ onto its ith coordinate set X_i which is defined by $p_i(x) = x_i$—that is, the mapping whose value at an arbitrary element of the product is the ith coordinate of that element —is called the *projection* onto the ith coordinate set. The projection p_i selects the ith coordinate of each element in its domain. There is clearly one projection for each element of the index set I, and the set of all projections plays an important role in the general theory of topological spaces.

Problems

1. The graph of a mapping $f : X \to Y$ is a subset of the product $X \times Y$. What properties characterize the graphs of mappings among all subsets of $X \times Y$?

2. Let X and Y be non-empty sets. If A_1 and A_2 are subsets of X, and B_1 and B_2 subsets of Y, show the following:

$$(A_1 \times B_1) \cap (A_2 \times B_2) = (A_1 \cap A_2) \times (B_1 \cap B_2);$$
$$(A_1 \times B_1) - (A_2 \times B_2) = (A_1 - A_2) \times (B_1 - B_2)$$
$$\cup (A_1 \cap A_2) \times (B_1 - B_2)$$
$$\cup (A_1 - A_2) \times (B_1 \cap B_2).$$

3. Let X and Y be non-empty sets, and let **A** and **B** be rings of subsets of X and Y, respectively. Show that the class of all finite unions of sets of the form $A \times B$ with $A \, \varepsilon \, \mathbf{A}$ and $B \, \varepsilon \, \mathbf{B}$ is a ring of subsets of $X \times Y$.

5. PARTITIONS AND EQUIVALENCE RELATIONS

In the first part of this section we consider a non-empty set X, and we study decompositions of X into non-empty subsets which fill it out and have no elements in common with one another. We give special attention to the tools (equivalence relations) which are normally used to generate such decompositions.

A *partition* of X is a disjoint class $\{X_i\}$ of non-empty subsets of X whose union is the full set X itself. The X_i's are called the *partition sets*. Expressed somewhat differently, a partition of X is the result of splitting it, or subdividing it, into non-empty subsets in such a way that each element of X belongs to one and only one of the given subsets.

If X is the set $\{1, 2, 3, 4, 5\}$, then $\{1, 3, 5\}$, $\{2, 4\}$ and $\{1, 2, 3\}$, $\{4, 5\}$ are two different partitions of X. If X is the set R of all real numbers, then we can partition X into the set of all rationals and the set of all irrationals, or into the infinitely many closed-open intervals of the form $[n, n + 1)$ where n is an integer. If X is the set of all points in the coordinate plane, then we can partition X in such a way that each partition set consists of all points with the same x coordinate (vertical lines), or so that each partition set consists of all points with the same y coordinate (horizontal lines).

Other partitions of each of these sets will readily occur to the reader. In general, there are many different ways in which any given set can be partitioned. These manufactured examples are admittedly rather uninspiring and serve only to make our ideas more concrete. Later in this section we consider some others which are more germane to our present purposes.

A *binary relation* in the set X is a mathematical symbol or verbal phrase, which we denote by R in this paragraph, such that for each ordered pair (x,y) of elements of X the statement $x \, R \, y$ is meaningful, in the sense that it can be classified definitely as true or false. For such a binary relation, $x \, R \, y$ symbolizes the assertion that x *is* related by R to y, and $x \, \not{R} \, y$ the negation of this, namely, the assertion that x *is not* related by R to y. Many examples of binary relations can be given, some familiar and others less so, some mathematical and others not. For instance, if X is the set of all integers and R is interpreted to mean "is less than," which of course is usually denoted by the symbol $<$, then we clearly have $4 < 7$ and $5 \not< 2$. We have been speaking of binary relations, which are so named because they apply only to ordered pairs of elements, rather than to ordered triples, etc. In our work we drop the qualifying adjective and speak simply of a *relation* in X, since we shall have occasion to consider only relations of this kind.[1]

We now assume that a partition of our non-empty set X is given,

---

[1] Some writers prefer to regard a relation R in X as a subset R of $X \times X$. From this point of view, $x \, R \, y$ and $x \, \not{R} \, y$ are simply equivalent ways of writing $(x,y) \, \varepsilon$ R and $(x,y) \notin$ R. This definition has the advantage of being more tangible than ours, and the disadvantage that few people really think of a relation in this way.

and we associate with this partition a relation in X. This relation is defined in the following way: we say that x is *equivalent* to y and write this $x \sim y$ (the symbol $\sim$ is pronounced "wiggle"), if x and y belong to the same partition set. It is obvious that the relation $\sim$ has the following properties:

(1) $x \sim x$ for every x *(reflexivity)*;

(2) $x \sim y \Rightarrow y \sim x$ *(symmetry)*;

(3) $x \sim y$ and $y \sim z \Rightarrow x \sim z$ *(transitivity)*.

This particular relation in X arose in a special way, in connection with a given partition of X, and its properties are immediate consequences of its definition. Any relation whatever in X which possesses these three properties is called an *equivalence relation* in X.

We have just seen that each partition of X has associated with it a natural equivalence relation in X. We now reverse the situation and show that a given equivalence relation in X determines a natural partition of X.

Let $\sim$ be an equivalence relation in X; that is, assume that it is reflexive, symmetric, and transitive in the sense described above. If x is an element of X, the subset of X defined by $[x] = \{y : y \sim x\}$ is called the *equivalence set* of x. The equivalence set of x is thus the set of all elements which are equivalent to x. We show that the class of all distinct equivalence sets forms a partition of X. By reflexivity, $x \varepsilon [x]$ for each element x in X, so each equivalence set is non-empty and their union is X. It remains to be shown that any two equivalence sets $[x_1]$ and $[x_2]$ are either disjoint or identical. We prove this by showing that if $[x_1]$ and $[x_2]$ are not disjoint, then they must be identical. Suppose that $[x_1]$ and $[x_2]$ are not disjoint; that is, suppose that they have a common element z. Since z belongs to both equivalence sets, $z \sim x_1$ and $z \sim x_2$, and by symmetry, $x_1 \sim z$. Let y be any element of $[x_1]$, so that $y \sim x_1$. Since $y \sim x_1$ and $x_1 \sim z$, transitivity shows that $y \sim z$. By another application of transitivity, $y \sim z$ and $z \sim x_2$ imply that $y \sim x_2$, so that y is in $[x_2]$. Since y was chosen arbitrarily in $[x_1]$, we see by this that $[x_1] \subseteq [x_2]$. The same reasoning shows that $[x_2] \subseteq [x_1]$, and from this we conclude (see the last paragraph of Sec. 1) that $[x_1] = [x_2]$.

The above discussion demonstrates that there is no real distinction (other than a difference in language) between partitions of a set and equivalence relations in the set. If we start with a partition, we get an equivalence relation by regarding elements as equivalent if they belong to the same partition set, and if we start with an equivalence relation, we get a partition by grouping together into subsets all elements which are equivalent to one another. We have here a single mathematical idea, which we have been considering from two different points of view, and the approach we choose in any particular application depends entirely on our own convenience. In practice, it is almost invariably the case that we use equivalence relations (which are usually easy to define) to obtain partitions (which are sometimes difficult to describe fully).

We now turn to several of the more important simple examples of equivalence relations.

Let I be the set of all integers. If a and b are elements of this set, we write $a = b$ (and say that a *equals* b) if a and b are the same integer. Thus $2 + 3 = 5$ means that the expressions on the left and right are simply different ways of writing the same integer. It is apparent that $=$ used in this sense is an equivalence relation in the set I:

(1) $a = a$ for every a;

(2) $a = b \Rightarrow b = a$;

(3) $a = b$ and $b = c \Rightarrow a = c$.

Clearly, each equivalence set consists of precisely one integer.

Another familiar example is the relation of equality commonly used for fractions. We remind the reader that, strictly speaking, a fraction is merely a symbol of the form a/b, where a and b are integers and b is not zero. The fractions $\frac{2}{3}$ and $\frac{4}{6}$ are obviously not identical, but nevertheless we consider them to be equal. In general, we say that two fractions a/b and c/d are *equal*, written $a/b = c/d$, if ad and bc are equal as integers in the usual sense (see the above paragraph). We leave it to the reader to show that this is an equivalence relation in the set of all fractions. An equivalence set of fractions is what we call a *rational number*. Everyday usage ignores the distinction between fractions and rational numbers, but it is important to recognize that from the strict point of view it is the rational numbers (and not the fractions) which form part of the real number system.

Our final example has a deeper significance, for it provides us with the basic tool for our work of the next two sections.

For the remainder of this section we consider a relation between pairs of non-empty sets, and each set mentioned (whether we say so explicitly or not) is assumed to be non-empty. If X and Y are two sets, we say that X is *numerically equivalent* to Y if there exists a one-to-one correspondence between X and Y, i.e., if there exists a one-to-one mapping of X onto Y. This relation is reflexive, since the identity mapping $i_X : X \to X$ is one-to-one onto; it is symmetric, since if $f : X \to Y$ is one-to-one onto, then its inverse mapping $f^{-1} : Y \to X$ is also one-to-one onto; and it is transitive, since if $f : X \to Y$ and $g : Y \to Z$ are one-to-one onto, then $gf : X \to Z$ is also one-to-one onto. Numerical equivalence has all the properties of an equivalence relation, and if we consider it as an equivalence relation in the class of all non-empty subsets of some universal set U, it groups together into equivalence sets all those subsets of U which have the *same number of elements*. After we state and prove the following very useful but rather technical theorem, we shall continue in Secs. 6 and 7 with an exploration of the implications of these ideas.

The theorem we have in mind—the *Schroeder-Bernstein theorem*—is the following: *if X and Y are two sets each of which is numerically equivalent to a subset of the other, then all of X is numerically equivalent to all of Y.* There are several proofs of this classic theorem, some of which are quite difficult. The very elegant proof we give is essentially due to Birkhoff and MacLane.

Now for the proof. We assume that $f:X \to Y$ is a one-to-one mapping of X into Y, and that $g:Y \to X$ is a one-to-one mapping of Y into X. Our task is to produce a mapping $F:X \to Y$ which is one-to-one *onto*. We may assume that neither f nor g is onto, since if f is, we can define F to be f, and if g is, we can define F to be g^{-1}. Since both f and g are one-to-one, it is permissible to use the mappings f^{-1} and g^{-1} as long as we clearly understand that f^{-1} is defined only on $f(X)$ and g^{-1} only on $g(Y)$. We obtain the mapping F by splitting both X and Y into subsets which we characterize in terms of the ancestry of their elements. Let x be an element of X. We apply g^{-1} to it (if we can) to get the element $g^{-1}(x)$ in Y. If $g^{-1}(x)$ exists, we call it the first ancestor of x. The element x itself we call the zeroth ancestor of x. We now apply f^{-1} to $g^{-1}(x)$ if we can, and if $(f^{-1}g^{-1})(x)$ exists, we call it the second ancestor of x. We now apply g^{-1} to $(f^{-1}g^{-1})(x)$ if we can, and if $(g^{-1}f^{-1}g^{-1})(x)$ exists, we call it the third ancestor of x. As we continue this process of tracing back the ancestry of x, it becomes apparent that there are three possibilities. (1) x has infinitely many ancestors. We denote by X_i the subset of X which consists of all elements with infinitely many ancestors. (2) x has an even number of ancestors; this means that x has a last ancestor (that is, one which itself has no first ancestor) in X. We denote by X_e the subset of X consisting of all elements with an even number of ancestors. (3) x has an odd number of ancestors; this means that x has a last ancestor in Y. We denote by X_o the subset of X which consists of all elements with an odd number of ancestors. The three sets X_i, X_e, X_o form a disjoint class whose union is X. We decompose Y in just the same way into three subsets Y_i, Y_e, Y_o. It is easy to see that f maps X_i onto Y_i and X_e onto Y_o, and that g^{-1} maps X_o onto Y_e; and we complete the proof by defining F in the following piecemeal manner:

$$F(x) = \begin{cases} f(x) & \text{if } x \in X_i \cup X_e, \\ g^{-1}(x) & \text{if } x \in X_o. \end{cases}$$

We attempt to illustrate these ideas in Fig. 12. Here we present two replicas of the situation: on the left, X and Y are represented by the vertical lines, and f and g by the lines slanting down to the right and left; and on the right, we schematically trace the ancestry of three elements in X, of which x_1 has no first ancestor, x_2 has a first and second ancestor, and x_3 has a first, second, and third ancestor.

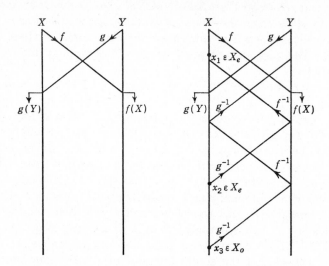

Fig. 12.   The proof of the Schroeder-Bernstein theorem.

The Schroeder-Bernstein theorem has great theoretical and practical significance.   Its main value for us lies in its role as a tool by means of which we can prove numerical equivalence with a minimum of effort for many specific sets.   We put it to work in Sec. 7.

Problems

1.   Let $f: X \to Y$ be an arbitrary mapping.   Define a relation in X as follows: $x_1 \sim x_2$ means that $f(x_1) = f(x_2)$.   Show that this is an equivalence relation and describe the equivalence sets.

2.   In the set R of all real numbers, let $x \sim y$ mean that $x - y$ is an integer.   Show that this is an equivalence relation and describe the equivalence sets.

3.   Let I be the set of all integers, and let m be a fixed positive integer. Two integers a and b are said to be *congruent* modulo m—symbolized by $a \equiv b \pmod{m}$—if $a - b$ is exactly divisible by m, i.e., if $a - b$ is an integral multiple of m.   Show that this is an equivalence relation, describe the equivalence sets, and state the number of distinct equivalence sets.

4.   Decide which ones of the three properties of reflexivity, symmetry, and transitivity are true for each of the following relations in the set

of all positive integers: $m \leq n$, $m < n$, m divides n. Are any of these equivalence relations?

5. Give an example of a relation which is (a) reflexive but not symmetric or transitive; (b) symmetric but not reflexive or transitive; (c) transitive but not reflexive or symmetric; (d) reflexive and symmetric but not transitive; (e) reflexive and transitive but not symmetric; (f) symmetric and transitive but not reflexive.

6. Let X be a non-empty set and $\sim$ a relation in X. The following purports to be a proof of the statement that if this relation is symmetric and transitive, then it is necessarily reflexive: $x \sim y \Rightarrow y \sim x$; $x \sim y$ and $y \sim x \Rightarrow x \sim x$; therefore $x \sim x$ for every x. In view of Problem 5f, this cannot be a valid proof. What is the flaw in the reasoning?

7. Let X be a non-empty set. A relation $\sim$ in X is called *circular* if $x \sim y$ and $y \sim z \Rightarrow z \sim x$, and *triangular* if $x \sim y$ and $x \sim z \Rightarrow y \sim z$. Prove that a relation in X is an equivalence relation $\Leftrightarrow$ it is reflexive and circular $\Leftrightarrow$ it is reflexive and triangular.

6. COUNTABLE SETS

The subject of this section and the next—*infinite cardinal numbers*—lies at the very foundation of modern mathematics. It is a vital instrument in the day-to-day work of many mathematicians, and we shall make extensive use of it ourselves. This theory, which was created by the German mathematician Cantor, also has great aesthetic appeal, for it begins with ideas of extreme simplicity and develops through natural stages into an elaborate and beautiful structure of thought. In the course of our discussion we shall answer questions which no one before Cantor's time thought to ask, and we shall ask a question which no one can answer to this day.

Without further ado, we can say that *cardinal numbers* are those used in counting, such as the positive integers (or natural numbers) 1, 2, 3, . . . familiar to us all. But there is much more to the story than this.

The act of counting is undoubtedly one of the oldest of human activities. Men probably learned to count in a crude way at about the same time as they began to develop articulate speech. The earliest men who lived in communities and domesticated animals must have found it necessary to record the number of goats in the village herd by means of a pile of stones or some similar device. If the herd was counted in each night by removing one stone from the pile for each goat accounted for, then stones left over would have indicated strays, and herdsmen would have gone out to search for them. Names for numbers and symbols for them, like our 1, 2, 3, . . . , would have been superfluous. The simple and yet profound idea of a one-to-one correspondence between the stones and the goats would have fully met the needs of the situation.

In a manner of speaking, we ourselves use the infinite set

$$N = \{1, 2, 3, \ldots\}$$

of all positive integers as a "pile of stones." We carry this set around with us as part of our intellectual equipment. Whenever we want to count a set, say, a stack of dollar bills, we start through the set N and tally off one bill against each positive integer as we come to it. The last number we reach, corresponding to the last bill, is what we call the number of bills in the stack. If this last number happens to be 10, then "10" is our symbol for the number of bills in the stack, as it also is for the number of our fingers, and for the number of our toes, and for the number of elements in any set which can be put into one-to-one correspondence with the finite set $\{1, 2, \ldots, 10\}$. Our procedure is slightly more sophisticated than that of the primitive savage. We have the symbols 1, 2, 3, . . . for the numbers which arise in counting; we can record them for future use, and communicate them to other people, and manipulate them by the operations of arithmetic. But the underlying idea, that of the one-to-one correspondence, remains the same for us as it probably was for him.

The positive integers are adequate for the purpose of counting any non-empty finite set, and since outside of mathematics all sets appear to be of this kind, they suffice for all non-mathematical counting. But in the world of mathematics we are obliged to consider many infinite sets, such as the set of all positive integers itself, the set of all integers, the set of all rational numbers, the set of all real numbers, the set of all points in a plane, and so on. It is often important to be able to count such sets, and it was Cantor's idea to do this, and to develop a theory of infinite cardinal numbers, by means of one-to-one correspondences.

In comparing the sizes of two sets, the basic concept is that of numerical equivalence as defined in the previous section. We recall that two non-empty sets X and Y are said to be numerically equivalent if there exists a one-to-one mapping of one onto the other, or—and this amounts to the same thing—if there can be found a one-to-one correspondence between them. To say that two non-empty finite sets are numerically equivalent is of course to say that they have the *same number of elements* in the ordinary sense. If we count one of them, we simply establish a one-to-one correspondence between its elements and a set of positive integers of the form $\{1, 2, \ldots, n\}$, and we then say that n is the *number of elements possessed by both*, or the *cardinal number of both*. The positive integers are the *finite cardinal numbers*. We encounter many surprises as we follow Cantor and consider numerical equivalence for infinite sets.

The set $N = \{1, 2, 3, \ldots\}$ of all positive integers is obviously "larger" than the set $\{2, 4, 6, \ldots\}$ of all even positive integers, for it contains this set as a proper subset. It appears on the surface that N has "more" elements. But it is very important to avoid jumping to conclusions when dealing with infinite sets, and we must remember that our criterion in these matters is whether there exists a one-to-one correspondence between the sets (not whether one set is or is not a proper subset of the other). As a matter of fact, the pairing

$$1, 2, 3, \ldots, n, \ldots$$
$$2, 4, 6, \ldots, 2n, \ldots$$

serves to establish a one-to-one correspondence between these sets, in which each positive integer in the upper row is matched with the even positive integer (its double) directly below it, and these two sets must therefore be regarded as having the *same number of elements*. This is a very remarkable circumstance, for it seems to contradict our intuition and yet is based only on solid common sense. We shall see below, in Problems 6 and 7-4, that every infinite set is numerically equivalent to a proper subset of itself. Since this property is clearly not possessed by any finite set, some writers even use it as the definition of an infinite set.

In much the same way as above, we can show that N is numerically equivalent to the set of *all* even integers:

$$1, 2, \quad 3, 4, \quad 5, 6, \quad 7, \ldots$$
$$0, 2, \; -2, 4, \; -4, 6, \; -6, \ldots$$

Here our device is to start with 0 and follow each even positive integer as we come to it by its negative. Similarly, N is numerically equivalent to the set of all integers:

$$1, 2, \quad 3, 4, \quad 5, 6, \quad 7, \ldots$$
$$0, 1, \; -1, 2, \; -2, 3, \; -3, \ldots$$

It is of considerable historical interest to note that Galileo observed in the early seventeenth century that there are precisely as many perfect squares (1, 4, 9, 16, 25, etc.) among the positive integers as there are positive integers altogether. This is clear from the pairing

$$1, \quad 2, \quad 3, \quad 4, \quad 5, \ldots$$
$$1^2, \quad 2^2, \quad 3^2, \quad 4^2, \quad 5^2, \ldots$$

It struck him as very strange that this should be true, considering how sparsely strewn the squares are among all the positive integers. But the time appears not to have been ripe for the exploration of this phenomenon, or perhaps he had other things on his mind; in any case, he did not follow up his idea.

These examples should make it clear that all that is really necessary in showing that an infinite set X is numerically equivalent to N is that we be able to list the elements of X, with a first, a second, a third, and so on, in such a way that it is completely exhausted by this counting off of its elements. It is for this reason that any infinite set which is numerically equivalent to N is said to be *countably infinite*. We say that a set is *countable* if it is non-empty and finite (in which case it can obviously be counted) or if it is countably infinite.

One of Cantor's earliest discoveries in his study of infinite sets was that the set of all positive rational numbers (which is very large: it contains N and a great many other numbers besides) is actually countable. We cannot list the positive rational numbers in order of size, as we can the positive integers, beginning with the smallest, then the next smallest, and so on, for there is no smallest, and between any two there are infinitely many others. We must find some other way of counting them, and following Cantor, we arrange them not in order of size, but according to the size of the sum of the numerator and denominator. We begin with all positive rationals whose numerator and denominator add up to 2: there is only one, $\frac{1}{1} = 1$. Next we list (with increasing numerators) all those for which this sum is $3:\frac{1}{2}$, $\frac{2}{1} = 2$. Next, all those for which this sum is $4:\frac{1}{3}$, $\frac{2}{2} = 1$, $\frac{3}{1} = 3$. Next, all those for which this sum is $5:\frac{1}{4}$, $\frac{2}{3}$, $\frac{3}{2}$, $\frac{4}{1} = 4$. Next, all those for which this sum is $6:\frac{1}{5}$, $\frac{2}{4} = \frac{1}{2}$, $\frac{3}{3} = 1$, $\frac{4}{2} = 2$, $\frac{5}{1} = 5$. And so on. If we now list all these together from the beginning, omitting those already listed when we come to them, we get a sequence

$$1, \tfrac{1}{2}, 2, \tfrac{1}{3}, 3, \tfrac{1}{4}, \tfrac{2}{3}, \tfrac{3}{2}, 4, \tfrac{1}{5}, 5, \ldots$$

which contains each positive rational number once and only once. Figure 13 gives a schematic representation of this manner of listing the positive rationals. In this figure the first row contains all positive rationals with numerator 1, the second all with numerator 2, etc.; and the first column contains all with denominator 1, the second all with denominator 2, and so on. Our listing amounts to traversing this array of numbers as the arrows indicate, where of course all those numbers already encountered are left out.

It's high time that we christened the infinite cardinal number we've been discussing, and for this purpose we use the first letter of the Hebrew alphabet ($\aleph$, pronounced "aleph") with 0 as a subscript. We say that $\aleph_0$ is the number of elements in any countably infinite set. Our complete list of cardinal numbers so far is

$$1, 2, 3, \ldots , \aleph_0.$$

We expand this list in the next section.

Suppose now that m and n are two cardinal numbers (finite or infinite). The statement that m is *less than* n (written $m < n$) is defined to mean the following: if X and Y are sets with m and n elements, then

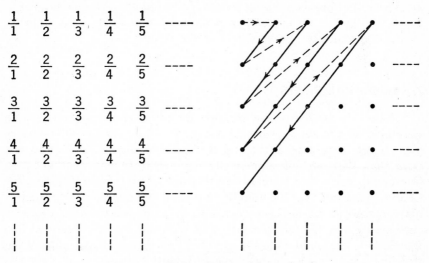

Fig. 13.   A listing of the positive rationals.

(1) there exists a one-to-one mapping of X into Y, and (2) there does not exist a one-to-one mapping of X onto Y. Using this concept, it is easy to relate our cardinal numbers to one another by means of

$$1 < 2 < 3 < \cdots < \aleph_0.$$

With respect to the finite cardinal numbers, this ordering corresponds to their usual ordering as real numbers.

Problems

1. Prove that the set of all rational numbers (positive, negative, and zero) is countable. (*Hint:* see our method of showing that the set of all integers is countable.)
2. Use the idea behind Fig. 13 to prove that if $\{X_i\}$ is a countable class of countable sets, then $\cup_i X_i$ is also countable. We usually express this by saying that *any countable union of countable sets is countable.*

3. Prove that the set of all rational points in the coordinate plane R^2 (i.e., all points whose coordinates are both rational) is countable.
4. Prove that if X_1 and X_2 are countable, then $X_1 \times X_2$ is also countable.
5. Prove that if $X_1, X_2, \ldots, X_n$ are countable, where n is any positive integer, then $X_1 \times X_2 \times \cdots \times X_n$ is also countable.
6. Prove that every countably infinite set is numerically equivalent to a proper subset of itself.
7. Prove that any non-empty subset of a countable set is countable.
8. Let X and Y be non-empty sets, and f a mapping of X onto Y. If X is countable, prove that Y is also countable.

7. UNCOUNTABLE SETS

All the infinite sets we considered in the previous section were countable, so it might appear at this stage that *every* infinite set is countable. If this were true, if the end result of the analysis of infinite sets were that they are all numerically equivalent to one another, then Cantor's theory would be relatively trivial. But this is not the case, for Cantor discovered that the infinite set R of all real numbers is *not* countable—or, as we phrase it, R is *uncountable* or *uncountably infinite*. Since we customarily identify the elements of R with the points of the real line (see Sec. 4), this amounts to the assertion that the set of *all* points on the real line represents a "higher type of infinity" than that of only the integral points or only the rational points.

Cantor's proof of this is very ingenious, but it is actually quite simple. In outline the procedure is as follows: we assume that all the real numbers (in decimal form) can be listed, and in fact have been listed; then we produce a real number which cannot be in this list—thus contradicting our initial assumption that a complete listing is possible. In representing real numbers by decimals, we use the scheme of decimal expansion in which infinite chains of 9's are avoided; for instance, we write $\frac{1}{2}$ as .5000 . . . and not as .4999 . . . . In this way we guarantee that each real number has one and only one decimal representation. Suppose now that we can list all the real numbers, and that they have been listed in a column like the one below (where we use particular numbers for the purpose of illustration).

$$
\begin{array}{ll}
\text{1st number} & 13 + .712983 \ldots \\
\text{2nd number} & -4 + .913572 \ldots \\
\text{3rd number} & 0 + .843265 \ldots \\
\quad \cdot \ \cdot \ \cdot \ \cdot \ \cdot \ \cdot \ \cdot & \quad \cdot \ \cdot \ \cdot \ \cdot \ \cdot \ \cdot \ \cdot \ \cdot \ \cdot \ \cdot \ \cdot \ \cdot
\end{array}
$$

Since it is impossible actually to write down this infinite list of decimals, our assumption that all the real numbers can be listed in this way means that we assume that we have available some general rule according to which the list is constructed, similar to that used for listing the positive rationals, and that every conceivable real number occurs somewhere in this list. We now demonstrate that this assumption is false by exhibiting a decimal $.a_1a_2a_3 \ldots$ which is constructed in such a way that it is not in the list. We choose a_1 to be 1 unless the first digit after the decimal point of the first number in our list is 1, in which case we choose a_1 to be 2. Clearly, our new decimal will differ from the first number in our list regardless of how we choose its remaining digits. Next, we choose a_2 to be 1 unless the second digit after the decimal point of the second number in our list is 1, in which case we choose a_2 to be 2. Just as above, our new decimal will necessarily differ from the second number in our list. We continue building up the decimal $.a_1a_2a_3 \ldots$ in this way, and since the process can be continued indefinitely, it defines a real number in decimal form ($.121 \ldots$ in the case of our illustrative example) which is different from each number in our list. This contradicts our assumption that we can list all the real numbers and completes our proof of the fact that the set R of all real numbers is uncountable.

We have seen (in Problem 6-1) that the set of all rational points on the real line is countable, and we have just proved that the set of *all* points on the real line is uncountable. We conclude at once from this that irrational points on the real line (i.e., irrational numbers) must exist. In fact, it is very easy to see by means of Problem 6-2 that the set of all irrational numbers is uncountably infinite. To vary slightly a striking metaphor coined by E. T. Bell, the rational numbers are spotted along the real line like stars against a black sky, and the dense blackness of the background is the firmament of the irrationals. The reader is probably familiar with a proof of the fact that the square root of 2 is irrational. This proof demonstrates the existence of irrational numbers by exhibiting a specimen. Our remarks, on the other hand, do not show that this or that particular number is irrational; they merely show that such numbers must exist, and moreover must exist in overwhelming abundance.

If the reader supposes that the set of all points on the real line R is uncountable because R is infinitely long, then we can disillusion him by the following argument, which shows that any open interval on R, no matter how short it may be, has precisely as many points as R itself. Let a and b be any two real numbers with $a < b$, and consider the open interval (a,b). Figure 14 shows how to establish a one-to-one correspondence between the points P of (a,b) and the points P' of R: we bend (a,b) into a semicircle; we rest this semicircle tangentially on the real line R as shown in the figure; and we link P and P' by projecting from its center. If formulas are preferred over geometric reasoning of this kind, we observe that $y = a + (b - a)x$ is a numerical equivalence between real numbers $x \, \varepsilon \, (0,1)$ and $y \, \varepsilon \, (a,b)$, and that $z = \tan \pi(x - \frac{1}{2})$ is another numerical equivalence between $(0,1)$ and all of R. It now follows that (a,b) and R are numerically equivalent to one another.

We are now in a position to show that any subset X of the real line R which contains an open interval I is numerically equivalent to R, no matter how complicated the structure of X may be. The proof of this fact is very simple, and it uses only the Schroeder-Bernstein theorem and our above result that I is numerically equivalent to R. The argument can be given in two sentences. Since X is numerically equivalent to itself, it is obviously numerically equivalent to a subset of R; and R is numerically equivalent to a subset of X, namely, to I. It is now a direct consequence of the Schroeder-Bernstein theorem that X and R are numerically equivalent to one another. We point out that all numerical equivalences up to this point have been established by actually exhibiting one-to-one correspondences between the sets concerned. In the present situation, however, it is not feasible to do this, for very little has been assumed about the specific nature of the set X. Without the help of the Schroeder-Bernstein theorem it would be very difficult to prove theorems of this type.

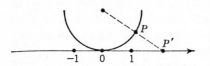

Fig. 14. A one-to-one correspondence between an open interval and the real line.

We give another interesting application of the Schroeder-Bernstein theorem. Consider the coordinate plane R^2 and the subset X of R^2 defined by $X = \{(x,y) : 0 \leq x < 1 \text{ and } 0 \leq y < 1\}$. We show that X is numerically equivalent to the closed-open interval

$$I = \{(x,y) : 0 \leq x < 1 \text{ and } y = 0\}$$

which forms its base (see Fig. 15). Since I is numerically equivalent to a subset of X, namely, to I itself, our conclusion will follow at once from the Schroeder-Bernstein theorem if we can establish a one-to-one mapping of X into I. This we now do. Let (x,y) be an arbitrary point of X. Each of the coordinates x and y has a unique decimal expansion which does not end in an infinite chain of 9's. We form another decimal z from these by alternating their digits; for example, if $x = .327 \ldots$ and $y = .614 \ldots$, then $z = .362174 \ldots$. We now identify z (which cannot end in an infinite chain of 9's) with a point of I. This gives the required one-to-one mapping of X into I and yields the somewhat startling result that there are no more points inside a square than there are on one of its sides.

In Sec. 6 we introduced the symbol $\aleph_0$ for the number of elements in any countably infinite set. At the beginning of this section we proved that the set R of all real numbers (or of all points on the real line) is uncountably infinite. We now introduce the symbol c (called the *cardinal number of the continuum*) for the number of elements in R. c is the cardinal number of R and of any set which is numerically equivalent to R. In the above three paragraphs we have demonstrated that c is the cardinal number of any open interval, of any subset of R which contains an open interval, and

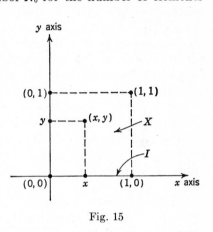

Fig. 15

of the subset X of the coordinate plane which is illustrated in Fig. 15. Our list of cardinal numbers has now grown to

$$1, 2, 3, \ldots , \aleph_0, c,$$

and they are related to each other by

$$1 < 2 < 3 < \cdots < \aleph_0 < c.$$

At this point we encounter one of the most famous unsolved problems of mathematics. Is there a cardinal number greater than $\aleph_0$ and less than c? No one knows the answer to this question. Cantor himself thought that there is no such number, or in other words, that c is the next infinite cardinal number greater than $\aleph_0$, and his guess has come to be known as *Cantor's continuum hypothesis*. The continuum hypothesis can also be expressed by the assertion that every uncountable set of real numbers has c as its cardinal number.[1]

There is another question which arises naturally at this stage, and this one we are fortunately able to answer. Are there any infinite cardinal numbers greater than c? Yes, there are; for example, the cardinal number of the class of all subsets of R. This answer depends on the following fact: if X is any non-empty set, then the cardinal number of X is less than the cardinal number of the class of all subsets of X.

We prove this statement as follows. In accordance with the definition given in the last paragraph of the previous section, we must show

[1] For further information about the continuum hypothesis, see Wilder [42, p. 125] and Gödel [12].

(1) that there exists a one-to-one mapping of X into the class of all its subsets, and (2) that there does not exist such a mapping of X onto this class. To prove (1), we have only to point to the mapping $x \rightarrow \{x\}$, which makes correspond to each element x that set $\{x\}$ which consists of the element x alone. We prove (2) indirectly. Let us assume that there does exist a one-to-one mapping f of X onto the class of all its subsets. We now deduce a contradiction from the assumed existence of such a mapping. Let A be the subset of X defined by $A = \{x : x \notin f(x)\}$. Since our mapping f is onto, there must exist an element a in X such that $f(a) = A$. Where is the element a? If a is in A, then by the definition of A we have $a \notin f(a)$, and since $f(a) = A$, $a \notin A$. This is a contradiction, so a cannot belong to A. But if a is not in A, then again by the definition of A we have $a \varepsilon f(a)$ or $a \varepsilon A$, which is another contradiction. The situation is impossible, so our assumption that such a mapping exists must be false.

This result guarantees that given any cardinal number, there always exists a greater one. If we start with a set $X_1 = \{1\}$ containing one element, then there are two subsets, the empty set $\emptyset$ and the set $\{1\}$ itself. If $X_2 = \{1,2\}$ is a set containing two elements, then there are four subsets: $\emptyset$, $\{1\}$, $\{2\}$, $\{1,2\}$. If $X_3 = \{1, 2, 3\}$ is a set containing three elements, then there are eight subsets: $\emptyset$, $\{1\}$, $\{2\}$, $\{3\}$, $\{1,2\}$, $\{1,3\}$, $\{2,3\}$, $\{1, 2, 3\}$. In general, if X_n is a set with n elements, where n is any finite cardinal number, then X_n has 2^n subsets. If we now take n to be any infinite cardinal number, the above facts suggest that we *define* 2^n to be the number of subsets of any set with n elements. If n is the first infinite cardinal number, namely, $\aleph_0$, then it can be shown that

$$2^{\aleph_0} = c.$$

The simplest proof of this fact depends on the ideas developed in the following paragraph.

Consider the closed-open unit interval $[0,1)$ and a real number x in this set. Our concern is with the meaning of the *decimal, binary,* and *ternary expansions* of x. For the sake of clarity, let us take x to be $\frac{1}{4}$. How do we arrive at the decimal expansion of $\frac{1}{4}$? First, we split $[0,1)$ into the 10 closed-open intervals

$$[0,\tfrac{1}{10}), [\tfrac{1}{10},\tfrac{2}{10}), \ldots, [\tfrac{9}{10},1),$$

and we use the 10 digits $0, 1, \ldots, 9$ to number them in order. Our number $\frac{1}{4}$ belongs to exactly one of these intervals, namely, to $[\tfrac{2}{10},\tfrac{3}{10})$. We have labeled this interval with the digit 2, so 2 is the first digit after the decimal point in the decimal expansion of $\frac{1}{4}$:

$$\tfrac{1}{4} = .2 \ldots$$

Next, we split the interval $[\frac{2}{10}, \frac{3}{10})$ into the 10 closed-open intervals

$$[\frac{2}{10}, \frac{21}{100}), [\frac{21}{100}, \frac{22}{100}), \ldots , [\frac{29}{100}, \frac{3}{10}),$$

and we use the 10 digits to number these in order. Our number $\frac{1}{4}$ belongs to $[\frac{25}{100}, \frac{26}{100})$, which is labeled with the digit 5, so 5 is the second number after the decimal point in the decimal expansion of $\frac{1}{4}$:

$$\frac{1}{4} = .25 \ldots$$

If we continue this process exactly as we started it, we can obtain the decimal expansion of $\frac{1}{4}$ to as many places as we wish. As a matter of fact, if we do continue, we get 0 at each stage from this point on:

$$\frac{1}{4} = .25000 \ldots$$

The reader should notice that there is no ambiguity in this system as we have explained it: contrary to customary usage, .24999 . . . is *not* to be regarded as another decimal expansion of $\frac{1}{4}$ which is "equivalent" to .25000 . . . . In this system, each real number x in $[0,1)$ has *one and only one* decimal expansion which cannot end in an infinite chain of 9's. There is nothing magical about the role of the number 10 in the above discussion. If at each stage we split our closed-open interval into two equal closed-open intervals, and if we use the two digits 0 and 1 to number them, we obtain the binary expansion of any real number x in $[0,1)$. The binary expansion of $\frac{1}{4}$ is easily seen to be .01000 . . . . The ternary expansion of x is found similarly: at each stage we split our closed-open interval into three equal closed-open intervals, and we use the three digits 0, 1, and 2 to number them. A moment's thought should convince the reader that the ternary expansion of $\frac{1}{4}$ is .020202 . . . . Just as (in our system) the decimal expansion of a number in $[0,1)$ cannot end in an infinite chain of 9's, so also its binary expansion cannot end in an infinite chain of 1's, and its ternary expansion cannot end in an infinite chain of 2's.

We now use this machinery to give a proof of the fact that

$$2^{\aleph_0} = c.$$

Consider the two sets $N = \{1, 2, 3, \ldots\}$ and $I = [0,1)$, the first with cardinal number $\aleph_0$ and the second with cardinal number c. If $\mathbf{N}$ denotes the class of all subsets of N, then by definition $\mathbf{N}$ has cardinal number $2^{\aleph_0}$. Our proof amounts to showing that there exists a one-to-one correspondence between $\mathbf{N}$ and I. We begin by establishing a one-to-one mapping f of $\mathbf{N}$ into I. If A is a subset of N, then $f(A)$ is that real number x in I whose decimal expansion $x = .d_1 d_2 d_3 \ldots$ is defined by the condition that d_n is 3 or 5 according as n is or is not in A. Any other two digits can be used here, as long as neither of them is 9. Next, we construct a one-to-one mapping g of I into $\mathbf{N}$.   If x is a real number in I, and if $x = .b_1b_2b_3 \ldots$ is its binary expansion (so that each b_n is either 0 or 1), then $g(x)$ is that subset A of N defined by $A = \{n : b_n = 1\}$.   We conclude the proof with an appeal to the Schroeder-Bernstein theorem, which guarantees that under these conditions $\mathbf{N}$ and I are numerically equivalent to one another.

If we follow up the hint contained in the fact that $2^{\aleph_0} = c$, and successively form 2^c, 2^{2^c}, and so on, we get a chain of cardinal numbers

$$1 < 2 < 3 < \cdots < \aleph_0 < c < 2^c < 2^{2^c} < \cdots$$

in which there are infinitely many infinite cardinal numbers.   Clearly, there is only one kind of countable infinity, symbolized by $\aleph_0$, and beyond this there is an infinite hierarchy of uncountable infinities which are all distinct from one another.

At this point we bring our discussion of these matters to a close. We have barely touched on Cantor's theory and have left entirely to one side, for instance, all questions relating to the addition and multiplication of infinite cardinal numbers and the rules of arithmetic which apply to these operations.   We have developed these ideas, not for their own sake, but for the sake of their applications in algebra and topology, and our main purpose throughout the last two sections has been to give the reader some of the necessary insight into countable and uncountable sets and the distinction between them.[1]

Problems

1. Show geometrically that the set of all points in the coordinate plane R^2 is numerically equivalent to the subset X of R^2 illustrated in Fig. 15 and defined by $X = \{(x,y) : 0 \leq x < 1 \text{ and } 0 \leq y < 1\}$, and that therefore R^2 has cardinal number c. [*Hint:* rest an open hemispherical surface ($=$ a hemispherical surface minus its boundary) tangentially on the center of X, project from various points on the line through its center and perpendicular to R^2, and use the Schroeder-Bernstein theorem.]

2. Show that the subset X of R^3 defined by

$$X = \{(x_1, x_2, x_3) : 0 \leq x_i < 1 \text{ for } i = 1, 2, 3\}$$

has cardinal number c.

[1] For the reader who wishes to learn something about the arithmetic of infinite cardinal numbers, we recommend Halmos [16, sec. 24], Kamke [24, chap. 2], Sierpinski [37, chaps. 7–10], or Fraenkel [9, chap. 2].

3. Let n be a positive integer and consider a polynomial equation of the form

$$a_nx^n + a_{n-1}x^{n-1} + \cdots + a_0 = 0,$$

with integral coefficients and $a_n \neq 0$. Such an equation has precisely n complex roots (some of which, of course, may be real). An *algebraic number* is a complex number which is a root of such an equation. The set of all algebraic numbers contains the set of all rational numbers (e.g., $\frac{2}{3}$ is the root of $3x - 2 = 0$) and many other numbers besides (the square root of 2 is a root of $x^2 - 2 = 0$, and $1 + i$ is a root of $x^2 - 2x + 2 = 0$). Complex numbers which are not algebraic are called *transcendental*. The numbers e and π are the best known transcendental numbers, though the fact that they are transcendental is quite difficult to prove (see Niven [33, chap. 9]). Prove that real transcendental numbers exist (*hint:* see Problem 6-5). Prove also that the set of all real transcendental numbers is uncountably infinite.

4. Prove that every infinite set is numerically equivalent to a proper subset of itself (*hint:* see Problem 6-6).

5. Prove that the set of all real functions defined on the closed unit interval has cardinal number 2^c. [*Hint:* there are at least as many such functions as there are *characteristic functions* (i.e., functions whose values are 0 or 1) defined on the closed unit interval.]

8. PARTIALLY ORDERED SETS AND LATTICES

There are two types of relations which often arise in mathematics: order relations and equivalence relations. We touched briefly on order relations in Problem 1-2, and in Section 5 we discussed equivalence relations in some detail. We now return to the topic of order relations and develop those parts of this subject which are necessary for our later work. The reader will find it helpful to keep in mind that a partial order relation (as we define it below) is a generalization of both set inclusion and the order relation on the real line.

Let P be a non-empty set. A *partial order relation* in P is a relation which is symbolized by $\leq$ and assumed to have the following properties:

(1) $x \leq x$ for every x (*reflexivity*);
(2) $x \leq y$ and $y \leq x \Rightarrow x = y$ (*antisymmetry*);
(3) $x \leq y$ and $y \leq z \Rightarrow x \leq z$ (*transitivity*).

We sometimes write $x \leq y$ in the equivalent form $y \geq x$. A non-empty set P in which there is defined a partial order relation is called a *partially ordered set.* It is clear that any non-empty subset of a partially ordered set is a partially ordered set in its own right.

Partially ordered sets are abundant in all branches of mathematics. Some are simple and easy to grasp, while others are complex and rather inaccessible. We give four examples which are quite different in nature but possess in common the virtues of being both important and easily described.

Example 1. Let P be the set of all positive integers, and let $m \leq n$ mean that m divides n.

Example 2. Let P be the set R of all real numbers, and let $x \leq y$ have its usual meaning (see Problem 1-2).

Example 3. Let P be the class of all subsets of some universal set U, and let $A \leq B$ mean that A is a subset of B.

Example 4. Let P be the set of all real functions defined on a non-empty set X, and let $f \leq g$ mean that $f(x) \leq g(x)$ for every x.

Two elements x and y in a partially ordered set are called *comparable* if one of them is less than or equal to the other, that is, if either $x \leq y$ or $y \leq x$. The word "partially" in the phrase "partially ordered set" is intended to emphasize that there may be pairs of elements in the set which are not comparable. In Example 1, for instance, the integers 4 and 6 are not comparable, because neither divides the other; and in Example 3, if the universal set U has more than one element, it is always possible to find two subsets of U neither of which is a subset of the other.

Some partial order relations possess a fourth property in addition to the three required by the definition:

(4)   any two elements are comparable.

A partial order relation with property (4) is called a *total* (or *linear*) *order relation,* and a partially ordered set whose relation satisfies condition (4) is called a *totally ordered set,* or a *linearly ordered set,* or, most frequently, a *chain.* Example 2 is a chain, as is the subset $\{2, 4, 8, \ldots, 2^n, \ldots\}$ of Example 1.

Let P be a partially ordered set. An element x in P is said to be *maximal* if $y \geq x \Rightarrow y = x$, that is, if no element other than x itself is greater than or equal to x. A maximal element in P is thus an element of P which is not less than or equal to any other element of P. Examples 1, 2, and 4 have no maximal elements. Example 3 has a single maximal element: the set U itself.

Let A be a non-empty subset of a partially ordered set P. An element x in P is called a *lower bound* of A if $x \leq a$ for each $a \in A$; and a lower bound of A is called a *greatest lower bound* of A if it is greater than or equal to every lower bound of A.   Similarly, an element y in P is said to be an *upper bound* of A if $a \leq y$ for every $a \, \varepsilon \, A$; and a *least upper bound* of A is an upper bound of A which is less than or equal to every upper bound of A.   In general, A may have many lower bounds and many upper bounds, but it is easy to prove (see Problem 1) that a greatest lower bound (or least upper bound) is unique if it exists.   It is therefore legitimate to speak of *the* greatest lower bound and *the* least upper bound if they exist.

We illustrate these concepts in some of the partially ordered sets mentioned above.

In Example 1, let the subset A consist of the integers 4 and 6.   An upper bound of $\{4,6\}$ is any positive integer divisible by both 4 and 6. 12, 24, 36, and so on, are all upper bounds of $\{4,6\}$.   12 is clearly its least upper bound, for it is less than or equal to (i.e., it divides) every upper bound.   The greatest lower bound of any pair of integers in this example is their greatest common divisor, and their least upper bound is their least common multiple—both of which are familiar notions from elementary arithmetic.

We now consider Example 2, the real line with its natural order relation.   The reader will doubtless recall from his study of calculus that 3 is an upper bound of the set $\{(1 + 1/n)^n : n = 1, 2, 3, \ldots\}$ and that its least upper bound is the fundamental constant $e = 2.7182 \ldots$. As we have stated before, it is a basic property of the real line that every non-empty subset of it which has a lower bound (or upper bound) has a greatest lower bound (or least upper bound).   There are several items of standard notation and terminology which must be mentioned in connection with this example.   Let A be any non-empty set of real numbers. If A has a lower bound, then its greatest lower bound is usually called its *infimum* and denoted by inf A.   Correspondingly, if A has an upper bound, then its least upper bound is called its *supremum* and written sup A.   If A happens to be finite, then inf A and sup A both exist and belong to A.   In this case, they are often called the *minimum* and *maximum* of A and are denoted by min A and max A.   If A consists of two real numbers a_1 and a_2, then min A is the smaller of a_1 and a_2, and max A is the larger.

Finally, consider Example 3, and let **A** be any non-empty class of subsets of U.   A lower bound of **A** is any subset of U which is contained in every set in **A**, and the greatest lower bound of **A** is the intersection of all its sets.   Similarly, the least upper bound of **A** is the union of all its sets.

One of our main aims in this section is to state *Zorn's lemma*, an exceedingly powerful tool of proof which is almost indispensable in many parts of modern pure mathematics.   Zorn's lemma asserts that *if P is a partially ordered set in which every chain has an upper bound, then P possesses a maximal element.* It is not possible to prove this in the usual sense of the word. However, it can be shown that Zorn's lemma is logically equivalent to the *axiom of choice*, which states: given any non-empty class of disjoint non-empty sets, a set can be formed which contains precisely one element taken from each set in the given class. The axiom of choice may strike the reader as being intuitively obvious, and in fact, either this axiom itself or some other principle equivalent to

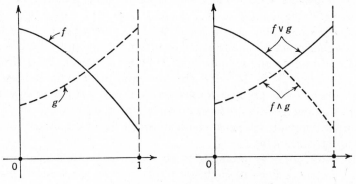

Fig. 16. The geometric meaning of $f \wedge g$ and $f \vee g$.

it is usually postulated in the logic with which we operate. We therefore assume Zorn's lemma as an axiom of logic. Any reader who is interested in these matters is urged to explore them further in the literature.[1]

A *lattice* is a partially ordered set L in which each pair of elements has a greatest lower bound and a least upper bound. If x and y are two elements in L, we denote their greatest lower bound and least upper bound by $x \wedge y$ and $x \vee y$. These notations are analogous to (and are intended to suggest) the notations for the intersection and union of two sets. We pursue this analogy even further, and call $x \wedge y$ and $x \vee y$ the *meet* and *join* of x and y. It is tempting to assume that all properties of intersections and unions in the algebra of sets carry over to lattices, but this is not a valid assumption. Some properties do carry over (see Problem 5), but others, for instance the distributive laws, are false in some lattices.

It is easy to see that all four of our examples are lattices. In Example 1, $m \wedge n$ is the greatest common divisor of m and n, and $m \vee n$ is their least common multiple; and in Example 3, $A \wedge B = A \cap B$ and $A \vee B = A \cup B$. In Example 2, if x and y are any two real numbers, then $x \wedge y$ is min $\{x,y\}$ and $x \vee y$ is max $\{x,y\}$. In Example 4, $f \wedge g$ is

[1] See, for example, Wilder [42, pp. 129–132], Halmos [16, secs. 15–16], Birkhoff [4, p. 42], Sierpinski [37, chap. 6], or Fraenkel and Bar-Hillel [10, p. 44].

the real function defined on X by $(f \wedge g)(x) = \min \{f(x),g(x)\}$, and $f \vee g$ is that defined by $(f \vee g)(x) = \max \{f(x),g(x)\}$. Figure 16 illustrates the geometric meaning of $f \wedge g$ and $f \vee g$ for two real functions f and g defined on the closed unit interval $[0,1]$.

Let L be a lattice. A *sublattice* of L is a non-empty subset L_1 of L with the property that if x and y are in L_1, then $x \wedge y$ and $x \vee y$ are also in L_1. If L is the lattice of all real functions defined on the closed unit interval, and if L_1 is the set of all continuous functions in L, then L_1 is easily seen to be a sublattice of L.

If a lattice has the additional property that every non-empty subset has a greatest lower bound and a least upper bound, then it is called a *complete lattice*. Example 3 is the only complete lattice in our list.

There are many distinct types of lattices, and the theory of these systems has a wide variety of interesting and significant applications (see Birkhoff [4]). We discuss some of these types in our Appendix on Boolean algebras.

Problems

1. Let A be a non-empty subset of a partially ordered set P. Show that A has at most one greatest lower bound and at most one least upper bound.

2. Consider the set $\{1, 2, 3, 4, 5\}$. What elements are maximal if it is ordered as Example 1? If it is ordered as Example 2?

3. Under what circumstances is Example 4 a chain?

4. Give an example of a partially ordered set which is not a lattice.

5. Let L be a lattice. If x, y, and z are elements of L, verify the following: $x \wedge x = x$, $x \vee x = x$, $x \wedge y = y \wedge x$, $x \vee y = y \vee x$,

$$x \wedge (y \wedge z) = (x \wedge y) \wedge z,$$

$x \vee (y \vee z) = (x \vee y) \vee z$, $(x \wedge y) \vee x = x$, $(x \vee y) \wedge x = x$.

6. Let A be a class of subsets of some non-empty universal set U. We say that A has the *finite intersection property* if every finite subclass of A has non-empty intersection. Use Zorn's lemma to prove that if A has the finite intersection property, then it is contained in some maximal class B with this property (to say that B is a *maximal* class with this property is to say that any class which properly contains B fails to have this property). (*Hint:* consider the family of all classes which contain A and have the finite intersection property, order this family by class inclusion, and show that any chain in the family has an upper bound in the family.)

7. Prove that if X and Y are any two non-empty sets, then there exists a one-to-one mapping of one into the other. (*Hint:* choose an

element x in X and an element y in Y, and establish the obvious one-to-one correspondence between the two single-element sets $\{x\}$ and $\{y\}$; define an *extension* to be a pair of subsets A of X and B of Y such that $\{x\} \subseteq A$ and $\{y\} \subseteq B$, together with a one-to-one correspondence between them under which x and y correspond with one another; order the set of all extensions in the natural way; and apply Zorn's lemma.)

8. Let m and n be any two cardinal numbers (finite or infinite).   The statement that m is *less than or equal to* n (written $m \leq n$) is defined to mean the following: if X and Y are sets with m and n elements, then there exists a one-to-one mapping of X into Y.   Prove that any non-empty set of cardinal numbers forms a chain when it is ordered in this way.   The fact that for any two cardinal numbers one is less than or equal to the other is usually called the *comparability theorem for cardinal numbers*.

9. Let X and Y be non-empty sets, and show that the cardinal number of X is less than or equal to the cardinal number of $Y \Leftrightarrow$ there exists a mapping of Y onto X.

10. Let $\{X_i\}$ be any infinite class of countable sets indexed by the elements i of an index set I, and show that the cardinal number of $\cup_i X_i$ is less than or equal to the cardinal number of I.   (*Hint:* if I is only countably infinite, this follows from Problem 6-2, and if I is uncountable, Zorn's lemma can be applied to represent it as the union of a disjoint class of countably infinite subsets.)

CHAPTER TWO

Metric Spaces

Classical analysis can be described as that part of mathematics which begins with calculus and, in essentially the same spirit, develops similar subject matter much further in many directions. It is a great nation in the world of mathematics, with many provinces, a few of which are ordinary and partial differential equations, infinite series (especially power series and Fourier series), and analytic functions of a complex variable. Each of these has experienced enormous growth over a long history, and each is rich enough in content to merit a lifetime of study.

In the course of its development, classical analysis became so complex and varied that even an expert could find his way around in it only with difficulty. Under these circumstances, some mathematicians became interested in trying to uncover the fundamental principles on which all analysis rests. This movement had associated with it many of the great names in mathematics of the last century: Riemann, Weierstrass, Cantor, Lebesgue, Hilbert, Riesz, and others. It played a large part in the rise to prominence of topology, modern algebra, and the theory of measure and integration; and when these new ideas began to percolate back through classical analysis, the brew which resulted was modern analysis.

As modern analysis developed in the hands of its creators, many a major theorem was given a simpler proof in a more general setting, in an effort to lay bare its inner meaning. Much thought was devoted to analyzing the texture of the real and complex number systems, which are the context of analysis. It was hoped—and these hopes were well founded —that analysis could be clarified and simplified, and that stripping away superfluous underbrush would give new emphasis to what really mattered from the point of view of the underlying theory.[1]

Analysis is primarily concerned with limit processes and continuity, so it is not surprising that mathematicians thinking along these lines soon found themselves studying (and generalizing) two elementary concepts: that of a convergent sequence of real or complex numbers, and that of a continuous function of a real or complex variable.

We remind the reader of the definitions.   First, a sequence

$$\{x_n\} = \{x_1, x_2, \ldots, x_n, \ldots\}$$

of real numbers is said to be *convergent* if there exists a real number x (called the *limit* of the sequence) such that, given $\epsilon > 0$, a positive integer n_0 can be found with the property that

$$n \geq n_0 \Rightarrow |x_n - x| < \epsilon.$$

This condition means that x_n must be "close" to x for all "sufficiently large" n, and it is usually symbolized by

$$x_n \to x \qquad \text{or} \qquad \lim x_n = x$$

and expressed by saying that x_n *approaches* x or x_n *converges to* x.   Second, a real function f defined on a non-empty subset X of the real line is said to be *continuous at x_0 in X* if for each $\epsilon > 0$ there exists $\delta > 0$ such that

$$x \text{ in } X \text{ and } |x - x_0| < \delta \Rightarrow |f(x) - f(x_0)| < \epsilon,$$

and f is said to be *continuous* if it is continuous at each point of X. When X is an interval, this definition gives precise expression to the intuitive requirement that f have a graph without breaks or gaps.   The corresponding definitions for sequences of complex numbers and complex functions of a complex variable are word for word the same.

Our purpose in giving these definitions in detail here is a simple one. We wish to point out explicitly that each is dependent for its meaning on the concept of the absolute value of the difference between two real or complex numbers.   We wish to observe also that this absolute value is the *distance between the numbers* when they are regarded as points on the real line or in the complex plane.

In many branches of mathematics—in geometry as well as analysis— it has been found extremely convenient to have available a notion of distance which is applicable to the elements of abstract sets.   A *metric space* (as we define it below) is nothing more than a non-empty set

---

[1] We illustrate these points in Appendix 1, where one of the basic existence theorems in the theory of differential equations is given a brief and uncluttered proof which depends only on the ideas of this chapter.

equipped with a concept of distance which is suitable for the treatment of convergent sequences in the set and continuous functions defined on the set.   Our purpose in this chapter is to develop in a systematic manner the main elementary facts about metric spaces.   These facts are important for their own sake, and also for the sake of the motivation they provide for our later work on topological spaces.

9. THE DEFINITION AND SOME EXAMPLES

Let X be a non-empty set.   A *metric* on X is a real function d of ordered pairs of elements of X which satisfies the following three conditions:

(1) $d(x,y) \geq 0$, and $d(x,y) = 0 \Leftrightarrow x = y$;

(2) $d(x,y) = d(y,x)$ (*symmetry*);

(3) $d(x,y) \leq d(x,z) + d(z,y)$ (the *triangle inequality*).

The function d assigns to each pair (x,y) of elements of X a non-negative real number $d(x,y)$, which by symmetry does not depend on the order of the elements; $d(x,y)$ is called the *distance* between x and y.   A *metric space* consists of two objects: a non-empty set X and a metric d on X. The elements of X are called the *points* of the metric space (X,d).   Whenever it can be done without causing confusion, we denote the metric space (X,d) by the symbol X which is used for the underlying set of points.   One should always keep in mind, however, that a metric space is not merely a non-empty set: it is a non-empty set together with a metric.   It often happens that several different metrics can be defined on a single given non-empty set, and in this case distinct metrics make the set into distinct metric spaces.

There are many different kinds of metric spaces, some of which play very significant roles in geometry and analysis.   Our first example is rather trivial, but it is often useful in showing that certain statements we might wish to make are not true.   It also shows that every non-empty set can be regarded as a metric space.

Example 1.   Let X be an arbitrary non-empty set, and define d by

$$d(x,y) = \begin{cases} 0 & \text{if } x = y, \\ 1 & \text{if } x \neq y. \end{cases}$$

The reader can easily see for himself that this definition yields a metric on X.

Our next two examples are the fundamental number systems of mathematics.

Example 2.   Consider the real line R and the real function $|x|$ defined on R.   Three elementary properties of this absolute value function are important for our purposes:

(i) $$|x| \geq 0, \text{ and } |x| = 0 \Leftrightarrow x = 0;$$
(ii) $$|-x| = |x|;$$
(iii) $$|x + y| \leq |x| + |y|.$$

We now define a metric on R by

$$d(x,y) = |x - y|.$$

This is called the *usual metric* on R, and the real line, as a metric space, is always understood to have this as its metric.   The fact that d actually is a metric follows from the three properties stated above.   This is a piece of reasoning which occurs frequently in our work, so we give the details.   By (i), $d(x,y) = |x - y|$ is a non-negative real number which equals $0 \Leftrightarrow x - y = 0 \Leftrightarrow x = y$.   By (ii),

$$d(x,y) = |x - y| = |-(y - x)| = |y - x| = d(y,x).$$

And by (iii),

$$d(x,y) = |x - y| = |(x - z) + (z - y)| \leq |x - z| + |z - y|$$
$$= d(x,z) + d(z,y).$$

Example 3.   Consider the complex plane C.   We mentioned C briefly in Sec. 4, and we described the sense in which it can be identified as a set with the coordinate plane R^2.   We now give a somewhat fuller discussion.   If z is a complex number, and if $z = a + ib$ where a and b are real numbers, then a and b are called the *real part* and the *imaginary part* of z and are denoted by $R(z)$ and $I(z)$.   Two complex numbers are said to be *equal* if their real and imaginary parts are equal:

$$a + ib = c + id \Leftrightarrow a = c \text{ and } b = d.$$

We add (or subtract) two complex numbers by adding (or subtracting) their real and imaginary parts, and we multiply them by multiplying them out as in elementary algebra and replacing i^2 by -1 wherever it appears:

$$(a + ib) \pm (c + id) = (a \pm c) + i(b \pm d),$$
and
$$(a + ib)(c + id) = ac + iad + ibc + i^2bd$$
$$= (ac - bd) + i(ad + bc).$$

Division is carried out in accordance with

$$\frac{a + ib}{c + id} = \frac{(a + ib)(c - id)}{(c + id)(c - id)} = \frac{(ac + bd) + i(bc - ad)}{c^2 + d^2}$$

$$= \frac{ac + bd}{c^2 + d^2} + i\,\frac{bc - ad}{c^2 + d^2},$$

where $c^2 + d^2$ is required to be non-zero. If $z = a + ib$ is a complex number, then its *negative* $-z$ and its *conjugate* $\bar{z}$ are defined by

$$-z = (-a) + i(-b)$$

and $\bar{z} = a + i(-b)$, which are usually written more informally as $-z = -a - ib$ and $\bar{z} = a - ib$. It is easy to see that

$$R(z) = \frac{z + \bar{z}}{2} \quad \text{and} \quad I(z) = \frac{z - \bar{z}}{2i}.$$

The real line R is usually regarded as part of the complex plane:

$$R = \{z : I(z) = 0\} = \{z : \bar{z} = z\}.$$

Simple calculations show directly that

$$\overline{z_1 + z_2} = \bar{z}_1 + \bar{z}_2, \qquad \overline{z_1 z_2} = \bar{z}_1 \cdot \bar{z}_2, \qquad \text{and} \qquad \bar{\bar{z}} = z.$$

The *origin*, or *zero*, is the complex number $0 = 0 + i0$. The ordinary distance from $z = a + ib$ to the origin is defined by

$$|z| = (a^2 + b^2)^{1/2}.$$

$|z|$ is called the *absolute value* of z, and it is easy to see that

$$|\bar{z}| = |z| \quad \text{and} \quad |z|^2 = z\bar{z}.$$

The *usual metric* on C is defined by

$$d(z_1, z_2) = |z_1 - z_2|.$$

Exactly as in Example 2, the fact that this is a metric is a consequence of the following properties of the real function $|z|$:

(i) $|z| \geq 0$, and $|z| = 0 \Leftrightarrow z = 0$;

(ii) $|-z| = |z|$;

(iii) $|z_1 + z_2| \leq |z_1| + |z_2|$.

Properties (i) and (ii) are obvious. Since $-z = (-1)z$, property (ii) is also a special case of the fact that

$$|z_1 z_2| = |z_1|\,|z_2|,$$

which we prove by means of

$$|z_1 z_2|^2 = z_1 z_2 \overline{z_1 z_2} = z_1 \overline{z_1} z_2 \overline{z_2} = |z_1|^2 |z_2|^2 = (|z_1| \, |z_2|)^2.$$

If we use the fact that $|R(z)| \le |z|$ for any z, property (iii) follows directly from

$$
\begin{aligned}
|z_1 + z_2|^2 = (z_1 + z_2)(\overline{z_1 + z_2}) &= (z_1 + z_2)(\overline{z_1} + \overline{z_2}) \\
&= z_1 \overline{z_1} + z_2 \overline{z_2} + z_1 \overline{z_2} + \overline{z_1} z_2 \\
&= |z_1|^2 + |z_2|^2 + (z_1 \overline{z_2} + \overline{z_1 \overline{z_2}}) \\
&= |z_1|^2 + |z_2|^2 + 2R(z_1 \overline{z_2}) \\
&\le |z_1|^2 + |z_2|^2 + 2|z_1 \overline{z_2}| \\
&= |z_1|^2 + |z_2|^2 + 2|z_1| \, |\overline{z_2}| \\
&= |z_1|^2 + |z_2|^2 + 2|z_1| \, |z_2| \\
&= (|z_1| + |z_2|)^2.
\end{aligned}
$$

Whenever the complex plane C is mentioned as a metric space, its metric is always assumed to be the usual metric defined above.

The remaining examples to be given in this section fit a common pattern, which we have tried to exhibit in our discussion of Examples 2 and 3. We now point out several major features of this pattern, so that the reader can see clearly how it applies in the slightly more complicated examples that follow.

I. The elements of each space can be added and subtracted in a natural way, and every element has a negative. Each space contains a special element, denoted by 0 and called the *origin*, or *zero element*.

II. In each space there is defined a notion of the distance from an arbitrary element to the origin, that is, a notion of the "size" of an arbitrary element. The size of an element x is a real number denoted below by $\|x\|$ and called its *norm*. Our use of the double vertical bars is intended to emphasize that the norm is a generalization of the absolute value functions in Examples 2 and 3, in the sense that it satisfies the following three conditions: (i) $\|x\| \ge 0$, and $\|x\| = 0 \Leftrightarrow x = 0$; (ii) $\|-x\| = \|x\|$; (iii) $\|x + y\| \le \|x\| + \|y\|$.

III. Finally, each metric arises as the norm of the difference between two elements: $d(x,y) = \|x - y\|$. As in Example 2, the fact that this is a metric follows from the properties of the norm listed in II. This metric is called the metric *induced by* the norm.

The knowledgeable reader will see at once that we are describing here (though incompletely and imprecisely) the concept of a *normed linear space*. Most of the metric spaces of major importance in analysis are of this type.

Example 4.   Let f be a real function defined on the closed unit interval
[0,1]. We say that f is a *bounded function* if there is a real number K
such that $|f(x)| \leq K$ for every $x \in [0,1]$. This concept is familiar to
the reader from elementary analysis, as is that of the continuity of f as
defined in the introduction to this chapter. The underlying set of points
in this example is the set of all bounded continuous real functions defined
on the closed unit interval. Actually, the boundedness of such a function
is a consequence of its other properties, but at this stage we assume it
explicitly. If f and g are two such functions, we add and subtract
them, and form negatives, pointwise:

$$(f + g)(x) = f(x) + g(x);$$
$$(f - g)(x) = f(x) - g(x);$$
$$(-f)(x) = -f(x).$$

The origin (denoted by 0) is the constant function which is identically
zero:

$$0(x) = 0$$

for all $x \in [0,1]$. We define the norm of a function f by

$$\|f\| = \int_0^1 |f(x)| \, dx,$$

and the induced metric by

$$d(f,g) = \|f - g\| = \int_0^1 |f(x) - g(x)| \, dx.$$

The integral involved in this definition is the Riemann integral of elementary calculus. Properties (i) and (ii) of the norm are easy to prove,
and (iii) follows from

$$\|f + g\| = \int_0^1 |f(x) + g(x)| \, dx \leq \int_0^1 (|f(x)| + |g(x)|) \, dx$$
$$= \int_0^1 |f(x)| \, dx + \int_0^1 |g(x)| \, dx$$
$$= \|f\| + \|g\|.$$

Example 5.   The set of points in the preceding example—that is, the
set of all bounded continuous real functions defined on the closed unit
interval—has another metric which is far more important for our purposes. It is defined by means of

$$\|f\| = \sup \, \{|f(x)| : x \in [0,1]\},$$

which we usually write more briefly as

$$\|f\| = \sup |f(x)|,$$

and $$d(f,g) = \|f - g\| = \sup |f(x) - g(x)|.$$

Properties (i) and (ii) of the norm are obvious, and in Problem 5 we ask the reader to prove (iii) in a slightly more general form. This example is typical of a large class of metric spaces which will play a major role in all our work throughout the rest of this book. We denote this space by $\mathcal{C}[0,1]$.

So much for the present for specific examples. We now turn to several fundamental principles relating to metric spaces in general.

Let X be a metric space with metric d. Let Y be an arbitrary non-empty subset of X. If the function d is considered to be defined only for points in Y, then (Y,d) is evidently itself a metric space. Y, with d restricted in this way, is called a *subspace* of X. This technique of forming subspaces of a given metric space enables us to obtain an infinity of further examples from the handful described above. For instance, the closed unit interval $[0,1]$ is a subspace of the real line, as is the set consisting of all the rational points; and the unit circle, the closed unit disc, and the open unit disc are subspaces of the complex plane. Also, the real line itself is a subspace of the complex plane.

It is desirable at this stage to introduce the *extended real number system*, by which we mean the ordinary real number system R with the symbols

$$-\infty \quad \text{and} \quad +\infty$$

adjoined. An *extended real number* is thus a real number or one of these symbols. We say (by definition) that

$$-\infty < +\infty;$$

also, if x is any real number, then

$$-\infty < x < +\infty.$$

The symbols $-\infty$ and $+\infty$ add nothing to our understanding of the real numbers. They are used mainly as a notational convenience, as we see below.

Let A be a non-empty set of real numbers which has an upper bound. In Sec. 8 we defined what is meant by the least upper bound (or supremum) of A: sup A is the smallest upper bound of A, that is, it is the smallest real number y such that $a \leq y$ for every a in A. With the stated assumptions about A, sup A always exists and is a real number. If A is a non-empty set of real numbers which has no upper bound, and therefore no least upper bound in R, we express this by writing

$$\text{sup } A = +\infty;$$

and if A is the empty subset of R, we put

$$\text{sup } A = -\infty,$$

The greatest lower bound (or infimum) of A is defined similarly: if A is non-empty and has a lower bound, inf A is the largest real number x such that $x \leq a$ for every a in A; if A is non-empty and has no lower bound, we put

$$\text{inf } A = -\infty;$$

and if A is empty, we put

$$\text{inf } A = +\infty.$$

These remarks illustrate one advantage of the extended real number system: it enables us to speak of sup A and inf A for subsets A of the real line without any restrictions whatever on the nature of A.

Another advantage of having available the symbols $-\infty$ and $+\infty$ is that they make convenient a reasonable extension of our concept of an interval on the real line. The reader should refer to the definitions given in Sec. 1 of the various kinds of intervals, for these are the definitions whose scope we are now widening. Let a and b be any two real numbers such that $a \leq b$; then the *closed interval* from a to b is the subset of the real line R defined by

$$[a,b] = \{x : a \leq x \leq b\}.$$

This extends our previous notion in that a closed interval may now consist of a single point (if $a = b$). If b is a real number and a is an extended real number such that $a < b$, then the *open-closed interval* from a to b is

$$(a,b] = \{x : a < x \leq b\}.$$

This allows open-closed intervals of the form $(-\infty,b]$. If a is a real number and b is an extended real number such that $a < b$, then the *closed-open interval* from a to b is

$$[a,b) = \{x : a \leq x < b\}.$$

This permits $[a, +\infty)$ to be considered a closed-open interval. If a and b are extended real numbers such that $a < b$, then the *open interval* from a to b is

$$(a,b) = \{x : a < x < b\}.$$

This adds to the previously defined open intervals those of the form $(-\infty,b)$ where b is real, $(a, +\infty)$ where a is real, and $(-\infty, +\infty)$. Throughout the rest of this book, the term *interval* will always signify one of the four types defined in this paragraph. The extended real numbers a and b are called the *end-points* of these intervals. We have used the symbols $-\infty$ and $+\infty$ with considerable freedom, and it therefore seems desirable to emphasize that an interval in our present sense is always a non-empty subset of the real number system: it never actually contains either of these symbols.

The very definition of a metric space presents us with the concept of the distance from one point to another. We now define the distance from a point to a set and the diameter of a set.

Let X be a metric space with metric d, and let A be a subset of X. If x is a point of X, then the *distance from x to A* is defined by

$$d(x,A) = \inf \{d(x,a) : a \in A\};$$

that is, it is the greatest lower bound of the distances from x to the points of A. The *diameter* of the set A is defined by

$$d(A) = \sup \{d(a_1,a_2) : a_1 \text{ and } a_2 \in A\}.$$

The diameter of A is thus the least upper bound of the distances between pairs of its points. A is said to have *finite diameter* or *infinite diameter* according as $d(A)$ is a real number or $\pm \infty$. We observe that the empty set has infinite diameter, since $d(\emptyset) = -\infty$. A *bounded set* is one whose diameter is finite. A mapping of a non-empty set into a metric space is called a *bounded mapping* if its range is a bounded set. Several of the simpler facts about these concepts are brought out in the following problems.

Problems

1. Let X be a metric space with metric d. Show that d_1, defined by $d_1(x,y) = d(x,y)/[1 + d(x,y)]$, is also a metric on X. Observe that X itself is a bounded set in the metric space (X,d_1).

2. Let X be a non-empty set, and let d be a real function of ordered pairs of elements of X which satisfies the following two conditions: $d(x,y) = 0 \Leftrightarrow x = y$, and $d(x,y) \leq d(x,z) + d(y,z)$. Show that d is a metric on X.

3. Let X be a non-empty set, and let d be a real function of ordered pairs of elements of X which satisfies the following three conditions: $d(x,y) \geq 0$, and $x = y \Rightarrow d(x,y) = 0$; $d(x,y) = d(y,x)$; and $d(x,y) \leq d(x,z) + d(z,y)$. A function d with these properties is called a *pseudo-metric* on X. A metric is obviously a pseudo-metric. Give an example of a pseudo-metric which is not a metric. Let d be a pseudo-metric on X, define a relation $\sim$ in X by means of

$$x \sim y \Leftrightarrow d(x,y) = 0,$$

and show that this is an equivalence relation whose corresponding class of equivalence sets can be made into a metric space in a natural way.

4. Let $X_1, X_2, \ldots, X_n$ be a finite class of metric spaces with metrics $d_1, d_2, \ldots, d_n$. Show that each of the functions d and $\bar{d}$ defined

as follows is a metric on the product $X_1 \times X_2 \times \cdots \times X_n$: $d(\{x_i\},\{y_i\}) = \max d_i(x_i,y_i); \bar{d}(\{x_i\},\{y_i\}) = \Sigma_{i=1}^n d_i(x_i,y_i)$.

5. Let X be a non-empty set and f a real function defined on X. Show that f is bounded in the sense of the definition given in the last paragraph of the text $\Leftrightarrow$ there exists a real number K such that $|f(x)| \leq K$ for every $x \in X \Leftrightarrow \sup |f(x)| < +\infty$. Consider the set of all bounded real functions defined on X, and define the norm of a function f in this set by

$$\|f\| = \sup |f(x)|.$$

It is obvious that $\|f\|$ is a non-negative real number such that $\|f\| = 0 \Leftrightarrow f = 0$, and that $\|-f\| = \|f\|$. Prove in detail that $\|f + g\| \leq \|f\| + \|g\|$.

6. Let I be a subset of the real line. Show that I is an interval $\Leftrightarrow$ it is non-empty and contains each point between any two of its points (in the sense that if x and z are in I and $x \leq y \leq z$, then y is in I). If $\{I_i\}$ is a non-empty class of intervals on the real line such that $\cap_i I_i$ is non-empty, show that $\cup_i I_i$ is an interval.

7. Let X be a metric space with metric d. If x is a point of X and A a subset of X, show the following: if A is non-empty, $d(x,A)$ is a non-negative real number; and $d(x,A) = +\infty \Leftrightarrow A$ is empty.

8. Let X be a metric space with metric d and A a subset of X. Show the following: if A is non-empty, $d(A)$ is a non-negative extended real number; $d(A) = -\infty \Leftrightarrow A$ is empty; and if A is bounded, it is non-empty.

10. OPEN SETS

Let X be a metric space with metric d. If x_0 is a point of X and r is a positive real number, the *open sphere* $S_r(x_0)$ with *center* x_0 and *radius* r is the subset of X defined by

$$S_r(x_0) = \{x : d(x,x_0) < r\}.$$

An open sphere is always non-empty, for it contains its center. In Example 9-1, an open sphere with radius 1 contains only its center. $S_r(x_0)$ is often called the open sphere with radius r *centered on* x_0; intuitively, it consists of all points in X which are "close" to x_0, with the degree of closeness given by r.

A few concrete examples are in order. It should be easy to visualize the open sphere $S_r(x_0)$ on the real line: it is the bounded open interval $(x_0 - r, x_0 + r)$ with mid-point x_0 and total length $2r$. Conversely, it is clear that any bounded open interval on the real line is an open sphere, so the open spheres on the real line are precisely the bounded open intervals.   The open sphere $S_r(z_0)$ in the complex plane (see Fig. 17) is the inside of the circle with center z_0 and radius r.   Figure 18 illustrates an open sphere in the space $\mathcal{C}[0,1]$: $S_r(f_0)$ consists of all functions f in $\mathcal{C}[0,1]$ whose graphs lie within the shaded band of vertical width $2r$ centered on the graph of f_0.

A subset G of the metric space X is called an *open set* if, given any point x in G, there exists a positive real number r such that $S_r(x) \subseteq G$,

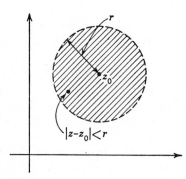

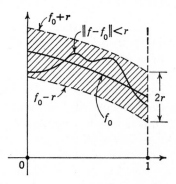

Fig. 17.   An open sphere in the complex plane.

Fig. 18.   An open sphere in $\mathcal{C}$ [0,1].

that is, if each point of G is the center of some open sphere contained in G.   Loosely speaking, a set is open if each of its points is "inside" the set, in the sense made precise by the definition.   On the real line, a set consisting of a single point is not open, for each bounded open interval centered on the point contains points not in the set.   Similarly, the subset [0,1) of the real line is not open, because the point 0 in [0,1) has the property that each bounded open interval centered on it (no matter how small it may be) contains points not in [0,1), e.g., negative points. If we omit the offending point 0, the resulting bounded open interval (0,1) is an open set (this is very easy to prove and is a special case of Theorem B below).   Further, it is quite clear that *any* open interval— bounded or not—is an open set, and also that the open intervals are the only intervals which are open sets.

Theorem A.   *In any metric space X, the empty set $\emptyset$ and the full space X are open sets.*

PROOF.   To show that $\emptyset$ is open, we must show that each point in $\emptyset$ is the center of an open sphere contained in $\emptyset$; but since there are no points in $\emptyset$, this requirement is automatically satisfied. X is clearly open, since every open sphere centered on each of its points is contained in X.

We have seen that $[0,1)$ is not open as a subset of the real line. However, if we consider $[0,1)$ as a metric space X in its own right, as a subspace of the real line, then $[0,1)$ is open as a subset of X, since from this point of view it is the full space. This apparent paradox disappears when we realize that points outside of a given metric space have no relevance to any discussion taking place within the context of that space. A set is open or not open only with respect to a specific metric space containing it, never on its own.

Our next theorem justifies the adjective in the expression "open sphere."

Theorem B. *In any metric space X, each open sphere is an open set.*

PROOF. Let $S_r(x_0)$ be an open sphere in X, and let x be a point in $S_r(x_0)$. We must produce an open sphere centered on x and contained in $S_r(x_0)$. Since $d(x,x_0) < r$, $r_1 = r - d(x,x_0)$ is a positive real number. We show that $S_{r_1}(x) \subseteq S_r(x_0)$. If y is a point in $S_{r_1}(x)$, so that $d(y,x) < r_1$, then $d(y,x_0) \le d(y,x) + d(x,x_0) < r_1 + d(x,x_0) = [r - d(x,x_0)] + d(x,x_0) = r$ shows that y is in $S_r(x_0)$.

The following characterization of open sets in terms of open spheres is a useful tool.

Theorem C. *Let X be a metric space. A subset G of X is open $\Leftrightarrow$ it is a union of open spheres.*

PROOF. We assume first that G is open, and we show that it is a union of open spheres. If G is empty, it is the union of the empty class of open spheres. If G is non-empty, then since it is open, each of its points is the center of an open sphere contained in it, and it is the union of all the open spheres contained in it.

We now assume that G is the union of a class S of open spheres. We must show that G is open. If S is empty, then G is also empty, and by Theorem A, G is open. Suppose that S is non-empty. G is also non-empty. Let x be a point in G. Since G is the union of the open spheres in S, x belongs to an open sphere $S_r(x_0)$ in S. By Theorem B, x is the center of an open sphere $S_{r_1}(x) \subseteq S_r(x_0)$. Since $S_r(x_0) \subseteq G$, $S_{r_1}(x) \subseteq G$ and we have an open sphere centered on x and contained in G. G is therefore open.

The fundamental properties of the open sets in a metric space are those stated in

Theorem D. *Let X be a metric space. Then (1) any union of open sets in X is open; and (2) any finite intersection of open sets in X is open.*

PROOF. To prove (1), let $\{G_i\}$ be an arbitrary class of open sets in X. We must show that $G = \cup_i G_i$ is open. If $\{G_i\}$ is empty, then G is empty, and by Theorem A, G is open.   Suppose that $\{G_i\}$ is non-empty. By Theorem C, each G_i (being an open set) is a union of open spheres; G is the union of all the open spheres which arise in this way; and by another application of Theorem C, G is open.

To prove (2), let $\{G_i\}$ be a finite class of open sets in X.   We must show that $G = \cap_i G_i$ is open.   If $\{G_i\}$ is empty, then $G = X$; and by Theorem A, G is open.   Suppose that $\{G_i\}$ is non-empty and that $\{G_i\} = \{G_1, G_2, \ldots, G_n\}$ for some positive integer n.   If G happens to be empty, then it is open by Theorem A, so we may assume that G is non-empty.   Let x be a point in G.   Since x is in each G_i, and each G_i is open, for each i there is a positive real number r_i such that $S_{r_i}(x) \subseteq G_i$. Let r be the smallest number in the set $\{r_1, r_2, \ldots, r_n\}$.   This number r is a positive real number such that $S_r(x) \subseteq S_{r_i}(x)$ for each i, so $S_r(x) \subseteq G_i$ for each i, and therefore $S_r(x) \subseteq G$.   Since $S_r(x)$ is an open sphere centered on x and contained in G, G is open.

The above theorem says that the class of all open sets in a metric space is closed under the formation of arbitrary unions and finite intersections.   The reader should clearly understand that Theorem A is an immediate consequence of this statement, since the empty set is the union of the empty class of open sets and the full space is its intersection. The limitation to *finite* intersections in this theorem is essential.   To see this, it suffices to consider the following sequence of open intervals on the real line:

$$(-1,1), \ (-\tfrac{1}{2},\tfrac{1}{2}), \ (-\tfrac{1}{3},\tfrac{1}{3}), \ \ldots$$

The intersection of these open sets is the set $\{0\}$ consisting of the single point 0, and this set is not open.

In an arbitrary metric space, the structure of the open sets can be very complicated indeed.   Theorem C contains the best information available in the general case: each open set is a union of open spheres. In the case of the real line, however, a description can be given of the open sets which is fairly explicit and reasonably satisfying to the intuition.

Theorem E.   *Every non-empty open set on the real line is the union of a countable disjoint class of open intervals.*

PROOF.   Let G be a non-empty open subset of the real line.   Let x be a point of G.   Since G is open, x is the center of a bounded open interval contained in G.   Define I_x to be the union of all the open intervals which contain x and are contained in G.   The following three facts are easily proved: I_x is an open interval (by Theorem D and Problem 9-6) which contains x and is contained in G; I_x contains every open interval which contains x and is contained in G; and if y is another point in I_x, then $I_x = I_y$. We next observe that if x and y are any two distinct points of G, then I_x and I_y are either disjoint or identical; for if they have a common point z, then $I_x = I_z$ and $I_y = I_z$, so $I_x = I_y$. Consider the class I of all distinct sets of the form I_x for points x in G. This is a disjoint class of open intervals, and G is obviously its union. It remains to be proved that I is countable. Let G_r be the set of rational points in G. G_r is clearly non-empty. We define a mapping f of G_r onto I as follows: for each r in G_r, let $f(r)$ be that unique interval in I which contains r. G_r is countable by Problem 6-7, and the fact that I is countable follows from Problem 6-8.

A firm grasp of the ideas involved in the theory of metric spaces depends on one's capacity to "see" these spaces with the mind's eye. The complex plane is perhaps the best metric space to use as a model from which to absorb this necessary intuitive understanding. When we consider an unspecified set A of complex numbers, we usually imagine it as a region bounded by a curve, as in Fig. 19. We think of the point z_1, which is completely surrounded by points of A, as being "inside" the set A, or in its "interior," while z_2 is on the "boundary" of A. More precisely, z_1 is the center of some open sphere contained in A, and each open sphere centered on z_2 in-

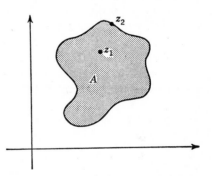

Fig. 19. A set A of complex numbers with interior point z_1 and boundary point z_2.

tersects both A and its complement A'. We formulate these ideas for a general metric space in the next paragraph and at the end of the next section.

Let X be an arbitrary metric space, and let A be a subset of X. A point in A is called an *interior point* of A if it is the center of some open sphere contained in A; and the *interior* of A, denoted by $\text{Int}(A)$, is the set of all its interior points. Symbolically,

$$\text{Int}(A) = \{x : x \in A \text{ and } S_r(x) \subseteq A \text{ for some } r\}.$$

The basic properties of interiors are the following:

(1) $\text{Int}(A)$ is an open subset of A which contains every open subset of A (this is often expressed by saying that the interior of A is the largest open subset of A);

(2) A is open $\Leftrightarrow A = \text{Int}(A)$;

(3) $\text{Int}(A)$ equals the union of all open subsets of A.

The proofs of these facts are quite easy, and we ask the reader to fill in the details as an exercise (see Problem 8).

Problems

1. Let X be a metric space, and show that any two distinct points of X can be separated by open spheres in the following sense: if x and y are distinct points in X, then there exists a disjoint pair of open spheres each of which is centered on one of the points.

2. Let X be a metric space. If $\{x\}$ is a subset of X consisting of a single point, show that its complement $\{x\}'$ is open. More generally, show that A' is open if A is any finite subset of X.

3. Let X be a metric space and $S_r(x)$ the open sphere in X with center x and radius r. Let A be a subset of X with diameter less than r which intersects $S_r(x)$. Prove that $A \subseteq S_{2r}(x)$.

4. Let X be a metric space. Show that every subset of X is open $\Leftrightarrow$ each subset of X which consists of a single point is open.

5. Let X be a metric space with metric d, and let d_1 be the metric defined in Problem 9-1. Show that the two metric spaces (X,d) and (X,d_1) have precisely the same open sets. (*Hint:* show that they have the same open spheres with one exception. What is this exception?)

6. If $X = X_1 \times X_2 \times \cdots \times X_n$ is the product in Problem 9-4, and if d and $\bar{d}$ are the metrics on X defined in that problem, show that the two metric spaces (X,d) and $(X,\bar{d})$ have precisely the same open sets. Observe that in this case the spaces do not have the same open spheres.

7. Let Y be a subspace of a metric space X, and let A be a subset of the metric space Y. Show that A is open as a subset of $Y \Leftrightarrow$ it is the intersection with Y of a set which is open in X.

8. Prove the statements made in the text about interiors.

9. Describe the interior of each of the following subsets of the real line: the set of all integers; the set of all rationals; the set of all irrationals; $(0,1)$; $[0,1]$; $(0,1) \cup \{1,2\}$. Do the same for each of the following subsets of the complex plane: $\{z:|z| < 1\}$; $\{z:|z| \leq 1\}$; $\{z:I(z) = 0\}$; $\{z:R(z)$ is rational$\}$.

10. Let A and B be two subsets of a metric space X, and prove the following:

    (*a*) $\text{Int}(A) \cup \text{Int}(B) \subseteq \text{Int}(A \cup B)$;

    (*b*) $\text{Int}(A) \cap \text{Int}(B) = \text{Int}(A \cap B)$.

    Give an example of two subsets A and B of the real line such that $\text{Int}(A) \cup \text{Int}(B) \neq \text{Int}(A \cup B)$.

11. CLOSED SETS

Let X be a metric space with metric d. If A is a subset of X, a point x in X is called a *limit point* of A if each open sphere centered on x contains at least one point of A different from x. The essential idea here is that the points of A different from x get "arbitrarily close" to x, or "pile up" at x.

The subset $\{1, \frac{1}{2}, \frac{1}{3}, \ldots\}$ of the real line has 0 as a limit point; in fact, 0 is its only limit point. The closed-open interval $[0,1)$ has 0 as a limit point which is in the set and 1 as a limit point which is not in the set; further, every real number x such that $0 < x < 1$ is also a limit point of this set. The set of all integral points on the real line has no limit points at all, whereas every real number is a limit point of the set of all rationals. In Example 9-1, every open sphere of radius less than 1 consists only of its center, so no subset of this space has any limit points.

A subset F of the metric space X is called a *closed set* if it contains each of its limit points. In rough terms, a set is closed if its points do not get arbitrarily close to any point outside of it. Among the subsets of the real line mentioned in the preceding paragraph, only the set of integral points is closed. In Example 9-1, every subset is closed.

Theorem A. *In any metric space X, the empty set $\emptyset$ and the full space X are closed sets.*

PROOF. The empty set has no limit points, so it contains them all and is therefore closed. Since the full space X contains all points, it automatically contains its own limit points and thus is closed.

The following theorem characterizes closed sets in terms of open sets. We already know a good deal about open sets, so this characterization provides us with a useful tool for establishing properties of closed sets.

Theorem B. *Let X be a metric space. A subset F of X is closed $\Leftrightarrow$ its complement F' is open.*

PROOF. Assume first that F is closed. We show that F' is open. If F' is empty, it is open by Theorem 10-A, so we may suppose that F' is non-empty. Let x be a point in F'. Since F is closed and x is not in F, x is not a limit point of F. Since x is not in F and is not a limit point of F, there exists an open sphere $S_r(x)$ which is disjoint from F. $S_r(x)$ is an open sphere centered on x and contained in F', and since x was taken to be any point of F', F' is open.

We now assume that F' is open and show that F is closed. The only way F can fail to be closed is to have a limit point in F'. This cannot happen, for since F' is open, each of its points is the center of an open sphere disjoint from F, and no such point can be a limit point of F.

If x_0 is a point in our metric space X, and r is a non-negative real number, the *closed sphere* $S_r[x_0]$ with *center* x_0 and *radius* r is the subset of X defined by

$$S_r[x_0] = \{x : d(x,x_0) \leq r\}.$$

$S_r[x_0]$ contains its center, and when $r = 0$ it contains only its center. The closed spheres on the real line are precisely the closed intervals. In this connection, we observe that though open spheres on the real line are open intervals, there are open intervals which are not open spheres, e.g., $(-\infty, +\infty)$.

The following theorem justifies the adjective in the phrase "closed sphere."

Theorem C.   *In any metric space X, each closed sphere is a closed set.*

PROOF.   Let $S_r[x_0]$ be a closed sphere in X. By Theorem B, it suffices to show that its complement $S_r[x_0]'$ is open. $S_r[x_0]'$ is open if it is empty, so we may assume that it is non-empty. Let x be a point in $S_r[x_0]'$. Since $d(x,x_0) > r$, $r_1 = d(x,x_0) - r$ is a positive real number. We take r_1 as the radius of an open sphere $S_{r_1}(x)$ centered on x, and we show that $S_r[x_0]'$ is open by showing that $S_{r_1}(x) \subseteq S_r[x_0]'$. Let y be a point in $S_{r_1}(x)$, so that $d(y,x) < r_1$. On the basis of this and the fact that $d(x_0,x) \leq d(x_0,y) + d(y,x)$, we see that

$$d(y,x_0) \geq d(x,x_0) - d(y,x) > d(x,x_0) - r_1 = d(x,x_0) - [d(x,x_0) - r] = r,$$

so that y is in $S_r[x_0]'$.

The main general facts about closed sets are those given in our next theorem.

Theorem D.   *Let X be a metric space. Then (1) any intersection of closed sets in X is closed; and (2) any finite union of closed sets in X is closed.*

PROOF.   By virtue of Eqs. 2-(2) and Theorem B above, this theorem is an immediate consequence of Theorem 10-D. We prove (1) as follows. If $\{F_i\}$ is an arbitrary class of closed subsets of X and $F = \cap_i F_i$, then by Theorem B, F is closed if F' is open; but $F' = \cup_i F_i'$ is open by Theorem 10-D, since by Theorem B each F_i' is open. The second statement is proved similarly.

In Theorem E of the previous section, we gave an explicit characterization of the open sets on the real line. We now consider the structure of its closed sets. Among the simplest closed sets on the real line are the closed intervals (which are the closed spheres) and finite unions of closed intervals. Finite sets are included among these, since a set consisting of a single point is a closed interval with equal end-points. What is the character of the most general closed set on the real line? Since closed sets are the complements of open sets, Theorem 10-E gives a complete answer to this question: the most general proper closed subset of the real line is obtained by removing a countable disjoint class of open intervals. This process sounds innocent enough, but in fact it leads to some rather curious and complicated examples. One of these examples is of particular importance. It was studied by Cantor and is usually called the *Cantor set*.

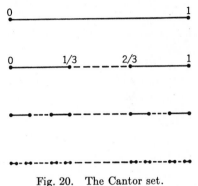

To construct the Cantor set, we proceed as follows (see Fig. 20).

Fig. 20.   The Cantor set.

First, denote the closed unit interval [0,1] by F_1. Next, delete from F_1 the open interval $(\frac{1}{3},\frac{2}{3})$ which is its middle third, and denote the remaining closed set by F_2. Clearly,

$$F_2 = [0,\tfrac{1}{3}] \cup [\tfrac{2}{3},1].$$

Next, delete from F_2 the open intervals $(\frac{1}{9},\frac{2}{9})$ and $(\frac{7}{9},\frac{8}{9})$, which are the middle thirds of its two pieces, and denote the remaining closed set by F_3. It is easy to see that

$$F_3 = [0,\tfrac{1}{9}] \cup [\tfrac{2}{9},\tfrac{1}{3}] \cup [\tfrac{2}{3},\tfrac{7}{9}] \cup [\tfrac{8}{9},1].$$

If we continue this process, at each stage deleting the open middle third of each closed interval remaining from the previous stage, we obtain a sequence of closed sets F_n, each of which contains all its successors. The Cantor set F is defined by

$$F = \bigcap_{n=1}^{\infty} F_n,$$

and it is closed by Theorem D. F consists of those points in the closed unit interval [0,1] which "ultimately remain" after the removal of all the open intervals $(\frac{1}{3},\frac{2}{3})$, $(\frac{1}{9},\frac{2}{9})$, $(\frac{7}{9},\frac{8}{9})$, ... What points do remain? F clearly contains the end-points of the closed intervals which make up each set F_n:

$$0,\ 1,\ \tfrac{1}{3},\ \tfrac{2}{3},\ \tfrac{1}{9},\ \tfrac{2}{9},\ \tfrac{7}{9},\ \tfrac{8}{9},\ \cdots$$

Does F contain any other points? We leave it to the reader to verify that $\frac{1}{4}$ is in F and is not an end-point. Actually, F contains a multitude of points other than the above end-points, for the set of these end-points is clearly countable, while the cardinal number of F itself is c, the cardinal number of the continuum.   To prove this, it suffices to exhibit a one-to-one mapping f of $[0,1)$ into F.   We construct such a mapping as follows. Let x be a point in $[0,1)$, and let $x = .b_1b_2b_3 \ . \ . \ .$ be its binary expansion (see Sec. 7).   Each b_n is either 0 or 1.   Let $t_n = 2b_n$, and regard $.t_1t_2t_3 \ . \ . \ .$ as the ternary expansion of a real number $f(x)$ in $[0,1)$.   The reader will easily convince himself that $f(x)$ is in the Cantor set F: since t_1 is 0 or 2, $f(x)$ is not in $[\frac{1}{3},\frac{2}{3})$; since t_2 is 0 or 2, $f(x)$ is not in $[\frac{1}{9},\frac{2}{9})$ or $[\frac{7}{9},\frac{8}{9})$; etc. Also, it is easy to see that the mapping $f: [0,1) \to F$ is one-to-one. According to this, F contains exactly as many points as the entire closed unit interval $[0,1]$.   It is interesting to compare this conclusion with the fact that the sum of the lengths of all the open intervals removed is precisely 1, since

$$\tfrac{1}{3} + \tfrac{2}{9} + \tfrac{4}{27} + \cdot \ \cdot \ \cdot \ = 1.$$

It is also interesting to observe (by doing a little arithmetic) that F_{25} is the union of 16,777,216 disjoint closed intervals of the same length which are rather irregularly distributed along $[0,1]$.   These facts may suffice to indicate that the Cantor set is a very intricate mathematical object and is just the sort of thing mathematicians delight in.   We shall encounter this set again from time to time, for its properties illustrate several phenomena discussed in later sections.

   We conclude this section by defining two additional concepts which are often useful.

   Let X be an arbitrary metric space, and let A be a subset of X.   The *closure* of A, denoted by $\bar{A}$, is the union of A and the set of all its limit points.   Intuitively, $\bar{A}$ is A itself together with all other points in X which are arbitrarily close to A.   As an example, if A is the open unit disc $\{z:|z| < 1\}$ in the complex plane, then $\bar{A}$ is the closed unit disc $\{z:|z| \leq 1\}$.   The main facts about closures are the following:

   (1) $\bar{A}$ is a closed superset of A which is contained in every closed superset of A (we express this by saying that $\bar{A}$ is the smallest closed superset of A);

   (2) A is closed $\Leftrightarrow A = \bar{A}$;

   (3) $\bar{A}$ equals the intersection of all closed supersets of A.

It is a routine exercise to prove these statements, and we leave this task to the reader (in Problem 6).

   Our second concept relates to the discussion of Fig. 19 given at the end of the previous section.   Again, let X be a metric space and A a subset of X.   A point in X is called a *boundary point* of A if each open sphere centered on the point intersects both A and A', and the *boundary* of A is the set of all its boundary points.   This concept possesses the following properties:

(1)   the boundary of A equals $\bar{A} \cap \overline{A'}$;
(2)   the boundary of A is a closed set;
(3) A is closed $\Leftrightarrow$ it contains its boundary.
We ask the reader to give the proofs in Problem 11.

Problems

1. Let X be a metric space, and extend Problem 10-1 by proving the following statements:

    (a)   any point and disjoint closed set in X can be separated by open sets, in the sense that if x is a point and F a closed set which does not contain x, then there exists a disjoint pair of open sets G_1 and G_2 such that $x \in G_1$ and $F \subseteq G_2$;

    (b)   any disjoint pair of closed sets in X can be separated by open sets, in the sense that if F_1 and F_2 are disjoint closed sets, then there exists a disjoint pair of open sets G_1 and G_2 such that $F_1 \subseteq G_1$ and $F_2 \subseteq G_2$.

2. Let X be a metric space, and let A be a subset of X. If x is a limit point of A, show that each open sphere centered on x contains an infinite number of distinct points of A. Use this result to show that a finite subset of X is closed.

3. Show that a subset of a metric space is bounded $\Leftrightarrow$ it is non-empty and is contained in some closed sphere.

4. Give an example of an infinite class of closed sets whose union is not closed. Give an example of a set which (a) is both open and closed; (b) is neither open nor closed; (c) contains a point which is not a limit point of the set; and (d) contains no point which is not a limit point of the set.

5. Describe the interior of the Cantor set.

6. Prove the statements made in the text about closures.

7. Let X be a metric space and A a subset of X. Prove the following facts:
    (a) $\bar{A}' = \text{Int}(A')$;
    (b) $\bar{A} = \{x : d(x,A) = 0\}$.

8. Describe the closure of each of the following subsets of the real line: the integers; the rationals; the Cantor set; $(0, +\infty)$; $(-1,0) \cup (0,1)$. Do the same for each of the following subsets of the complex plane: $\{z : |z| \text{ is rational}\}$; $\{z : 1/R(z) \text{ is an integer}\}$; $\{z : |z| < 1 \text{ and } I(z) < 0\}$.

9. Let X be a metric space, let x be a point of X, and let r be a positive real number. One is inclined to believe that the closure of $S_r(x)$ must equal $S_r[x]$. Give an example to show that this is not necessarily true. (*Hint:* see Example 9-1.)

10. Let X be a metric space, and let G be an open set in X. Prove that G is disjoint from a set $A \Leftrightarrow G$ is disjoint from $\bar{A}$.

11. Prove the facts about boundaries stated in the text.

12. Describe the boundary of each of the following subsets of the real line: the integers; the rationals; [0,1]; (0,1). Do the same for each of the following subsets of the complex plane: $\{z : |z| < 1\}$; $\{z : |z| \leq 1\}$; $\{z : I(z) > 0\}$.

13. Let X be a metric space and A a subset of X. A is said to be *dense* (or *everywhere dense*) if $\bar{A} = X$. Prove that A is dense $\Leftrightarrow$ the only closed superset of A is $X \Leftrightarrow$ the only open set disjoint from A is $\emptyset \Leftrightarrow A$ intersects every non-empty open set $\Leftrightarrow A$ intersects every open sphere.

12. CONVERGENCE, COMPLETENESS, AND BAIRE'S THEOREM

As we emphasized in the introduction to this chapter, one of our main aims in considering metric spaces is to study convergent sequences in a context more general than that of classical analysis. The fruits of this study are many, and among them is the added insight gained into ordinary convergence as it is used in analysis.

Let X be a metric space with metric d, and let

$$\{x_n\} = \{x_1, x_2, \ldots, x_n, \ldots\}$$

be a sequence of points in X. We say that $\{x_n\}$ is *convergent* if there exists a point x in X such that either

(1)   for each $\epsilon > 0$, there exists a positive integer n_0 such that $n \geq n_0 \Rightarrow d(x_n, x) < \epsilon$; or equivalently,

(2)   for each open sphere $S_\epsilon(x)$ centered on x, there exists a positive integer n_0 such that x_n is in $S_\epsilon(x)$ for all $n \geq n_0$.

The reader should observe that the first condition is a direct generalization of convergence for sequences of numbers as defined in the introduction, and that the second can be thought of as saying that each open sphere centered on x contains all points of the sequence from some place on. If we rely on our knowledge of what is meant by a convergent sequence of real numbers, the statement that $\{x_n\}$ is convergent can equally well be defined as follows: there exists a point x in X such that $d(x_n, x) \to 0$. We usually symbolize this by writing

$$x_n \to x,$$

and we express it verbally by saying that x_n *approaches* x, or that x_n *converges to* x. It is easily seen from condition (2) and Problem 10-1 that the point x in this discussion is unique, that is, that $x_n \to y$ with $y \neq x$

is impossible.   The point x is called the *limit* of the sequence $\{x_n\}$, and we sometimes write $x_n \to x$ in the form

$$\lim x_n = x.$$

The statements $\qquad x_n \to x \qquad$ and $\qquad \lim x_n = x$

mean exactly the same thing, namely, that $\{x_n\}$ is a convergent sequence with limit x.

Every convergent sequence $\{x_n\}$ has the following property: for each $\epsilon > 0$, there exists a positive integer n_0 such that $m, n \geq n_0 \Rightarrow d(x_m, x_n) < \epsilon$.   For if $x_n \to x$, then there exists a positive integer n_0 such that $n \geq n_0 \Rightarrow d(x_n, x) < \epsilon/2$, and from this we see that

$$m, n \geq n_0 \Rightarrow d(x_m, x_n) \leq d(x_m, x) + d(x, x_n) < \epsilon/2 + \epsilon/2 = \epsilon.$$

A sequence with this property is called a *Cauchy sequence*, and we have just shown that every convergent sequence is a Cauchy sequence. Loosely speaking, this amounts to the statement that if the terms of a sequence approach a limit, then they get close to one another.   It is of basic importance to understand that the converse of this need not be true, that is, that a Cauchy sequence is not necessarily convergent.   As an example, consider the subspace $X = (0,1]$ of the real line.   The sequence defined by $x_n = 1/n$ is easily seen to be a Cauchy sequence in this space, but it is not convergent, since the point 0 (which it wants to converge to) is not a point of the space.   The difficulty which arises in this example stems from the fact that the notion of a convergent sequence is not intrinsic to the sequence itself, but also depends on the structure of the space in which it lies.   A convergent sequence is not convergent "on its own"; it must converge to some point in the space.   Some writers emphasize the distinction between convergent sequences and Cauchy sequences by calling the latter "intrinsically convergent" sequences.

A *complete metric space* is a metric space in which every Cauchy sequence is convergent.   In rough terms, a metric space is complete if every sequence in it which tries to converge is successful, in the sense that it finds a point in the space to converge to.   The space $(0,1]$ mentioned above is not complete, but it evidently can be made so by adjoining the point 0 to it to form the slightly larger space $[0,1]$.   As a matter of fact, any metric space, if it isn't already complete, can be made so by suitably adjoining additional points.   We outline this process in a problem at the end of Sec. 14.

It is a fundamental fact of elementary analysis that the real line is a complete metric space.   The complex plane is also complete, as we see from the following argument.   Let $\{z_n\}$, where $z_n = a_n + ib_n$, be a Cauchy sequence of complex numbers.   Then $\{a_n\}$ and $\{b_n\}$ are themselves Cauchy sequences of real numbers, since

$$|a_m - a_n| \leq |z_m - z_n|$$
and
$$|b_m - b_n| \leq |z_m - z_n|.$$

By the completeness of the real line, there exist real numbers a and b such that $a_n \to a$ and $b_n \to b$. If we now put $z = a + ib$, then we see that $z_n \to z$ by means of

$$\begin{aligned} |z_n - z| &= |(a_n + ib_n) - (a + ib)| \\ &= |(a_n - a) + i(b_n - b)| \\ &\leq |a_n - a| + |b_n - b| \end{aligned}$$

and the fact that both final terms on the right approach 0. The completeness of the complex plane thus depends directly on the completeness of the real line. The metric space defined in Example 9-1 is also complete; for in this space a Cauchy sequence must be constant (i.e., it must consist of a single point repeated) from some place on, and it converges with that point as its limit.

The first three of the five metric spaces given as examples in Sec. 9 are therefore complete. What about the last two?

We ask the reader to show in Problem 5 that Example 9-4 is not complete. The problem of completing this space leads to the modern theory of Lebesgue integration, and it would carry us too far afield to pursue this matter to its natural conclusion.

On the other hand, the space $\mathbb{C}[0,1]$ defined in Example 9-5 is complete. We prove this in a more general form in Sec. 14. The completeness of this space, and of others similar to it, is one of the major focal points of topology and modern analysis.

The terms *limit* and *limit point* are often a source of confusion for people not thoroughly accustomed to them. On the real line, for instance, the constant sequence $\{1, 1, \ldots, 1, \ldots\}$ is convergent with limit 1; but the set of points of this sequence is the set consisting of the single element 1, and by Problem 11-2, the point 1 is not a limit point of this set. The essence of the matter is that a sequence of points in a set is not a subset of the set: it is a function defined on the positive integers with values in the set, and is usually specified by listing its values, as in $\{x_n\} = \{x_1, x_2, \ldots, x_n, \ldots\}$, where x_n is the value of the function at the integer n. A sequence may have a limit, but cannot have a limit point; and the set of points of a sequence may have a limit point, but cannot have a limit. The following theorem relates these concepts to one another and is a useful tool for some of our later work.

Theorem A. *If a convergent sequence in a metric space has infinitely many distinct points, then its limit is a limit point of the set of points of the sequence.*

PROOF.   Let X be a metric space, and let $\{x_n\}$ be a convergent sequence in X with limit x.   We assume that x is not a limit point of the set of points of the sequence, and we show that it follows from this that the sequence has only finitely many distinct points.   Our assumption implies that there exists an open sphere $S_r(x)$ centered on x which contains no point of the sequence different from x.   However, since x is the limit of the sequence, all x_n's from some place on must lie in $S_r(x)$, hence must coincide with x.   From this we see that there are only finitely many distinct points in the sequence.

Our next theorem guarantees the completeness of many metric spaces which arise as subspaces of complete metric spaces.

Theorem B.   *Let X be a complete metric space, and let Y be a subspace of X.   Then Y is complete $\Leftrightarrow$ it is closed.*

PROOF.   We assume first that Y is complete as a subspace of X, and we show that it is closed.   Let y be a limit point of Y.   For each positive integer n, $S_{1/n}(y)$ contains a point y_n in Y.   It is clear that $\{y_n\}$ converges to y in X and is a Cauchy sequence in Y, and since Y is complete, y is in Y. Y is therefore closed.

We now assume that Y is closed, and we show that it is complete. Let $\{y_n\}$ be a Cauchy sequence in Y.   It is also a Cauchy sequence in X, and since X is complete, $\{y_n\}$ converges to a point x in X.   We show that x is in Y.   If $\{y_n\}$ has only finitely many distinct points, then x is that point infinitely repeated and is thus in Y.   On the other hand, if $\{y_n\}$ has infinitely many distinct points, then, by Theorem A, x is a limit point of the set of points of the sequence; it is therefore also a limit point of Y, and since Y is closed, x is in Y.

A sequence $\{A_n\}$ of subsets of a metric space is called a *decreasing sequence* if

$$A_1 \supseteq A_2 \supseteq A_3 \supseteq \cdots .$$

The following theorem gives conditions under which the intersection of such a sequence is non-empty.

Theorem C (Cantor's Intersection Theorem).   *Let X be a complete metric space, and let $\{F_n\}$ be a decreasing sequence of non-empty closed subsets of X such that $d(F_n) \to 0$.   Then $F = \bigcap_{n=1}^{\infty} F_n$ contains exactly one point.*

PROOF.   It is first of all evident from the assumption $d(F_n) \to 0$ that F cannot contain more than one point, so it suffices to show that F is non-empty.   Let x_n be a point in F_n.   Since $d(F_n) \to 0$, $\{x_n\}$ is a Cauchy sequence.   Since X is complete, $\{x_n\}$ has a limit x.   We show that x is in F, and for this it suffices to show that x is in F_{n_0} for a fixed but arbitrary n_0.   If $\{x_n\}$ has only finitely many distinct points, then x is that point infinitely repeated, and is therefore in F_{n_0}. If $\{x_n\}$ has infinitely many distinct points, then x is a limit point of the set of points of the sequence, it is a limit point of the subset $\{x_n : n \geq n_0\}$ of the set of points of the sequence, it is a limit point of F_{n_0}, and thus (since F_{n_0} is closed) it is in F_{n_0}.

A subset A of a metric space is said to be *nowhere dense* if its closure has empty interior. It is easy to see that A is nowhere dense $\Leftrightarrow \bar{A}$ does not contain any non-empty open set $\Leftrightarrow$ each non-empty open set has a non-empty open subset disjoint from $\bar{A} \Leftrightarrow$ each non-empty open set has a non-empty open subset disjoint from $A \Leftrightarrow$ each non-empty open set contains an open sphere disjoint from A. If a nowhere dense set is thought of as a set which doesn't cover very much of the space, then our next theorem says that a complete metric space cannot be covered by any sequence of such sets.

Theorem D.   *If $\{A_n\}$ is a sequence of nowhere dense sets in a complete metric space X, then there exists a point in X which is not in any of the A_n's.*

PROOF.   For the duration of this proof, we abandon our usual notations for open spheres and closed spheres. Since X is open and A_1 is nowhere dense, there is an open sphere S_1 of radius less than 1 which is disjoint from A_1. Let F_1 be the concentric closed sphere whose radius is one-half that of S_1, and consider its interior. Since A_2 is nowhere dense, $\text{Int}(F_1)$ contains an open sphere S_2 of radius less than $\frac{1}{2}$ which is disjoint from A_2. Let F_2 be the concentric closed sphere whose radius is one-half that of S_2, and consider its interior. Since A_3 is nowhere dense, $\text{Int}(F_2)$ contains an open sphere S_3 of radius less than $\frac{1}{4}$ which is disjoint from A_3. Let F_3 be the concentric closed sphere whose radius is one-half that of S_3. Continuing in this way, we get a decreasing sequence $\{F_n\}$ of non-empty closed subsets of X such that $d(F_n) \to 0$. Since X is complete, Theorem C guarantees that there exists a point x in X which is in all the F_n's. This point is clearly in all the S_n's, and therefore (since S_n is disjoint from A_n) it is not in any of the A_n's.

For our purposes, the following equivalent form of Theorem D is often more convenient.

Theorem E.   *If a complete metric space is the union of a sequence of its subsets, then the closure of at least one set in the sequence must have non-empty interior.*

Theorems D and E are really one theorem expressed in two different ways. We refer to both (or either) as *Baire's theorem*. This theorem is admittedly rather technical in nature, and the reader can hardly be expected to appreciate its significance at the present stage of our work.

He will find, however, that a need for it crops up from time to time, and when this need arises, Baire's theorem is an indispensable tool.[1]

Problems

1. Let X be a metric space. If $\{x_n\}$ and $\{y_n\}$ are sequences in X such that $x_n \to x$ and $y_n \to y$, show that $d(x_n, y_n) \to d(x, y)$.

2. Show that a Cauchy sequence is convergent $\Leftrightarrow$ it has a convergent subsequence.

3. If $X = X_1 \times X_2 \times \cdots \times X_n$ is the product in Problem 9-4, and if each of the coordinate spaces $X_1, X_2, \ldots, X_n$ is complete, show that X is complete with respect to each of the metrics d and $\bar{d}$ defined in that problem.

4. Let X be any non-empty set. By Problem 9-5, the set of all bounded real functions defined on X is a metric space with respect to the metric induced by the norm defined in that problem. Show that this metric space is complete. (*Hint*: if $\{f_n\}$ is a Cauchy sequence, then $\{f_n(x)\}$ is a Cauchy sequence of real numbers for each point x in X.)

5. In Example 9-4, show that the following functions f_n defined on $[0,1]$ form a Cauchy sequence in this space which is not convergent: $f_n(x) = 1$ if $0 \leq x \leq \frac{1}{2}$, $f_n(x) = -2^n(x - \frac{1}{2}) + 1$ if $\frac{1}{2} \leq x \leq \frac{1}{2} + (\frac{1}{2})^n$, and $f_n(x) = 0$ if $\frac{1}{2} + (\frac{1}{2})^n \leq x \leq 1$.

6. Give an example to show that the set F in Cantor's intersection theorem may be empty if the hypothesis $d(F_n) \to 0$ is dropped.

7. Show that a closed set is nowhere dense $\Leftrightarrow$ its complement is everywhere dense.

8. Show that the Cantor set is nowhere dense.

13. CONTINUOUS MAPPINGS

In the previous section we extended the idea of convergence to the context of a general metric space. We now do the same for continuity.

Let X and Y be metric spaces with metrics d_1 and d_2, and let f be a mapping of X into Y. f is said to be *continuous at a point x_0 in X* if either

---

[1] There is some rather undescriptive terminology which is often used in connection with Baire's theorem. We shall not make use of it ourselves, but the reader ought to be acquainted with it. A subset of a metric space is called a set of the *first category* if it can be represented as the union of a sequence of nowhere dense sets, and a set of the *second category* if it is not a set of the first category. Baire's theorem—sometimes called the *Baire category theorem*—can now be expressed as follows: any complete metric space (considered as a subset of itself) is a set of the second category.

of the following equivalent conditions is satisfied:

(1) for each $\epsilon > 0$ there exists $\delta > 0$ such that $d_1(x,x_0) < \delta \Rightarrow$ $d_2(f(x),f(x_0)) < \epsilon$;

(2) for each open sphere $S_\epsilon(f(x_0))$ centered on $f(x_0)$ there exists an open sphere $S_\delta(x_0)$ centered on x_0 such that $f(S_\delta(x_0)) \subseteq S_\epsilon(f(x_0))$.

The reader will notice that the first condition generalizes the elementary definition given in the introduction to this chapter, and that the second translates the first into the language of open spheres.

Our first theorem expresses continuity at a point in terms of sequences which converge to the point.

Theorem A. *Let X and Y be metric spaces and f a mapping of X into Y. Then f is continuous at x_0 if and only if $x_n \to x_0 \Rightarrow f(x_n) \to f(x_0)$.*

PROOF. We first assume that f is continuous at x_0. If $\{x_n\}$ is a sequence in X such that $x_n \to x_0$, we must show that $f(x_n) \to f(x_0)$. Let $S_\epsilon(f(x_0))$ be an open sphere centered on $f(x_0)$. By our assumption, there exists an open sphere $S_\delta(x_0)$ centered on x_0 such that $f(S_\delta(x_0)) \subseteq S_\epsilon(f(x_0))$. Since $x_n \to x_0$, all x_n's from some place on lie in $S_\delta(x_0)$. Since $f(S_\delta(x_0)) \subseteq S_\epsilon(f(x_0))$, all $f(x_n)$'s from some place on lie in $S_\epsilon(f(x_0))$. We see from this that $f(x_n) \to f(x_0)$.

To prove the other half of our theorem, we assume that f is not continuous at x_0, and we show that $x_n \to x_0$ does not imply $f(x_n) \to f(x_0)$. By this assumption, there exists an open sphere $S_\epsilon(f(x_0))$ with the property that the image under f of each open sphere centered on x_0 is not contained in it. Consider the sequence of open spheres $S_1(x_0)$, $S_{1/2}(x_0)$, $\ldots$, $S_{1/n}(x_0)$, $\ldots$. Form a sequence $\{x_n\}$ such that $x_n \in S_{1/n}(x_0)$ and $f(x_n) \notin S_\epsilon(f(x_0))$. It is clear that x_n converges to x_0 and that $f(x_n)$ does not converge to $f(x_0)$.

A mapping of one metric space into another is said to be *continuous* if it is continuous at each point in its domain. The following theorem is an immediate consequence of Theorem A and this definition.

Theorem B. *Let X and Y be metric spaces and f a mapping of X into Y. Then f is continuous if and only if $x_n \to x \Rightarrow f(x_n) \to f(x)$.*

This result shows that continuous mappings of one metric space into another are precisely those which send convergent sequences into convergent sequences, or, in other words, which *preserve convergence*. Our next theorem characterizes continuous mappings in terms of open sets.

Theorem C. *Let X and Y be metric spaces and f a mapping of X into Y. Then f is continuous $\Leftrightarrow f^{-1}(G)$ is open in X whenever G is open in Y.*

PROOF. We first assume that f is continuous. If G is an open set in Y, we must show that $f^{-1}(G)$ is open in X. $f^{-1}(G)$ is open if it is empty, so we may assume that it is non-empty.   Let x be a point in $f^{-1}(G)$.   Then $f(x)$ is in G, and since G is open, there exists an open sphere $S_\epsilon(f(x))$ centered on $f(x)$ and contained in G.   By the definition of continuity, there exists an open sphere $S_\delta(x)$ such that $f(S_\delta(x)) \subseteq S_\epsilon(f(x))$.   Since $S_\epsilon(f(x)) \subseteq G$, we also have $f(S_\delta(x)) \subseteq G$, and from this we see that $S_\delta(x) \subseteq f^{-1}(G)$. $S_\delta(x)$ is therefore an open sphere centered on x and contained in $f^{-1}(G)$, so $f^{-1}(G)$ is open.

We now assume that $f^{-1}(G)$ is open whenever G is, and we show that f is continuous.   We show that f is continuous at an arbitrary point x in X.   Let $S_\epsilon(f(x))$ be an open sphere centered on $f(x)$.   This open sphere is an open set, so its inverse image is an open set which contains x.   By this, there exists an open sphere $S_\delta(x)$ which is contained in this inverse image.   It is clear that $f(S_\delta(x))$ is contained in $S_\epsilon(f(x))$, so f is continuous at x.   Finally, since x was taken to be an arbitrary point in X, f is continuous.

The fact just established—that continuous mappings are precisely those which pull open sets back to open sets—will be of great importance for all our work from Chap. 3 on.

We now come to the useful concept of *uniform* continuity.   In order to explain what this is, we examine the definition of continuity expressed in condition (1) at the beginning of this section.   Let X and Y be metric spaces with metrics d_1 and d_2, and let f be a mapping of X into Y.   We assume that f is continuous, that is, that for each point x_0 in X the following is true: given $\epsilon > 0$, a number $\delta > 0$ can be found such that $d_1(x,x_0) < \delta \Rightarrow d_2(f(x),f(x_0)) < \epsilon$.   The reader is no doubt familiar with the idea that if x_0 is held fixed and ϵ is made smaller, then, in general, δ has to be made correspondingly smaller.   Thus, in the case of the real function f defined by $f(x) = 2x$, δ can always be chosen as any positive number $\leq \epsilon/2$, and no larger δ will do.   In general, therefore, δ depends on ϵ.   Let us return to our examination of the definition.   It says that for our given ϵ, a δ can be found which "works" in the above sense at the particular point x_0 under consideration.   But if we hold ϵ fixed and move to another point x_0, then it may happen that this δ no longer works; that is, it may be necessary to take a smaller δ to satisfy the requirement of the definition.   We see in this way that δ may well depend, in general, not only on ϵ but also on x_0.   Uniform continuity is essentially continuity plus the added condition that for each ϵ we can find a δ which works *uniformly over the entire space* X, in the sense that it does not depend on x_0.   The formal definition is as follows.   If X and Y are metric spaces with metrics d_1 and d_2, then a mapping f of X into Y is said to be *uniformly continuous* if for each $\epsilon > 0$ there exists $\delta > 0$ such that $d_1(x,x') < \delta \Rightarrow d_2(f(x),f(x')) < \epsilon$.   It is clear that any uniformly continuous mapping is automatically continuous. The reader will observe that the above real function f defined on the entire real line R by $f(x) = 2x$ is uniformly continuous. On the other hand, the function g defined on R by $g(x) = x^2$ is continuous but not uniformly continuous. Similarly, the continuous function h defined on $(0,1)$ by $h(x) = 1/x$ is not uniformly continuous.

Uniformly continuous mappings—as opposed to those which are merely continuous—are of particular significance in analysis. The following theorem expresses a property of these mappings which is often useful.

Theorem D. *Let X be a metric space, let Y be a complete metric space, and let A be a dense subspace of X. If f is a uniformly continuous mapping of A into Y, then f can be extended uniquely to a uniformly continuous mapping g of X into Y.*

PROOF. Let d_1 and d_2 be the metrics on X and Y. If $A = X$, the conclusion is obvious. We therefore assume that $A \neq X$. We begin by showing how to define the mapping g. If x is a point in A, we define $g(x)$ to be $f(x)$. Now let x be a point in $X - A$. Since A is dense, x is the limit of a convergent sequence $\{a_n\}$ in A. Since $\{a_n\}$ is a Cauchy sequence and f is uniformly continuous, $\{f(a_n)\}$ is a Cauchy sequence in Y (see Problem 8). Since Y is complete, there exists a point in Y—we call this point $g(x)$—such that $f(a_n) \to g(x)$. We must make sure that $g(x)$ depends only on x, and not on the sequence $\{a_n\}$. Let $\{b_n\}$ be another sequence in A such that $b_n \to x$. Then $d_1(a_n,b_n) \to 0$, and by the uniform continuity of f, $d_2(f(a_n),f(b_n)) \to 0$. It readily follows from this that $f(b_n) \to g(x)$.

We next show that g is uniformly continuous. Let $\epsilon > 0$ be given, and use the uniform continuity of f to find $\delta > 0$ such that for a and a' in A we have $d_1(a,a') < \delta \Rightarrow d_2(f(a),f(a')) < \epsilon$. Let x and x' be any points in X such that $d_1(x,x') < \delta$. It suffices to show that $d_2(g(x), g(x')) \leq \epsilon$. Let $\{a_n\}$ and $\{a_n'\}$ be sequences in A such that $a_n \to x$ and $a_n' \to x'$. By the triangle inequality, we see that $d_1(a_n,a_n') \leq d_1(a_n,x) + d_1(x,x') + d_1(x',a_n')$. This inequality, together with the facts that $d_1(a_n,x) \to 0$, $d_1(x,x') < \delta$, and $d_1(x',a_n') \to 0$, implies that $d_1(a_n,a_n') < \delta$ for all sufficiently large n. It now follows that $d_2(f(a_n),f(a_n')) < \epsilon$ for all sufficiently large n. By Problem 12-1,

$$d_2(g(x),g(x')) = \lim d_2(f(a_n),f(a_n')),$$

and from this and the previous statement we see that $d_2(g(x),g(x')) \leq \epsilon$.

All that remains is to show that g is unique, and this is easily proved by means of Problem 3 below.

There is an important type of uniformly continuous mapping which often arises in practice. If X and Y are metric spaces with metrics d_1 and d_2, a mapping f of X onto Y is called an *isometry* (or an *isometric mapping*) if $d_1(x,x') = d_2(f(x),f(x'))$ for all points x and x' in X; and if such a mapping exists, we say that X is *isometric* to Y. It is clear that an isometry is necessarily one-to-one. If X is isometric to Y, then the points of these spaces can be put into one-to-one correspondence in such a way that the distances between pairs of corresponding points are the same. The spaces therefore differ only in the nature of their points, and this is often unimportant. We usually consider isometric spaces to be identical with one another. It is often convenient to be able to use this terminology in the case of mappings which are not necessarily onto. If f is a mapping of X into Y which preserves distances in the above sense, then we call f an isometry of X *into* Y, or an isometry of X *onto* the subspace $f(X)$ of Y. In this situation, we often say that Y contains an *isometric image* of X, namely, the subspace $f(X)$.

Problems

1. Let X and Y be metric spaces and f a mapping of X into Y. If f is a constant mapping, show that f is continuous. Use this to show that a continuous mapping need not have the property that the image of every open set is open.

2. Let X be a metric space with metric d, and let x_0 be a fixed point in X. Show that the real function f_{x_0} defined on X by $f_{x_0}(x) = d(x,x_0)$ is continuous. Is it uniformly continuous?

3. Let X and Y be metric spaces and A a non-empty subset of X. If f and g are continuous mappings of X into Y such that $f(x) = g(x)$ for every x in A, show that $f(x) = g(x)$ for every x in $\bar{A}$.

4. Let X and Y be metric spaces and f a mapping of X into Y. Show that f is continuous $\Leftrightarrow f^{-1}(F)$ is closed in X whenever F is closed in $Y \Leftrightarrow f(\bar{A}) \subseteq \overline{f(A)}$ for every subset A of X.

5. Show that any mapping of the metric space defined in Example 9-1 into any other metric space is continuous.

6. Consider the real function f defined on the real line R by $f(x) = x^2$. If b is a given positive real number, show that the restriction of f to the closed interval $[0,b]$ is uniformly continuous by starting with an $\epsilon > 0$ and exhibiting a $\delta > 0$ which satisfies the requirement of the definition.

7. Determine which of the following functions are uniformly continuous on the open unit interval $(0,1)$: $1/(1 - x)$; $1/(2 - x)$; $\sin x$; $\sin (1/x)$; $x^{1/2}$; x^3. Which are uniformly continuous on the open interval $(0, +\infty)$?

8. In the proof of Theorem D we used the following fact: the image of a Cauchy sequence under a uniformly continuous mapping is again a Cauchy sequence. Give the details of the proof.

9. Let f be a continuous real function defined on R which satisfies the functional equation $f(x + y) = f(x) + f(y)$. Show that this function must have the form $f(x) = mx$ for some real number m. (*Hint:* the subspace of rational numbers is dense in the metric space R.)

14. SPACES OF CONTINUOUS FUNCTIONS

In Example 9-5 we gave a brief description of the metric space $\mathcal{C}[0,1]$. The reader will recall that the points of this space are the bounded continuous real functions defined on the closed unit interval $[0,1]$ and that its metric is defined by $d(f,g) = \sup |f(x) - g(x)|$. We have two aims in this section: to generalize this very important example by considering functions defined on an arbitrary metric space, and to place all function spaces of this type in their proper context by giving the details of the structural pattern (discussed briefly in Sec. 9) which they all have in common with one another. We begin with the second, and define the algebraic systems which are relevant to our present interests.

Let L be a non-empty set, and assume that each pair of elements x and y in L can be combined by a process called *addition* to yield an element z in L denoted by $z = x + y$. Assume also that this operation of addition satisfies the following conditions:

(1) $x + y = y + x$;

(2) $x + (y + z) = (x + y) + z$;

(3) there exists in L a unique element, denoted by 0 and called the *zero element*, or the *origin*, such that $x + 0 = x$ for every x;

(4) to each element x in L there corresponds a unique element in L, denoted by $-x$ and called the *negative* of x, such that $x + (-x) = 0$.

We adopt the device of referring to the system of real numbers or to the system of complex numbers as the *scalars*. We now assume that each scalar α and each element x in L can be combined by a process called *scalar multiplication* to yield an element y in L denoted by $y = \alpha x$ in such a way that

(5) $\alpha(x + y) = \alpha x + \alpha y$;

(6) $(\alpha + \beta)x = \alpha x + \beta x$;

(7) $(\alpha\beta)x = \alpha(\beta x)$;

(8) $1 \cdot x = x$.

The algebraic system L defined by these operations and axioms is called a *linear space*. Depending on the numbers admitted as scalars (only the real numbers, or all the complex numbers), we distinguish when necessary between *real linear spaces* and *complex linear spaces*. For geometric reasons discussed in the next section, a linear space is often called a *vector space*, and its elements are spoken of as *vectors*.

We are not concerned here with developing the algebraic theory of linear spaces. Our only interest is in making available some pertinent concepts and terminology which are useful as a background against which to view the metric spaces we wish to study. With this in mind, we mention a few simple facts which are quite easy to prove from the axioms for a linear space: $0 + x = x$ for every x; $x + z = y + z \Rightarrow x = y$ (*hint:* add $-z$ to both sides on the right); $\alpha \cdot 0 = 0$ (*hint:* $\alpha \cdot 0 + \alpha x = \alpha(0 + x) = \alpha x = 0 + \alpha x$); $0 \cdot x = 0$ (*hint:* $0 \cdot x + \alpha x = (0 + \alpha)x = \alpha x = 0 + \alpha x$); and $(-1)x = -x$ (*hint:* $x + (-1)x = 1 \cdot x + (-1)x = (1 + (-1))x = 0 \cdot x = 0$). The reader will notice that in the relation $0 \cdot x = 0$ we have used the symbol 0 with two different meanings: as a scalar on the left and as a vector on the right. Several other meanings will be given to this single symbol, but fortunately it is always possible to avoid confusion by attending closely to the context in which it occurs. It is convenient to introduce the operation of *subtraction* by using the symbol $x - y$ as an abbreviation for $x + (-y)$; $x - y$ is called the *difference* between x and y.

A non-empty subset M of a linear space L is called a *linear subspace* of L if $x + y$ is in M whenever x and y are and if αx is in M (for any scalar α) whenever x is. Since M is non-empty, $0 \cdot x = 0$ shows that 0 is in M. Since $-x = (-1)x$, $-x$ is in M whenever x is. It will be seen at once that a linear subspace of a linear space is itself a linear space with respect to the same operations.

A *normed linear space* is a linear space on which there is defined a *norm*, i.e., a function which assigns to each element x in the space a real number $\|x\|$ in such a manner that

(1) $\|x\| \geq 0$, and $\|x\| = 0 \Leftrightarrow x = 0$;

(2) $\|x + y\| \leq \|x\| + \|y\|$;

(3) $\|\alpha x\| = |\alpha|\, \|x\|$.

In general terms, a normed linear space is simply a linear space in which there is available a satisfactory notion of the distance from an arbitrary element to the origin. From (3) and the fact that $-x = (-1)x$, we obtain $\|-x\| = \|x\|$. As we saw in Sec. 9, a normed linear space is a metric space with respect to the induced metric defined by

$$d(x,y) = \|x - y\|.$$

A *Banach space* is a normed linear space which is complete as a metric space.   By Theorem 12-B, any closed linear subspace of a Banach space is itself a Banach space with respect to the same algebraic operations and the same norm.

So much for the technical framework.   We now turn to the metric spaces which really concern us.   They are all *function spaces,* in the sense that they are linear spaces whose elements are functions defined on some non-empty set X with addition and scalar multiplication defined pointwise, i.e., by $(f + g)(x) = f(x) + g(x)$ and $(\alpha f)(x) = \alpha f(x)$.   We note that the zero element in such a linear space is the constant function 0 whose only value is the scalar 0 and that $(-f)(x) = -f(x)$.

Suppose, then, that X is an arbitrary non-empty set, and consider the set L of all real functions defined on X.   It is clear that L is a real linear space with respect to the operations described above.   We now restrict ourselves to the subset B consisting of the bounded functions in L. B is obviously a linear subspace of L, so it is a linear space in its own right.   Even more, if we define a norm on B by $\|f\| = \sup |f(x)|$, then B is a Banach space (see Problems 9-5 and 12-4).

We next assume that the underlying set X is a metric space.   This enables us to consider the possible continuity of functions defined on X.   We define $\mathcal{C}(X,R)$ to be that subset of B which consists of continuous functions. $\mathcal{C}(X,R)$ is thus the set of all bounded continuous real functions defined on the metric space X, and it is non-empty by Problem 13-1.

Lemma.   *If f and g are continuous real functions defined on a metric space X, then $f + g$ and αf are also continuous, where α is any real number.*

PROOF.   Let d be the metric on X.   We show that $f + g$ is continuous by showing that it is continuous at an arbitrary point x_0 in X.   Let $\epsilon > 0$ be given.   Since f and g are continuous, and thus continuous at x_0, we can find $\delta_1 > 0$ and $\delta_2 > 0$ such that $d(x,x_0) < \delta_1 \Rightarrow |f(x) - f(x_0)| < \epsilon/2$ and $d(x,x_0) < \delta_2 \Rightarrow |g(x) - g(x_0)| < \epsilon/2$.   Let δ be the smaller of the numbers δ_1 and δ_2.   Then the continuity of $f + g$ at x_0 follows from

$$d(x,x_0) < \delta \Rightarrow |(f + g)(x) - (f + g)(x_0)| = |[f(x) + g(x)] - [f(x_0)$$
$$+ g(x_0)]| = |[f(x) - f(x_0)] + [g(x) - g(x_0)]| \le |f(x) - f(x_0)|$$
$$+ |g(x) - g(x_0)| < \epsilon/2 + \epsilon/2 = \epsilon.$$

We leave it to the reader to show similarly that αf is continuous.

This lemma implies that $\mathcal{C}(X,R)$ is a linear subspace of the linear space B.   We next prove that it is closed as a subset of the metric space B.

Lemma. $\mathcal{C}(X,R)$ *is a closed subset of the metric space B.*

PROOF. Let f be a function in B which is in the closure of $\mathcal{C}(X,R)$. We show that f is continuous, and therefore is in $\mathcal{C}(X,R)$, by showing that it is continuous at an arbitrary point x_0 in X. Since a set which equals its closure is closed, this will suffice to prove the lemma. Let d be the metric on X, and let $\epsilon > 0$ be given. Since f is in the closure of $\mathcal{C}(X,R)$, there exists a function f_0 in $\mathcal{C}(X,R)$ such that $\|f - f_0\| < \epsilon/3$, from which it follows that $|f(x) - f_0(x)| < \epsilon/3$ for every point x in X. Since f_0 is continuous, and hence continuous at x_0, we can find a $\delta > 0$ such that $d(x,x_0) < \delta \Rightarrow |f_0(x) - f_0(x_0)| < \epsilon/3$. The fact that f is continuous at x_0 now follows from

$$d(x,x_0) < \delta \Rightarrow |f(x) - f(x_0)| = |[f(x) - f_0(x)] + [f_0(x) - f_0(x_0)]$$
$$+ [f_0(x_0) - f(x_0)]| \leq |f(x) - f_0(x)| + |f_0(x) - f_0(x_0)|$$
$$+ |f_0(x_0) - f(x_0)| < \epsilon/3 + \epsilon/3 + \epsilon/3 = \epsilon.$$

Since a closed linear subspace of a Banach space is itself a Banach space, we can summarize the result of the above discussion and lemmas in the following theorem.

Theorem A. *The set $\mathcal{C}(X,R)$ of all bounded continuous real functions defined on a metric space X is a real Banach space with respect to pointwise addition and scalar multiplication and the norm defined by $\|f\| = \sup |f(x)|$.*

It is desirable at this stage to make a clear distinction between two types of convergence for sequences of functions. Let X be a metric space, and let $\{f_n\}$ be a sequence of real functions defined on X. If for each x in X it happens that $\{f_n(x)\}$ is a Cauchy sequence of real numbers, then by the completeness of the real number system we can define a limit function f by $f(x) = \lim f_n(x)$. We then say that f_n *converges pointwise* to f, or that f is the *pointwise limit* of f_n. It is often important to know what properties of the functions f_n carry over to the limit function f, but unless we strengthen the mode of convergence, very little can be said along these lines. The stronger type of convergence normally needed to conclude anything of interest is called *uniform convergence*. In order to explain what this is, we inspect a little more closely what is involved in pointwise convergence. To say that f_n converges pointwise to f is to say the following: for each point x in X, if $\epsilon > 0$ is given, then a positive integer n_0 can be found such that $|f_n(x) - f(x)| < \epsilon$ for all $n \geq n_0$. In general, the integer n_0 may depend on x as well as ϵ. If, however, for each given ϵ an integer n_0 can be found which serves for *all* points x, then we say that f_n *converges uniformly* to f, or that f is the *uniform limit* of f_n. The reader will observe that these concepts are quite independent of the assumption that X is a metric space and that they are meaningful for functions defined on an arbitrary non-empty set.

It will be seen at once that convergence in the function space $\mathcal{C}(X,R)$ is precisely uniform convergence as we have just defined it. The fact that $\mathcal{C}(X,R)$ is complete can be restated as follows in the language of uniform convergence: if a bounded real function f defined on X is the uniform limit of a sequence $\{f_n\}$ of bounded *continuous* real functions defined on X, then f is also continuous. In other words, in the presence of uniform convergence, continuity carries over from the f_n's to the limit function f.

A moment's thought will convince the reader that the entire discussion given above, beginning with our definition of the linear space L, could perfectly well have been based on complex functions. Without going again through all the details, we state the following theorem and consider it proved.

Theorem B. *The set $\mathcal{C}(X,C)$ of all bounded continuous complex functions defined on a metric space X is a complex Banach space with respect to point-wise addition and scalar multiplication and the norm defined by*

$$\|f\| = \sup |f(x)|.$$

In summary, we associate with each metric space X two linear spaces of continuous functions defined on X. The first—$\mathcal{C}(X,R)$—contains only real functions, and the second—$\mathcal{C}(X,C)$—consists of complex functions. Further, all functions considered are assumed to be bounded, so that the norm defined by $\|f\| = \sup |f(x)|$ is always a real number. In the special case in which X is a closed interval $[a,b]$ on the real line, we write $\mathcal{C}(X,R)$ in the simpler form $\mathcal{C}[a,b]$.

Problems

1. Show that a non-empty subset A of a Banach space is bounded $\Leftrightarrow$ there exists a real number K such that $\|x\| \leq K$ for every x in A.
2. Construct a sequence of continuous functions defined on $[0,1]$ which converges pointwise but *not* uniformly to a continuous limit.
3. Construct a sequence of continuous functions defined on $[0,1]$ which converges pointwise to a discontinuous limit.
4. Let X and Y be metric spaces with metrics d_1 and d_2, and let $\{f_n\}$ be a sequence of mappings of X into Y which converges pointwise to a mapping f of X into Y, in the sense that $f_n(x) \to f(x)$ for each x in X. Define what ought to be meant by the statement that f_n converges uniformly to f, and prove that under this assumption f is continuous if each f_n is continuous.
5. In this problem we give a procedure for constructing the *completion* X^* of an arbitrary metric space X. Denote by d the metric on X.

Let x_0 be a fixed point in X, and to each point x in X make correspond the real function f_x defined on X by $f_x(y) = d(y,x) - d(y,x_0)$.

(a)   Show that f_x is bounded.   (*Hint:* $|f_x(y)| \leq d(x,x_0)$.)

(b)   Show that f_x is continuous.   (*Hint:* $|f_x(y_1) - f_x(y_2)| \leq 2d(y_1,y_2)$.)
By (a) and (b), the mapping F defined by $F(x) = f_x$ is a mapping of X into $\mathfrak{C}(X,R)$.

(c)   Show that F is an isometry.   (*Hint:* $|f_{x_1}(y) - f_{x_2}(y)| \leq d(x_1,x_2)$.)

F is thus an isometry of X into the complete metric space $\mathfrak{C}(X,R)$. We define the *completion* X^* of X to be the closure of $F(X)$ in $\mathfrak{C}(X,R)$.

(d)   Show that X^* is a complete metric space which contains an isometric image of X.

(e)   Show that there is a natural isometry of X^* into any complete metric space Y which contains an isometric image of X (to say that an isometry of X^* into Y is "natural" means that the image of a point in X^* which corresponds to a point in X is the point in Y which corresponds to this same point in X).

(f)   Show that (d) and (e) characterize X^* in the following sense: if Z is a complete metric space which contains an isometric image of X, and if there is a natural isometry of Z into any complete metric space Y which contains an isometric image of X, then there is a natural isometry of Z onto X^*.

(g)   Show that if X occurs as a subspace of a complete metric space, then there is a natural isometry of the closure of X onto X^*.

(h)   Show that there is a natural isometry of any complete metric space which contains X as a dense subspace onto X^*.[1]

15. EUCLIDEAN AND UNITARY SPACES

Let n be a fixed positive integer, and consider the set R^n of all ordered n-tuples $x = (x_1, x_2, \ldots, x_n)$ of real numbers.[2] We promised in Sec. 4 to make this set into a space, and we are now in a position to do so.

---

[1] The construction outlined in (a) to (c) clearly depends on the initial choice of the fixed point x_0. If another fixed point x_0 is chosen, then another isometry F of X into $\mathfrak{C}(X,R)$ is determined. It would seem, therefore, that there is little justification for calling the particular X^* defined in this problem *the* completion of X. In practice, however, we usually pursue the reasonable course of regarding isometric spaces as essentially identical. From this point of view, the X^* defined here is a complete metric space which contains X as a dense subspace; and since by (h) it is the only complete metric space with this property, it is natural to call it *the* completion of X.

[2] From this point on, we omit the adjective "ordered." It is to be understood that an n-tuple is *always* ordered.

We begin by defining addition and scalar multiplication in R^n. If $x = (x_1, x_2, \ldots, x_n)$ and $y = (y_1, y_2, \ldots, y_n)$, then we define $x + y$ and αx (where α is any real number) by

$$x + y = (x_1 + y_1, x_2 + y_2, \ldots, x_n + y_n)$$

and

$$\alpha x = (\alpha x_1, \alpha x_2, \ldots, \alpha x_n).$$

With the algebraic operations defined *coordinatewise* in this way, R^n is a real linear space. The origin or zero element is clearly $0 = (0, 0, \ldots, 0)$, and the negative of an element $x = (x_1, x_2, \ldots, x_n)$ is

$$-x = (-x_1, -x_2, \ldots, -x_n).$$

When we speak of R^n as an *n-dimensional* space, all we mean at this stage is that each element $x = (x_1, x_2, \ldots, x_n)$ is the ordered array of its n *coordinates* $x_1, x_2, \ldots, x_n$.

The reader is probably familiar with vector algebra in the ordinary three-dimensional space of our physical intuition. If so, then he is

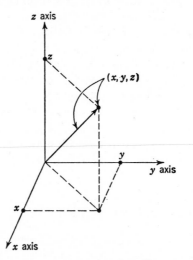

Fig. 21.   A vector (or point) in ordinary space.

accustomed to regarding a point in this space as being essentially identical with the arrow (or vector) from the origin to that point, in the sense that given the point, the vector is determined, and given the vector, the point is determined. This situation is illustrated in Fig. 21. The above definitions of addition and scalar multiplication in R^n correspond to vector addition and the multiplication of a vector by a real number. A word of warning must be given. In ordinary vector algebra, a vector is usually allowed to have its tail at any point in the space and its head at any other point. It should be clearly understood, however, that for us a vector always has its tail at the origin. In accordance with this intuitive picture, we may think of the elements of the real linear space R^n either as points or as generalized vectors from the origin to those points. The latter view is often more fruitful and illuminating.

There is yet a third interpretation of the elements of R^n, of great significance from the point of view of generalizations. An n-tuple $x = (x_1, x_2, \ldots , x_n)$ of real numbers can be thought of as a real function f defined on the set $\{1, 2, \ldots , n\}$ of the first n positive integers. The ith coordinate x_i of x is then just the value of this function at the integer i $(f(i) = x_i)$, and the coordinatewise operations defined above become pointwise operations. This way of thinking about the elements of R^n should help to allay any doubts which might be felt as to the feasibility of visualizing n-dimensional spaces for $n \geq 4$. The four-dimensional space R^4, for instance, is merely the space of all real functions defined on the set consisting of the first four positive integers, and there is surely nothing mysterious or incomprehensible about this. The advantages of the function notation are so great that we shall often (but not always) use it in preference to the n-tuple notation. The reader will find it profitable to keep in mind all three aspects of the elements of R^n—as points, as vectors, and as functions—and he will train himself to use that interpretation (and notation) which appears most natural in any given situation.

Our next task is to define a suitable norm on the linear space R^n. We recall that in solid analytic geometry the usual distance from a point (x, y, z) to the origin (see Fig. 21) is given by the expression

$$\sqrt{x^2 + y^2 + z^2}.$$

If $x = (x_1, x_2, \ldots , x_n)$ is an arbitrary element of R^n, then it is natural to define $\|x\|$—the distance from the point x to the origin, or the length of the vector x—by

$$\|x\| = \sqrt{|x_1|^2 + |x_2|^2 + \cdots + |x_n|^2}$$
$$= \left(\sum_{i=1}^{n} |x_i|^2 \right)^{\frac{1}{2}}.$$

If we think of R^n as composed of real functions f defined on $\{1, 2, \ldots , n\}$, then this definition becomes

$$\|f\| = \left(\sum_{i=1}^{n} |f(i)|^2 \right)^{\frac{1}{2}}.$$

This is called the *Euclidean norm* on R^n, and the real linear space R^n normed in this way is called *n-dimensional Euclidean space*. The

Euclidean plane is the real linear space R^2 with its Euclidean norm; that is, it is the coordinate plane equipped with the above algebraic operations and the above norm. For reasons which will appear a little later, we observe that our formula defining $\|x\|$ can be applied equally well to n-tuples of complex numbers.

We have not yet proved, of course, that the above expression for $\|x\|$ possesses the three properties required by the definition of a norm. The first and third of these conditions are clearly satisfied. The second, namely, that

$$\|x + y\| \leq \|x\| + \|y\|,$$

is another matter. We prove this by the following two lemmas, of which the first is essentially a tool used in the proof of the second.

Lemma (Cauchy's Inequality). *Let* $x = (x_1, x_2, \ldots, x_n)$ *and* $y = (y_1, y_2, \ldots, y_n)$ *be two n-tuples of real or complex numbers. Then*

$$\sum_{i=1}^{n} |x_i y_i| \leq \Big(\sum_{i=1}^{n} |x_i|^2 \Big)^{\frac{1}{2}} \Big(\sum_{i=1}^{n} |y_i|^2 \Big)^{\frac{1}{2}},$$

or, in our notation, $\Sigma_{i=1}^{n} |x_i y_i| \leq \|x\| \, \|y\|$.

PROOF. We first remark that if a and b are any two non-negative real numbers, then $a^{\frac{1}{2}} b^{\frac{1}{2}} \leq (a + b)/2$; for on squaring both sides and rearranging, this is equivalent to $0 \leq (a - b)^2$, which is obviously true. If $x = 0$ or $y = 0$, the assertion of the lemma is clear. We therefore assume that $x \neq 0$ and $y \neq 0$. We define a_i and b_i by $a_i = (|x_i|/\|x\|)^2$ and $b_i = (|y_i|/\|y\|)^2$. By the above remark, we obtain the following for each i:

$$\frac{|x_i y_i|}{\|x\| \, \|y\|} \leq \frac{|x_i|^2/\|x\|^2 + |y_i|^2/\|y\|^2}{2}.$$

Summing these inequalities as i varies from 1 to n yields

$$\frac{\sum_{i=1}^{n} |x_i y_i|}{\|x\| \, \|y\|} \leq \frac{1 + 1}{2} = 1,$$

from which our conclusion follows at once.

Lemma (Minkowski's Inequality). *Let* $x = (x_1, x_2, \ldots, x_n)$ *and* $y = (y_1, y_2, \ldots, y_n)$ *be two n-tuples of real or complex numbers. Then*

$$\Big(\sum_{i=1}^{n} |x_i + y_i|^2 \Big)^{\frac{1}{2}} \leq \Big(\sum_{i=1}^{n} |x_i|^2 \Big)^{\frac{1}{2}} + \Big(\sum_{i=1}^{n} |y_i|^2 \Big)^{\frac{1}{2}},$$

or, in our notation, $\|x + y\| \leq \|x\| + \|y\|$.

PROOF. Using Cauchy's inequality, we have the following chain of relations:

$$\|x+y\|^2 = \sum_{i=1}^{n} |x_i + y_i|\, |x_i + y_i|$$

$$\leq \sum_{i=1}^{n} |x_i + y_i|(|x_i| + |y_i|)$$

$$= \sum_{i=1}^{n} |x_i + y_i|\, |x_i| + \sum_{i=1}^{n} |x_i + y_i|\, |y_i|$$

$$\leq \|x+y\|\, \|x\| + \|x+y\|\, \|y\|$$

$$= \|x+y\|(\|x\| + \|y\|),$$

or summarizing,

$$\|x+y\|^2 \leq \|x+y\|(\|x\| + \|y\|).$$

If $\|x+y\| = 0$, our lemma is trivially true; otherwise, it follows from the inequality last written on dividing through by $\|x+y\|$.

We are now in a position to state

Theorem A. *The set R^n of all n-tuples $x = (x_1, x_2, \ldots, x_n)$ of real numbers is a real Banach space with respect to coordinatewise addition and scalar multiplication and the norm defined by $\|x\| = (\sum_{i=1}^{n} |x_i|^2)^{1/2}$.*

PROOF. In view of the above discussions, all that remains is to prove completeness. It will be convenient here to use the function notation, so that a typical element of our space is regarded as a real function defined on $\{1, 2, \ldots, n\}$. Let $\{f_m\}$ be a Cauchy sequence in R^n. If $\epsilon > 0$ is given, then for all sufficiently large m and m' we have $\|f_m - f_{m'}\| < \epsilon$, $\|f_m - f_{m'}\|^2 < \epsilon^2$, and $\sum_{i=1}^{n} |f_m(i) - f_{m'}(i)|^2 < \epsilon^2$; and from this we see that $|f_m(i) - f_{m'}(i)| < \epsilon$ for each i (and all sufficiently large m and m'). The sequence $\{f_m\}$ therefore converges pointwise to a limit function f defined by $f(i) = \lim f_m(i)$. Since the set $\{1, 2, \ldots, n\}$ is finite, this convergence is uniform. We can thus find a positive integer m_0 such that $|f_m(i) - f(i)| < \epsilon/n^{1/2}$ for all $m \geq m_0$ and every i. Squaring each of these inequalities and summing as i varies from 1 to n yields $\sum_{i=1}^{n} |f_m(i) - f(i)|^2 < \epsilon^2$ or $\|f_m - f\| < \epsilon$ for all $m \geq m_0$. This shows that the Cauchy sequence $\{f_m\}$ converges to the limit f, so R^n is complete.

Just as in the previous section, virtually every statement we have made about n-tuples of real numbers (or about real functions defined on $\{1, 2, \ldots, n\}$) has its complex analogue. We therefore consider the following theorem to be fully proved.

Theorem B. *The set C^n of all n-tuples $z = (z_1, z_2, \ldots, z_n)$ of complex numbers is a complex Banach space with respect to coordinatewise addition and scalar multiplication and the norm defined by $\|z\| = (\sum_{i=1}^{n} |z_i|^2)^{1/2}$.*

The space C^n, with these algebraic operations and this norm, is called *n-dimensional unitary space*. Needless to say, it can equally well be viewed as the set of all complex functions f defined on the set $\{1, 2, \ldots, n\}$, with addition and scalar multiplication (by complex scalars) defined pointwise and the norm defined by $\|f\| = (\Sigma_{i=1}^{n} |f(i)|^2)^{1/2}$.

The four spaces defined and discussed in this and the previous section—$\mathcal{C}(X,R)$ and $\mathcal{C}(X,C)$, and R^n and C^n—form the foundation for all our future work. In Chaps. 3 to 7 we generalize the first two by loosening the restrictions on the underlying space X. In Chaps. 9 to 11 we study all four from a wider point of view, with special emphasis on C^n. And in the last three chapters we pull these lines of development together in such a way that each aspect of our work sheds light on all the others.

Problems

1. Show that a non-empty subset A of R^n is bounded $\Leftrightarrow$ there exists a real number K such that for each $x = (x_1, x_2, \ldots, x_n)$ in A we have $|x_i| \leq K$ for every subscript i.

2. Let X be the set $\{1, 2, \ldots, n\}$, equipped with the metric defined in Example 9-1. Then $\mathcal{C}(X,R)$ and R^n are two Banach spaces which are essentially identical as real linear spaces but which have different norms. Show that they have the same open sets.

3. Prove the following extension of Minkowski's inequality. If $x = \{x_1, x_2, \ldots, x_n, \ldots\}$ and $y = \{y_1, y_2, \ldots, y_n, \ldots\}$ are two sequences of real or complex numbers such that $\Sigma_{n=1}^{\infty} |x_n|^2$ and $\Sigma_{n=1}^{\infty} |y_n|^2$ are convergent, then $\Sigma_{n=1}^{\infty} |x_n + y_n|^2$ is also convergent, and

$$\Big(\sum_{n=1}^{\infty} |x_n + y_n|^2 \Big)^{1/2} \leq \Big(\sum_{n=1}^{\infty} |x_n|^2 \Big)^{1/2} + \Big(\sum_{n=1}^{\infty} |y_n|^2 \Big)^{1/2}.$$

    This statement is also called Minkowski's inequality—for infinite sums.

4. The set of all sequences $x = \{x_1, x_2, \ldots, x_n, \ldots\}$ of real numbers such that $\Sigma_{n=1}^{\infty} |x_n|^2$ converges is denoted by R^∞. If addition and scalar multiplication are defined coordinatewise (or termwise), and if a norm is defined by $\|x\| = (\Sigma_{n=1}^{\infty} |x_n|^2)^{1/2}$, show that R^∞ is a real Banach space. R^∞ is called *infinite-dimensional Euclidean space*. The *infinite-dimensional unitary space* C^∞ is defined similarly, and is a complex Banach space.

CHAPTER THREE

Topological Spaces

In the previous chapter we defined the concept of a continuous mapping of one metric space into another, and this definition was formulated in terms of the metrics on the spaces involved. It often happens, however, that it is convenient—even essential—to be able to speak of continuous mappings in situations where no useful metrics are defined, readily definable, or capable of being defined. In order to deal effectively with circumstances of this kind, it is necessary for us to liberate our concept of continuity from its dependence on metric spaces.

Theorem 13-C shows that the continuity of a mapping of one metric space into another can be expressed solely in terms of open sets, without any direct reference to metrics. This suggests the possibility of discarding metrics altogether and of replacing them as the source of our theory by open sets. With this in mind, our attention is drawn to Theorem 10-D, which gives the main internal properties of the class of open sets in a metric space. These two theorems provide the leading hint on which we base our generalization of metric spaces to topological spaces—a *topological space* being simply a non-empty set in which there is given a class of subsets, called *open sets*, with the properties expressed in Theorem 10-D.

Our underlying purpose in this and the next four chapters is to study topological spaces and continuous mappings of topological spaces into one another. We shall see that these spaces provide the ideal context for a theory of continuity in its purest form.

This chapter is devoted primarily to explaining the concept of a general topological space.   We also construct some machinery which will be useful in the detailed study of these spaces.

Our main special interest in the four chapters that follow will be in continuous real or complex functions defined on particular types of topological spaces, and we shall develop the point of view that there is a constant illuminating interplay between the structure of these spaces and the properties of the continuous functions which they carry.

16. THE DEFINITION AND SOME EXAMPLES

Let X be a non-empty set.   A class **T** of subsets of X is called a *topology* on X if it satisfies the following two conditions:
   (1)   the union of every class of sets in **T** is a set in **T**;
   (2)   the intersection of every finite class of sets in **T** is a set in **T**.
A topology on X is thus a class of subsets of X which is closed under the formation of arbitrary unions and finite intersections.   A *topological space* consists of two objects: a non-empty set X and a topology **T** on X.   The sets in the class **T** are called the *open sets* of the topological space (X,T), and the elements of X are called its *points*.   It is customary to denote the topological space (X,T) by the symbol X which is used for its under-lying set of points.   No harm can come from this practice if one clearly understands that a topological space is more than merely a non-empty set: it is a non-empty set together with a specific topology on that set. We shall often be considering several topologies on a single given set, and in these circumstances distinct topologies make the set into distinct topological spaces.   We observe that the empty set and the full space are always open sets in every topological space, since they are the union and intersection of the empty class of sets, which is a subclass of every topology.

We now list several simple examples of topological spaces.   In order to exhibit a topological space, one must specify a non-empty set, tell which subsets are to be considered the open sets, and verify that this given class of sets satisfies conditions (1) and (2) above.   In the examples which follow, we leave this third step to the reader.

Example 1.   Let X be any metric space, and let the topology be the class of all subsets of X which are open in the sense of the definition in Sec. 10. This is called the *usual topology* on a metric space, and we say that these sets are the open sets *generated by* the metric on the space.   Metric spaces are the most important topological spaces, and whenever we speak of a metric space as a topological space, it is understood (unless we say something to the contrary) that its topology is the usual topology described here.

Example 2. Let X be any non-empty set, and let the topology be the class of all subsets of X. This is called the *discrete topology* on X, and any topological space whose topology is the discrete topology is called a *discrete space*.

Example 3. Let X be any non-empty set, and let the topology consist only of the empty set $\emptyset$ and the full space X. This topology is at the opposite extreme from that described in Example 2, but they coincide when X is a set with only one element.

Example 4. Let X be any infinite set, and let the topology consist of the empty set $\emptyset$ together with all subsets of X whose complements are finite.

Example 5. Let X be the three-element set $\{a, b, c\}$, and let the topology consist of the following subsets of X: $\emptyset$, $\{a\}$, $\{a,b\}$, $\{a,c\}$, X. Spaces of this type serve mainly to illustrate certain aspects of the theory which will emerge in later chapters.

A *metrizable space* is a topological space X with the property that there exists at least one metric on the set X whose class of generated open sets is precisely the given topology. A metrizable space is thus a topological space which is—so far as its open sets are concerned—essentially a metric space. We shall encounter many important topological spaces which are not metrizable, and it is the existence of such spaces which gives our present theory a wider scope than the theory of metric spaces. It is a problem of considerable interest to determine what types of topological spaces are metrizable, and we shall return to this question in Sec. 29.

Let X be a topological space, and let Y be a non-empty subset of X. Problem 10-7 suggests a natural way of making Y into a topological space. The *relative topology* on Y is defined to be the class of all intersections with Y of open sets in X; and when Y is equipped with its relative topology, it is called a *subspace* of X.

Let X and Y be topological spaces and f a mapping of X into Y. f is called a *continuous mapping* if $f^{-1}(G)$ is open in X whenever G is open in Y, and an *open mapping* if $f(G)$ is open in Y whenever G is open in X. A mapping is continuous if it pulls open sets back to open sets, and open if it carries open sets over to open sets. Any image $f(X)$ of a topological space X under a continuous mapping f is called a *continuous image* of X.

A *homeomorphism* is a one-to-one continuous mapping of one topological space onto another which is also an open mapping. Two topological spaces X and Y are said to be *homeomorphic* if there exists a homeomorphism of X onto Y (and in this case, Y is called a *homeomorphic image* of X). If X and Y are homeomorphic, then their points can be put into one-to-one correspondence in such a way that their open sets also correspond to one another. The two spaces therefore differ only in the nature of their points, and can, from the point of view of topology, be considered essentially identical.

We have just used the word *topology* in its primary sense, as the name of a branch of mathematics. This word derives from two Greek words, and its literal meaning is "the science of position." How can we account for this? Let us say that a *topological property* is a property which, if possessed by a topological space X, is also possessed by every homeomorphic image of X. The subject of *topology* can now be defined as the study of all topological properties of topological spaces. If, very roughly, we think of a topological space as a general type of geometric configuration, say, a diagram drawn on a sheet of rubber, then a homeomorphism may be thought of as any deformation of this diagram (by stretching, bending, etc.) which does not tear the sheet. A circle can be deformed in this way into an ellipse, a triangle, or a square, but not into a figure eight, a horseshoe, or a single point. A topological property would then be any property of the diagram which is invariant under (or unchanged by) such a deformation. Distances, angles, and the like, are not topological properties, because they can be altered by suitable "non-tearing" deformations. What sorts of properties are topological? In the case of the circle, the fact that it has one "inside" and one "outside" (a point has no inside, and a figure eight has two). Also, the fact that when two points are removed from a circle it falls into two pieces, whereas if only one point is removed, then the circle remains in one piece. These remarks may suffice to indicate why topology is often described to non-mathematicians as "rubber sheet geometry." For an excellent nontechnical discussion of topology from this geometric point of view, see Courant and Robbins [6, chap. 5].

Problems

1. Let T_1 and T_2 be two topologies on a non-empty set X, and show that $T_1 \cap T_2$ is also a topology on X.
2. Let X be a non-empty set, and consider the class of subsets of X consisting of the empty set $\emptyset$ and all sets whose complements are countable. Is this a topology on X?
3. Which topological spaces given as examples in the text are metrizable? (*Hint:* if a space is metrizable, then its open sets must have certain properties.)

4. Show that if a topological space is metrizable, then it is metrizable in an infinite number of different ways (i.e., by means of an infinite number of different metrics).
5. Show that a subspace of a topological space is itself a topological space.
6. Let X be a topological space, and let Y and Z be subspaces of X such that $Y \subseteq Z$. Show that the topology which Y has as a subspace of X is the same as that which it has as a subspace of Z.
7. Let f be a continuous mapping of a topological space X into a topological space Y. If Z is a subspace of X, show that the restriction of f to Z is continuous.
8. Let X and Y be topological spaces, and f a mapping of X into Y. Show that f is continuous $\Leftrightarrow$ it is continuous as a mapping of X onto the subspace $f(X)$ of Y.
9. Let X, Y, and Z be topological spaces. If $f:X \to Y$ and $g:Y \to Z$ are continuous mappings, show that $gf:X \to Z$ is also continuous.
10. Let f be a one-to-one mapping of one topological space onto another, and show that f is a homeomorphism $\Leftrightarrow$ both f and f^{-1} are continuous.
11. Give an example to show that a one-to-one continuous mapping of one topological space onto another need not be a homeomorphism. (*Hint:* consider Examples 2 and 3.)
12. Show that a topological space X is metrizable $\Leftrightarrow$ there exists a homeomorphism of X onto a subspace of some metric space Y.
13. If X and Y are topological spaces, let $X \sim Y$ mean that X and Y are homeomorphic. Show that this relation is reflexive, symmetric, and transitive.

17. ELEMENTARY CONCEPTS

We have taken open sets as the starting point in our development of topology, and we now define a number of other basic concepts in terms of open sets. Most of these will be familiar to the reader from the previous chapter, and he will observe that in every case the definition given here is a strict generalization of our earlier definition or some equivalent form of it.

A *closed set* in a topological space is a set whose complement is open. The following theorem is an immediate consequence of Eqs. 2-(2) and the assumed properties of open sets.

Theorem A. *Let X be a topological space. Then* (1) *any intersection of closed sets in X is closed; and* (2) *any finite union of closed sets in X is closed.*

By considering the empty class of closed sets, we see at once that the empty set and the full space—its union and intersection—are always closed sets in every topological space.

If A is a subset of a topological space, then its *closure* (denoted by $\bar{A}$) is the intersection of all closed supersets of A. It is easy to see that the closure of A is a closed superset of A which is contained in every closed superset of A, and that A is closed $\Leftrightarrow A = \bar{A}$. A subset A of a topological space X is said to be *dense* (or *everywhere dense*) if $\bar{A} = X$, and X is called a *separable space* if it has a countable dense subset. For reasons which will become clear at the end of this section, we summarize the main facts about the operation of forming closures in the following theorem. Its proof is a direct application of the above statements.

Theorem B. *Let X be a topological space. If A and B are arbitrary subsets of X, then the operation of forming closures has the following four properties*: (1) $\bar{\emptyset} = \emptyset$; (2) $A \subseteq \bar{A}$; (3) $\bar{\bar{A}} = \bar{A}$; *and* (4) $\overline{A \cup B} = \bar{A} \cup \bar{B}$.

A *neighborhood* of a point (or a set) in a topological space is an open set which contains the point (or the set). A class of neighborhoods of a point is called an *open base for the point* (or an *open base at the point*) if each neighborhood of the point contains a neighborhood in this class. In the case of a point in a metric space, an open sphere centered on the point is a neighborhood of the point, and the class of all such open spheres is an open base for the point. Our next theorem gives a useful characterization (in terms of neighborhoods) of the closure of a set.

Theorem C. *Let X be a topological space and A an arbitrary subset of X. Then $\bar{A} = \{x : each\ neighborhood\ of\ x\ intersects\ A\}$.*

PROOF. We begin by proving that $\bar{A}$ is contained in the given set (the set on the right) by showing that any point not in the given set is not in $\bar{A}$. Let x be a point with a neighborhood which does not intersect A. Then the complement of this neighborhood is a closed superset of A which does not contain x, and since $\bar{A}$ is the intersection of all closed supersets of A, x is not in $\bar{A}$. In the same way, it can easily be shown that $\bar{A}$ contains the given set.

Let X be a topological space and A a subset of X. A point in A is called an *isolated point* of A if it has a neighborhood which contains no other point of A. A point x in X is said to be a *limit point* of A if each of its neighborhoods contains a point of A different from x. The *derived set* of A—denoted by $D(A)$—is the set of all limit points of A.

Theorem D. *Let X be a topological space and A a subset of X. Then* (1) $\bar{A} = A \cup D(A)$; *and* (2) A *is closed* $\Leftrightarrow A \supseteq D(A)$.

PROOF. To prove (1), we use Theorem C to show that any point not in one side is also not in the other. If x is not in $\bar{A}$, then it has a neighborhood disjoint from A, so it is not in A or $D(A)$; and if x is not in A or $D(A)$, then it has a neighborhood disjoint from A, so it is not in $\bar{A}$.

We prove (2) as follows. If A is closed, so that $A = \bar{A}$, then by (1) $A = A \cup D(A)$, from which we see that $A \supseteq D(A)$; and if $A \supseteq D(A)$, so that $A \cup D(A) = A$, then by (1) we have $A = \bar{A}$, so A is closed.

By the above definitions, a point in a set is either an isolated point of the set or a limit point of the set, but not both. This fact leads to the following obvious but rather satisfying theorem.

Theorem E. *Let X be a topological space. Then any closed subset of X is the disjoint union of its set of isolated points and its set of limit points, in the sense that it contains these sets, they are disjoint, and it is their union.*

Let X be a topological space and A a subset of X. The *interior* of A [denoted by $\text{Int}(A)$] is the union of all open subsets of A, and a point in the interior of A is called an *interior point* of A. It is clear that the interior of A is an open subset of A which contains every open subset of A, and that A is open $\Leftrightarrow A = \text{Int}(A)$. Also, a point in A is an interior point of $A \Leftrightarrow$ it has a neighborhood which is contained in A. The *boundary* of A is $\bar{A} \cap \overline{A'}$, and a point in the boundary of A is called a *boundary point* of A. It follows at once from the definition that the boundary of A is a closed set, and that it consists of all points x in X with the property that each neighborhood of x intersects both A and A'.

It is easy to see from the neighborhood characterizations of the interior and boundary that a point in a set is an interior point of the set or a boundary point of the set, but not both. This immediately yields the following theorem, which serves to validate our feeling about the intuitive significance of interiors and boundaries.

Theorem F. *Let X be a topological space. Then any closed subset of X is the disjoint union of its interior and its boundary, in the sense that it contains these sets, they are disjoint, and it is their union.*

In defining a topological space, we chose "open set" as our primitive undefined term. Our next theorem shows that "closed set" would have served just as well.

Theorem G. *Let X be a non-empty set, and let there be given a class of subsets of X which is closed under the formation of arbitrary intersections and finite unions. Then the class of all complements of these sets is a topology on X whose closed sets are precisely those initially given.*

PROOF. This follows immediately from Eqs. 2-(2), the definition of a topology, and the definition of a closed set.

As the following theorem shows, we could even have taken the term "closure" as our undefined concept.

Theorem H. *Let X be a non-empty set, and let there be given a "closure" operation which assigns to each subset A of X a subset $\bar{A}$ of X in such a manner that* (1) $\bar{\emptyset} = \emptyset$, (2) $A \subseteq \bar{A}$, (3) $\bar{\bar{A}} = \bar{A}$, *and* (4) $\overline{A \cup B} = \bar{A} \cup \bar{B}$. *If a "closed" set A is defined to be one for which $A = \bar{A}$, then the class of all complements of such sets is a topology on X whose closure operation is precisely that initially given.*

PROOF. In view of Theorem G, it suffices to demonstrate two facts: that the class of all "closed" sets is closed under the formation of arbitrary intersections and finite unions; and that for any set A, $\bar{A}$ equals the intersection of all "closed" supersets of A.

By (1), the empty set is "closed," and from this and (4) we see that any finite union of "closed" sets is "closed." By (2), the full space X is "closed," so all that remains in the first part of our proof is to show that if $\{A_i\}$ is a non-empty class of sets such that $A_i = \overline{A_i}$ for every i, then $\bigcap_i A_i = \overline{\bigcap_i A_i}$. By (2), it suffices to prove that $\overline{\bigcap_i A_i} \subseteq \bigcap_i A_i$. For this, it suffices to show that $A \subseteq B \Rightarrow \bar{A} \subseteq \bar{B}$ (since $\bigcap_i A_i \subseteq A_i$ for each i, it will follow that $\overline{\bigcap_i A_i} \subseteq \overline{A_i} = A_i$ for each i, from which we see that $\overline{\bigcap_i A_i} \subseteq \bigcap_i A_i$). Assume that $A \subseteq B$. Then $B = A \cup B$, and by (4), $\bar{B} = \overline{A \cup B} = \bar{A} \cup \bar{B}$ or $\bar{A} \subseteq \bar{B}$.

We now let A be an arbitrary subset of X, and we show that $\bar{A}$ equals the intersection of all "closed" supersets of A. By (2) and (3), $\bar{A}$ is a "closed" superset of A, so it suffices to show that if $A \subseteq B$ and $B = \bar{B}$, then $\bar{A} \subseteq B$. Since $A \subseteq B$, $B = A \cup B$. By (4) and our assumption that $B = \bar{B}$, we obtain $B = \bar{B} = \overline{A \cup B} = \bar{A} \cup \bar{B} = \bar{A} \cup B$, so $\bar{A} \subseteq B$.

The four properties of the closure operation assumed in this theorem are called the *Kuratowski closure axioms.* The last two theorems show that it is possible to approach the subject of topological spaces by taking either closed sets or a closure operation as the basic undefined concept. A good deal of research was done along these lines in the early days of topology. It was found that there are many different ways of defining a topological space, all of which are equivalent to one another. Several decades of experience have convinced most mathematicians that the open set approach is the simplest, the smoothest, and the most natural.

Problems

1. Let $f: X \to Y$ be a mapping of one topological space into another. Show that f is continuous $\Leftrightarrow f^{-1}(F)$ is closed in X whenever F is closed in $Y \Leftrightarrow f(\bar{A}) \subseteq \overline{f(A)}$ for every subset A of X.

2. Let X be a topological space, Y a metric space, and A a subspace of X. If f is a continuous mapping of A into Y, show that f can be extended in at most one way to a continuous mapping of $\bar{A}$ into Y. (*Hint:* see Problem 13-3.)

3. Show that a subset of a topological space is dense $\Leftrightarrow$ it intersects every non-empty open set.
4. Let A be a non-empty subset of a topological space, and show that A is dense as a subset of the subspace $\bar{A}$.
5. A subset A of a topological space is called a *perfect set* if $A = D(A)$. Show that a set is perfect $\Leftrightarrow$ it is closed and has no isolated points. Show that the Cantor set is perfect.
6. Show that $\text{Int}(A') = \bar{A}'$ for every subset A of a topological space.
7. Show that a subset of a topological space is closed $\Leftrightarrow$ it contains its boundary.
8. Show that a subset of a topological space has empty boundary $\Leftrightarrow$ it is both open and closed. (Every topological space X has the property that the empty set $\emptyset$ and the full space X are both open and closed. In Chap. 6 we study the hypothesis that these are the *only* subsets of X which are both open and closed.)
9. A subset A of a topological space is said to be *nowhere dense* if $\bar{A}$ has empty interior.
    (a) Show that a set A is nowhere dense $\Leftrightarrow$ every non-empty open set has a non-empty open subset disjoint from A.
    (b) Show that a closed set is nowhere dense $\Leftrightarrow$ its complement is everywhere dense. Is this true for an arbitrary set?
    (c) Show that the boundary of a closed set is nowhere dense. Is this true for an arbitrary set?

18. OPEN BASES AND OPEN SUBBASES

A special role is played in the theory of metric spaces by the class of open spheres within the class of all open sets. The main feature of their relationship is that the open sets coincide with all unions of open spheres, and it follows from this that the continuity of a mapping can be expressed either in terms of open spheres or in terms of open sets, at our convenience. We now develop similar machinery for topological spaces.

Let X be a topological space. An *open base* for X is a class of open sets with the property that every open set is a union of sets in this class. This condition can also be expressed in the following equivalent form: if G is an arbitrary non-empty open set and x is a point in G, then there exists a set B in the open base such that $x \in B \subseteq G$. The sets in an open base are referred to as *basic open sets*. It is clear that the class of open spheres in a metric space is an open base, and also that any class of open sets which contains an open base is itself an open base.

Generally speaking, an open base is useful only if its sets are simple in form or few in number. For instance, a space which has a countable open base has many pleasant properties. A space of this kind is said to be a *second countable space*, or to satisfy the *second axiom of countability*.[1] It is easy to see that any subspace of a second countable space is also second countable, for the class of all intersections with the subspace of sets in an open base is evidently an open base for the subspace. The central fact about second countable spaces can be stated as follows.

Theorem A (Lindelöf's Theorem). *Let X be a second countable space. If a non-empty open set G in X is represented as the union of a class $\{G_i\}$ of open sets, then G can be represented as a countable union of G_i's.*

PROOF. Let $\{B_n\}$ be a countable open base for X. Let x be a point in G. The point x is in some G_i, and we can find a basic open set B_n such that $x \in B_n \subseteq G_i$. If we do this for each point x in G, we obtain a subclass of our countable open base whose union is G, and this subclass is necessarily countable. Further, for each basic open set in this subclass we can select a G_i which contains it. The class of G_i's which arises in this way is clearly countable, and its union is G.

Most applications of Lindelöf's theorem depend more directly on the following simple consequence of it.

Theorem B. *Let X be a second countable space. Then any open base for X has a countable subclass which is also an open base.*

PROOF. Let $\{B_n\}$ be a countable open base and $\{B_i\}$ an arbitrary open base. Since each B_n is a union of B_i's, we see by Lindelöf's theorem that each non-empty B_n is the union of a countable class of B_i's. In this way we obtain a countable family of countable classes of B_i's. The union of this family of classes is evidently an open base which is a countable subclass of the open base $\{B_i\}$.

If a topological space X has a countable open base $\{B_n\}$, then it also has a countable dense subset. To see this, we have only to select a point in each non-empty B_n and to note that the set of all these points is countable and dense in X. Thus every second countable space is separable. This simple result admits the following partial converse.

Theorem C. *Every separable metric space is second countable.*

PROOF. Let X be a separable metric space, and let A be a countable dense subset. If we consider the open spheres with rational radii centered on all the points of A, then the class of all these open spheres is a countable class of open sets. We show that it is an open base. Let G be an arbitrary non-empty open set and x a point in G. We must find an open sphere in our class which contains x and is contained in G. Let

---

[1] A *first countable space*—or a space which satisfies the *first axiom of countability*—is a topological space which has a countable open base at each of its points (see Sec. 17).

$S_r(x)$ be an open sphere centered on x and contained in G, and consider the concentric open sphere $S_{r/3}(x)$ with one-third its radius. Since A is dense, there exists a point a in A which is in $S_{r/3}(x)$. Let r_1 be a rational number such that $r/3 < r_1 < 2r/3$. We conclude the proof by observing that $x \varepsilon S_{r_1}(a) \subseteq S_r(x) \subseteq G$.

In order to form the simplest intuitive picture of our next concept, we give a brief discussion of rectangles and strips in the Euclidean plane R^2. Figure 22 is intended to illustrate our remarks. If (a_1,b_1) and (a_2,b_2) are bounded open intervals—one on the x_1 axis and the other on the x_2 axis—then their product

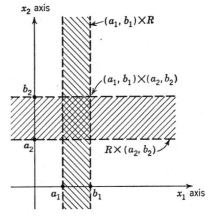

$$(a_1,b_1) \times (a_2,b_2) = \{(x_1,x_2) : a_i < x_i < b_i \text{ for } i = 1,2\}$$

is called an *open rectangle* in R^2. A *closed rectangle* is defined similarly, as a product of two closed intervals. It is easy to prove (see Problem 8)

Fig. 22. Open strips and an open rectangle.

that the class of all open rectangles is an open base for the Euclidean plane. We now observe that each open rectangle is the intersection of two open strips, in the following sense. We call sets of the form

$$(a_1,b_1) \times R = \{(x_1,x_2) : a_1 < x_1 < b_1, \ x_2 \text{ arbitrary}\}$$
$$\text{and} \qquad R \times (a_2,b_2) = \{(x_1,x_2) : a_2 < x_2 < b_2, \ x_1 \text{ arbitrary}\}$$

open strips in R^2. If we use closed intervals here, we get what we call *closed strips*. It is plain that

$$(a_1,b_1) \times (a_2,b_2) = [(a_1,b_1) \times R] \cap [R \times (a_2,b_2)].$$

Since every open strip in R^2 is clearly an open set, the class of all open strips is a class of open sets whose finite intersections form an open base, namely, the open base composed of the open strips, the open rectangles, the empty set, and the full space R^2.

Now let X be a topological space. An *open subbase* is a class of open subsets of X whose finite intersections form an open base. This open base is called the open base *generated by* the open subbase. We refer to the sets in an open subbase as *subbasic open sets*. It is easy to see that any class of open sets which contains an open subbase is also an open subbase. Since the bounded open intervals on the real line constitute an open base for this space, it is clear that all open intervals of the type

$(a, + \infty)$ and $(- \infty, b)$, where a and b are real numbers, form an open subbase. The open base generated by this open subbase consists of all open intervals of this kind, all bounded open intervals, the empty set, and the full space R. The ideas in the previous paragraph show at once that all open strips in the Euclidean plane form an open subbase for this space.

The practical value of open subbases rests mainly on the following theorem.

Theorem D. *Let X be any non-empty set, and let* S *be an arbitrary class of subsets of X. Then* S *can serve as an open subbase for a topology on X, in the sense that the class of all unions of finite intersections of sets in* S *is a topology.*

PROOF. If S is empty, then the class of all finite intersections of its sets is the single-element class $\{X\}$, and the class of all unions of sets in this class is the two-element class $\{\emptyset, X\}$. Since this is the topology described in Example 16-3, we may assume that S is non-empty. Let B be the class of all finite intersections of sets in S, and let T be the class of all unions of sets in B. We must show that T is a topology. T clearly contains $\emptyset$ and X, and is closed under the formation of arbitrary unions. All that remains is to show that if $\{G_1, G_2, \ldots, G_n\}$ is a non-empty finite class of sets in T, then $G = \bigcap_{i=1}^{n} G_i$ is also in T. Since the empty set is in T, we may assume that G is non-empty. Let x be a point in G. Then x is in each G_i, and by the definition of T, for each i there is a set B_i in B such that $x \in B_i \subseteq G_i$. Since each B_i is a finite intersection of sets in S, the intersection of all sets in S which arise in this way is a set in B which contains x and is contained in G. We conclude the proof by noting that this shows that G is a union of sets in B and is thus itself a set in T.

We speak of the topology in this theorem as the topology *generated by* the class S. As we shall see in later chapters, this theorem, though not particularly valuable as an end in itself, is quite a useful tool. It is normally used in the following manner. If X is a non-empty set, and if we have a class of subsets of X which we wish to regard as open sets, all we have to do is form the topology generated by this class in the sense of Theorem D.

Our next result often makes much lighter the task of proving that a given specific mapping is either continuous or open.

Theorem E. *Let $f: X \to Y$ be a mapping of one topological space into another, and let there be given an open base in X and an open subbase with its generated open base in Y. Then (1) f is continuous $\Leftrightarrow$ the inverse image* *of each basic open set is open ⇔ the inverse image of each subbasic open set is open; and (2) f is open ⇔ the image of each basic open set is open.*

PROOF. These statements are immediate consequences of the definitions and, respectively, Eqs. 3-(2) and 3-(3) and Eq. 3-(1).

We put these two theorems to work in the next section, where we develop a fragment of lattice theory which is very useful in the applications of topology to modern analysis.

Problems

1. Let X be a topological space, and **B** an open base with the property that each point in the space is contained in a basic open set different from X. Show that if $\emptyset$ and X happen to be in **B**, then the class which results when these two sets are dropped from **B** is still an open base.

2. Under what circumstances is the metric space defined in Example 9-1 separable?

3. Show that the real line and the complex plane are separable. Show also that R^n and C^n are separable. Show finally that R^∞ and C^∞ are separable.

4. Let X be the metric space whose points are the positive integers and whose metric is that defined in Example 9-1, and show that $\mathcal{C}(X,R)$ is not separable. (*Hint:* if $\{f_n\}$ is a sequence in $\mathcal{C}(X,R)$, and if f is the function in $\mathcal{C}(X,R)$ defined by $f(n) = 0$ if $|f_n(n)| \geq 1$ and $f(n) = |f_n(n)| + 1$ if $|f_n(n)| < 1$, then $\|f - f_n\| \geq 1$ for every n.)

5. Let X be any non-empty set with the metric defined in Example 9-1, and show that $\mathcal{C}(X,R)$ is separable ⇔ X is finite.

6. The following example demonstrates that a topological space with a countable dense subset need not be second countable. Let X be the set of all real numbers with the topology described in Example 16-4.
    (a) Show that any infinite subset of X is dense.
    (b) Show that X is not second countable. (*Hint:* assume that there exists a countable open base, let x_0 be a fixed point in X, show that the intersection of all basic open sets which contain x_0 is the single-element set $\{x_0\}$, and conclude from this that the complement of $\{x_0\}$ is countable.)

7. Show that the set of all isolated points of a second countable space is empty or countable. Show from this that any uncountable subset A of a second countable space must have at least one point which is a limit point of A.

8. Prove in detail that the open rectangles in the Euclidean plane form an open base.

9. Let $f: X \to Y$ be a mapping of one topological space into another. f is said to be *continuous at a point x_0 in X* if for each neighborhood H of $f(x_0)$ there exists a neighborhood G of x_0 such that $f(G) \subseteq H$.

(a) Show that f is continuous $\Leftrightarrow$ it is continuous at each point in X.

(b) If there is given an open base in Y, show that f is continuous at $x_0 \Leftrightarrow$ for each basic open set B which contains $f(x_0)$ there exists a neighborhood G of x_0 such that $f(G) \subseteq B$.

(c) If Y is a metric space, show that f is continuous at $x_0 \Leftrightarrow$ for each open sphere $S_r(f(x_0))$ centered on $f(x_0)$ there exists a neighborhood G of x_0 such that $f(G) \subseteq S_r(f(x_0))$.

19. WEAK TOPOLOGIES

Let X be a non-empty set. If T_1 and T_2 are topologies on X such that $\mathsf{T}_1 \subseteq \mathsf{T}_2$, we say that T_1 is *weaker* than T_2 (or T_2 is *stronger* than T_1). In rough terms, one topology is weaker than another if it has fewer open sets, and stronger than another if it has more open sets. The topology $\{\emptyset, X\}$ is the *weakest topology* on X, for it is weaker than every topology; and the discrete topology is the *strongest topology* on X, since it is stronger than every topology. It is clear that the family of all topologies on X is a partially ordered set with respect to the relation "is weaker than."

We next show that this partially ordered set is a complete lattice. In Problem 16-1 we asked the reader to prove that the intersection of any two topologies T_1 and T_2 on X is a topology on X. Since this topology is evidently weaker than both T_1 and T_2 and stronger than any topology which is weaker than both, it is the greatest lower bound of T_1 and T_2. It is equally easy to see that the intersection of any non-empty family of topologies on X is a topology on X; and since it is weaker than all these and stronger than any topology which is weaker than all these, it is the greatest lower bound of this family. What about least upper bounds? The situation here is a bit different, for the union of two topologies on X need not be a topology. However, if we have any non-empty family of topologies T_i, then the discrete topology is a topology stronger than each T_i. We can therefore appeal to our above remarks to conclude that the intersection of all topologies which are stronger than each T_i is a topology; and since it is stronger than each T_i and weaker than any topology which is stronger than each T_i, it is the least upper bound of our given family.

We summarize the results of this discussion in the following theorem.

Theorem A. *Let X be a non-empty set. Then the family of all topologies on X is a complete lattice with respect to the relation "is weaker than."*

Furthermore, this lattice has a least member (the weakest topology on X) and a greatest member (the discrete topology on X).

The reader will observe that if $\{T_i\}$ is a non-empty family of topologies on our set X, then the least upper bound of this family is precisely the topology generated by the class $\cup_i T_i$ in the sense of Theorem 18-D; that is, the class $\cup_i T_i$ is an open subbase for the least upper bound of the family $\{T_i\}$. In the present context, therefore, Theorem 18-D can be thought of as providing a mechanism for the direct construction of least upper bounds in our lattice of topologies.

Let X be a non-empty set, let $\{X_i\}$ be a non-empty class of topological spaces, and for each i let f_i be a mapping of X into X_i. It is clear that if X is given its discrete topology, then all the f_i's are continuous. If we look a little further, we may find other and weaker topologies on X which also have this property. There is, in fact, a unique weakest topology of this kind. The *weak topology generated by the f_i's* is defined to be the intersection of all topologies on X with respect to each of which all the f_i's are continuous mappings. This is clearly a topology on X which makes all the f_i's continuous, and it is weaker than any topology which has this property. It will appear in later chapters that many a topology which is used in practice is defined to be the weak topology generated by some set of mappings of particular interest in a given situation.

Problems

1. Let X be a non-empty set and $\{X_i\}$ a non-empty class of topological spaces. If for each i there is given a mapping f_i of X into X_i, denote by **T** the weak topology on X generated by the f_i's.

    (a) Show that **T** equals the topology generated by the class of all inverse images in X of open sets in the X_i's.

    (b) If an open subbase is given in each X_i, show that **T** equals the topology generated by the class of all inverse images in X of subbasic open sets in the X_i's.

    (c) If Y is a subspace of the topological space $(X, \mathbf{T})$, show that the relative topology on Y is the weak topology generated by the restrictions of the f_i's to Y.

2. In each of the following we specify a set $\{f_i\}$ of real functions defined on the real line R. In each case give a complete description of the weak topology on R generated by the f_i's.

    (a) $\{f_i\}$ consists of all constant functions.

    (b) $\{f_i\}$ consists of a single function f, defined by $f(x) = 0$ if $x \leq 0$ and $f(x) = 1$ if $x > 0$.

(c) $\{f_i\}$ consists of a single function f, defined by $f(x) = -1$ if $x < 0$, $f(0) = 0$, and $f(x) = 1$ if $x > 0$.

(d) $\{f_i\}$ consists of a single function f, defined by $f(x) = x$ for all x.

(e) $\{f_i\}$ consists of all bounded functions which are continuous with respect to the usual topology on R.

(f) $\{f_i\}$ consists of all functions which are continuous with respect to the usual topology on R.

20. THE FUNCTION ALGEBRAS $\mathcal{C}(X,R)$ AND $\mathcal{C}(X,C)$

Let X be an arbitrary topological space. We generalize the notations established in Sec. 14 by defining $\mathcal{C}(X,R)$ and $\mathcal{C}(X,C)$ to be the sets of all bounded continuous functions defined on X which are, respectively, real and complex.

It is desirable to extend our discussion of the algebraic structure of these sets beyond that given in Sec. 14 by introducing the following concepts. An *algebra* is a linear space whose vectors can be multiplied in such a way that

(1) $x(yz) = (xy)z$;

(2) $x(y + z) = xy + xz$ and $(x + y)z = xz + yz$;

(3) $\alpha(xy) = (\alpha x)y = x(\alpha y)$ for every scalar α.

We speak of a *real algebra* or a *complex algebra* according as the scalars are the real numbers or the complex numbers. A *commutative algebra* is an algebra whose multiplication satisfies the following condition:

(4) $xy = yx$.

In the case of a commutative algebra, the second part of (2) is clearly redundant. An *algebra with identity* is an algebra which possesses the following property:

(5) there exists a non-zero element in the algebra, denoted by 1 and called the *identity element* (or the *identity*), such that $1 \cdot x = x \cdot 1 = x$ for every x.

We speak of *the* identity because the identity in an algebra (if it has one) is unique; for if $1'$ is also an element such that $1' \cdot x = x \cdot 1' = x$ for every x, then $1' = 1' \cdot 1 = 1$. A *subalgebra* of an algebra is a linear subspace which contains the product of each pair of its elements. A subalgebra of an algebra is evidently an algebra in its own right.

In the case of a function space which is also an algebra, it is to be understood that multiplication is defined pointwise, that is, that the product fg of two functions in the space is defined by $(fg)(x) = f(x)g(x)$. This pointwise multiplication of functions should be clearly distinguished from the multiplication (or composition) of mappings discussed at the end of Sec. 3. If such an algebra has an identity element 1, then Problem

1 shows that in all cases of interest to us this identity is the constant function defined by $1(x) = 1$ for all x.

We prove two lemmas before going on to our main theorems.

Lemma. *If f and g are continuous real or complex functions defined on a topological space X, then $f + g$, αf, and fg are also continuous. Furthermore, if f and g are real, then $f \wedge g$ and $f \vee g$ are continuous.*

PROOF. We illustrate the method by showing that fg and $f \vee g$ are continuous.

We prove that fg is continuous by showing that it is continuous at an arbitrary point x_0 in X (see Problem 18-9). Let $\epsilon > 0$ be given, and find $\epsilon_1 > 0$ such that $\epsilon_1(|f(x_0)| + |g(x_0)|) + \epsilon_1{}^2 < \epsilon$. Since f is continuous, and thus continuous at x_0, there exists a neighborhood G_1 of x_0 such that $x \in G_1 \Rightarrow |f(x) - f(x_0)| < \epsilon_1$. Similarly, there exists a neighborhood G_2 of x_0 such that $x \in G_2 \Rightarrow |g(x) - g(x_0)| < \epsilon_1$. The continuity of fg at x_0 now follows from the fact that $G = G_1 \cap G_2$ is a neighborhood of x_0 such that

$$
\begin{aligned}
x \in G \Rightarrow |(fg)(x) - (fg)(x_0)| &= |f(x)g(x) - f(x_0)g(x_0)| \\
&= |[f(x)g(x) - f(x)g(x_0)] + [f(x)g(x_0) - f(x_0)g(x_0)]| \\
&\leq |f(x)|\,|g(x) - g(x_0)| + |g(x_0)|\,|f(x) - f(x_0)| < \epsilon_1|f(x)| + \epsilon_1|g(x_0)| \\
&= \epsilon_1|[f(x) - f(x_0)] + f(x_0)| + \epsilon_1|g(x_0)| \leq \epsilon_1|f(x) - f(x_0)| + \epsilon_1|f(x_0)| \\
&\quad + \epsilon_1|g(x_0)| < \epsilon_1(|f(x_0)| + |g(x_0)|) + \epsilon_1{}^2 < \epsilon.
\end{aligned}
$$

We prove that $f \vee g$ is continuous by recalling that all sets of the form $A = (a, +\infty)$ and $B = (-\infty, b)$ form an open subbase for the real line and by showing that the inverse image of any such set is open (see Theorem 18-E). All that is necessary is to observe that

$$
(f \vee g)^{-1}(A) = \{x : \max\{f(x), g(x)\} > a\} = \{x : f(x) > a\} \cup \{x : g(x) > a\},
$$

which is open since it is the union of two open sets, and that

$$
(f \vee g)^{-1}(B) = \{x : \max\{f(x), g(x)\} < b\} = \{x : f(x) < b\} \cap \{x : g(x) < b\},
$$

which is open since it is the intersection of two open sets.

Lemma. *Let X be a topological space, and let $\{f_n\}$ be a sequence of real or complex functions defined on X which converges uniformly to a function f defined on X. If all the f_n's are continuous, then f is also continuous.*

PROOF. We show that f is continuous by showing that it is continuous at an arbitrary point x_0 in X. Let $\epsilon > 0$ be given. Since f is the uniform limit of the f_n's, there exists a positive integer n_0 such that $|f(x) - f_{n_0}(x)| < \epsilon/3$ for all points x in X. Since f_{n_0} is continuous, and thus continuous at x_0, there exists a neighborhood G of x_0 such that $x \in G \Rightarrow |f_{n_0}(x) -$

$f_{n_0}(x_0)| < \epsilon/3$.   The continuity of f at x_0 now follows from the fact that

$$x \, \epsilon \, G \Rightarrow |f(x) - f(x_0)|$$
$$= |[f(x) - f_{n_0}(x)] + [f_{n_0}(x) - f_{n_0}(x_0)] + [f_{n_0}(x_0) - f(x_0)]|$$
$$\leq |f(x) - f_{n_0}(x)| + |f_{n_0}(x) - f_{n_0}(x_0)| + |f_{n_0}(x_0) - f(x_0)|$$
$$< \epsilon/3 + \epsilon/3 + \epsilon/3 = \epsilon.$$

This lemma is often stated more informally as follows: *any uniform limit of continuous functions is continuous.*

We are now in a position to give Theorem 14-A the following broader and richer form.

Theorem A.   *Let* $\mathcal{C}(X,R)$ *be the set of all bounded continuous real functions defined on a topological space* X. *Then* (1) $\mathcal{C}(X,R)$ *is a real Banach space with respect to pointwise addition and scalar multiplication and the norm defined by* $\|f\| = \sup |f(x)|$; (2) *if multiplication is defined pointwise,* $\mathcal{C}(X,R)$ *is a commutative real algebra with identity in which* $\|fg\| \leq \|f\| \, \|g\|$ *and* $\|1\| = 1$; *and* (3) *if* $f \leq g$ *is defined to mean that* $f(x) \leq g(x)$ *for all* x, $\mathcal{C}(X,R)$ *is a lattice in which the greatest lower bound and least upper bound of a pair of functions* f *and* g *are given by* $(f \wedge g)(x) = \min \{f(x),g(x)\}$ *and* $(f \vee g)(x) = \max \{f(x),g(x)\}$.

PROOF.   In view of the above lemmas, everything stated here is clear, except perhaps the fact that $\|fg\| \leq \|f\| \, \|g\|$; and this follows from

$$\|fg\| = \sup |(fg)(x)| = \sup |f(x)g(x)| = \sup |f(x)| \, |g(x)|$$
$$\leq (\sup |f(x)|)(\sup |g(x)|) = \|f\| \, \|g\|.$$

We also extend Theorem 14-B, but in a slightly different direction.

Theorem B.   *Let* $\mathcal{C}(X,C)$ *be the set of all bounded continuous complex functions defined on a topological space* X. *Then* (1) $\mathcal{C}(X,C)$ *is a complex Banach space with respect to pointwise addition and scalar multiplication and the norm defined by* $\|f\| = \sup |f(x)|$; (2) *if multiplication is defined pointwise,* $\mathcal{C}(X,C)$ *is a commutative complex algebra with identity in which* $\|fg\| \leq \|f\| \, \|g\|$ *and* $\|1\| = 1$; *and* (3) *if* $\bar{f}$ *is defined by* $\bar{f}(x) = \overline{f(x)}$, *then* $f \rightarrow \bar{f}$ *is a mapping of the algebra* $\mathcal{C}(X,C)$ *into itself which has the following properties:* $\overline{f + g} = \bar{f} + \bar{g}$, $\overline{\alpha f} = \bar{\alpha} \cdot \bar{f}$, $\overline{fg} = \bar{f} \cdot \bar{g}$, $\bar{\bar{f}} = f$, *and* $\|\bar{f}\| = \|f\|$.

PROOF.   This theorem is a direct consequence of the background provided above.   We do remark, however, that the fact that $\bar{f}$ is continuous when f is follows from $|\bar{f}(x) - \bar{f}(x_0)| = |f(x) - f(x_0)|$.

The function $\bar{f}$ defined in this theorem is called the *conjugate* of the function f, and the operation of forming $\bar{f}$ from f is called *conjugation.* It will become clear in the later chapters of this book that the operation of conjugation in the space $\mathcal{C}(X,C)$ is one of the chief supporting pillars of the theory we develop in those chapters.

We trust that the reader has noticed our insistence on assuming that a topological space always has at least one point. Our reason for this is that the empty set has no functions defined on it. If we were to allow a topological space X to be empty, then we would have to cope with the fact that its corresponding $\mathcal{C}(X,R)$ and $\mathcal{C}(X,C)$ are also empty, and so cannot be linear spaces, for a linear space must contain at least one vector (the zero vector). Since constant functions are always continuous, we avoid this difficulty by taking pains to assume that topological spaces are non-empty.

Problems

1. Let A be an algebra of real or complex functions defined on a non-empty set X, and assume that for each point x in X there is a function f in A such that $f(x) \neq 0$. Show that if A contains an identity element 1, than $1(x) = 1$ for all x.

2. Let f be a continuous real or complex function defined on a topological space X, and assume that f is not identically zero, i.e., that the set $Y = \{x : f(x) \neq 0\}$ is non-empty. Prove in detail that the function $1/f$ defined by $(1/f)(x) = 1/f(x)$ is continuous at each point of the subspace Y.

3. Let X be a topological space and A a subalgebra of $\mathcal{C}(X,R)$ or $\mathcal{C}(X,C)$. Show that its closure $\bar{A}$ is also a subalgebra. If A is a subalgebra of $\mathcal{C}(X,C)$ which contains the conjugate of each of its functions, show that $\bar{A}$ also contains the conjugate of each of its functions.

CHAPTER FOUR

Compactness

Like many other notions in topology, the concept of compactness for a topological space is an abstraction of an important property possessed by certain sets of real numbers. The property we have in mind is expressed by the *Heine-Borel theorem*, which asserts the following: if X is a closed and bounded subset of the real line R, then any class of open subsets of R whose union contains X has a finite subclass whose union also contains X. If we regard X as a topological space in its own right, as a subspace of R, then this theorem can be thought of as saying that any class of open subsets of X whose union is X has a finite subclass whose union is also X.

The Heine-Borel theorem has a number of profound and far-reaching applications in analysis. Many of these guarantee that continuous functions defined on closed and bounded sets of real numbers are well behaved. For instance, any such function is automatically bounded and uniformly continuous. In contrast to this satisfying behavior, we note that the function f defined on the open unit interval $(0,1)$ by $f(x) = 1/x$ is neither bounded nor uniformly continuous.

As is often the case with crucial theorems in analysis, the conclusion of the Heine-Borel theorem is converted into a definition in topology. This definition singles out for special attention what are called *compact topological spaces*. Our main business in this chapter is to develop the basic properties of these spaces and the continuous functions they carry, and, in the case of metric spaces, to establish several equivalent forms of compactness which are useful in applications.

110

21. COMPACT SPACES

Let X be a topological space.   A class $\{G_i\}$ of open subsets of X is said to be an *open cover* of X if each point in X belongs to at least one G_i, that is, if $\cup_i G_i = X$.   A subclass of an open cover which is itself an open cover is called a *subcover*.   A *compact space* is a topological space in which every open cover has a finite subcover.   A *compact subspace* of a topological space is a subspace which is compact as a topological space in its own right.   We begin by proving two simple but widely used theorems.

Theorem A.   *Any closed subspace of a compact space is compact.*

PROOF.   Let Y be a closed subspace of a compact space X, and let $\{G_i\}$ be an open cover of Y.   Each G_i, being open in the relative topology on Y, is the intersection with Y of an open subset H_i of X.   Since Y is closed, the class composed of Y' and all the H_i's is an open cover of X, and since X is compact, this open cover has a finite subcover.   If Y' occurs in this subcover, we discard it.   What remains is a finite class of H_i's whose union contains X.   Our conclusion that Y is compact now follows from the fact that the corresponding G_i's form a finite subcover of the original open cover of Y.

Theorem B.   *Any continuous image of a compact space is compact.*

PROOF.   Let $f\colon X \to Y$ be a continuous mapping of a compact space X into an arbitrary topological space Y.   We must show that $f(X)$ is a compact subspace of Y.   Let $\{G_i\}$ be an open cover of $f(X)$.   As in the above proof, each G_i is the intersection with $f(X)$ of an open subset H_i of Y.   It is clear that $\{f^{-1}(H_i)\}$ is an open cover of X, and by the compactness of X it has a finite subcover.   The union of the finite class of H_i's of which these are the inverse images clearly contains $f(X)$, so the class of corresponding G_i's is a finite subcover of the original open cover of $f(X)$, and $f(X)$ is compact.

It is sometimes quite difficult to prove that a given topological space is compact by appealing directly to the definition.   The following theorems give several equivalent forms of compactness which are often easier to apply.

Theorem C.   *A topological space is compact $\Leftrightarrow$ every class of closed sets with empty intersection has a finite subclass with empty intersection.*

PROOF.   This is a direct consequence of the fact that a class of open sets is an open cover $\Leftrightarrow$ the class of all their complements has empty intersection.

We recall from Problem 8-6 that a class of subsets of a non-empty set is said to have the *finite intersection property* if every finite subclass has non-empty intersection. This concept enables us to express Theorem C as follows.

Theorem D. *A topological space is compact $\Leftrightarrow$ every class of closed sets with the finite intersection property has non-empty intersection.*

Let X be a topological space. An open cover of X whose sets are all in some given open base is called a *basic open cover*, and if they all lie in some given open subbase, it is called a *subbasic open cover*. We observe the trivial fact that if X is compact, then every basic open cover has a finite subcover. Our next theorem asserts that compactness not only implies this property, but is also implied by it.

Theorem E. *A topological space is compact if every basic open cover has a finite subcover.*

PROOF. Let $\{G_i\}$ be an open cover and $\{B_j\}$ an open base. Each G_i is the union of certain B_j's, and the totality of all such B_j's is clearly a basic open cover. By our hypothesis, this class of B_j's has a finite subcover. For each set in this finite subcover we can select a G_i which contains it. The class of G_i's which arises in this way is evidently a finite subcover of the original open cover.

We go one more step in this direction and prove a similar (and much deeper) theorem relating to subbasic open covers. The proof is rather difficult, and we introduce the following concepts in an effort to make it as simple as possible. They also make some of its applications considerably easier to handle. Let X be a topological space. A class of closed subsets of X is called a *closed base* if the class of all complements of its sets is an open base, and a *closed subbase* if the class of all complements is an open subbase. Since the class of all finite intersections of sets in an open subbase is an open base, it follows that the class of all finite unions of sets in a closed subbase is a closed base. This is called the closed base *generated by* the closed subbase.

Theorem F. *A topological space is compact if every subbasic open cover has a finite subcover, or equivalently, if every class of subbasic closed sets with the finite intersection property has non-empty intersection.*

PROOF. The equivalence of the stated conditions is an easy consequence of Theorems C and D. Consider a closed subbase for our space, and let $\{B_i\}$ be its generated closed base, that is, the class of all finite unions of its sets. We assume that every class of subbasic closed sets with the finite intersection property has non-empty intersection, and we prove from this that every class of B_i's with the finite intersection property also has non-empty intersection. By Theorem E, this will suffice to prove our theorem.

Let $\{B_j\}$ be a class of B_i's with the finite intersection property. We must show that $\cap_j B_j$ is non-empty. We use Zorn's lemma to show that $\{B_j\}$ is contained in some class $\{B_k\}$ of B_i's which is maximal with respect to having the finite intersection property, in the sense that $\{B_k\}$ has this property and any class of B_i's which properly contains $\{B_k\}$ fails to have this property. The argument runs as follows. Consider the family of all classes of B_i's which contain $\{B_j\}$ and have the finite intersection property. This is a partially ordered set with respect to class inclusion. If we consider a chain in this partially ordered set, the union of all classes in it is a class of B_i's which contains every member of the chain and has the finite intersection property, as we see from the fact that every finite class of its sets is contained in some member of the chain, and that member has the finite intersection property. We conclude that every chain in our partially ordered set has an upper bound, so Zorn's lemma guarantees that the partially ordered set has a maximal element. This argument yields the existence of a class $\{B_k\}$ with the properties stated above. Since $\cap_k B_k \subseteq \cap_j B_j$, it now suffices to show that $\cap_k B_k$ is non-empty.

Each B_k is a finite union of sets in our closed subbase, for instance, $B_1 = S_1 \cup S_2 \cup \cdots \cup S_n$. It now suffices to show that at least one of the sets $S_1, S_2, \ldots, S_n$ belongs to the class $\{B_k\}$. For if we obtain such a set for each B_k, the resulting class of subbasic closed sets will have the finite intersection property (since it is contained in $\{B_k\}$), and therefore, by our hypothesis relating to the subbasic closed sets, it will have non-empty intersection; and since this non-empty intersection will be a subset of $\cap_k B_k$, we shall know that $\cap_k B_k$ is itself non-empty.

We finish the proof by showing that at least one of the sets $S_1, S_2, \ldots, S_n$ does in fact belong to the class $\{B_k\}$. We assume that each of these sets is not in this class, and we deduce a contradiction from this assumption. Since S_1 is a subbasic closed set, it is also a basic closed set; and since it is not in the class $\{B_k\}$, the class $\{B_k, S_1\}$ is a class of B_i's which properly contains $\{B_k\}$. By the maximality property of $\{B_k\}$, the class $\{B_k, S_1\}$ lacks the finite intersection property, so S_1 is disjoint from the intersection of some finite class of B_k's. If we do this for each of the sets $S_1, S_2, \ldots, S_n$, we see that B_1—the union of these sets—is disjoint from the intersection of the total finite class of all the B_k's which arise in this way. This contradicts the finite intersection property for the class $\{B_k\}$ and completes the proof.

The great power of this theorem can be surmised from the complexity of its proof. It is really a tool, and we illustrate the manner in which it can be used by applying it to give a simple proof of the classical Heine-Borel theorem stated in the introduction to this chapter.

Theorem G (the Heine-Borel Theorem). *Every closed and bounded subspace of the real line is compact.*

PROOF. A closed and bounded subspace of the real line is a closed subspace of some closed interval $[a,b]$, and by Theorem A it suffices to show that $[a,b]$ is compact. If $a = b$, this is clear, so we may assume that $a < b$. By Sec. 18, we know that the class of all intervals of the form $[a,d)$ and $(c,b]$, where c and d are any real numbers such that $a < c < b$ and $a < d < b$, is an open subbase for $[a,b]$; therefore the class of all $[a,c]$'s and all $[d,b]$'s is a closed subbase. Let $S = \{[a,c_i], [d_j,b]\}$ be a class of these subbasic closed sets with the finite intersection property. It suffices by Theorem F to show that the intersection of all sets in S is non-empty. We may assume that S is non-empty. If S contains only intervals of the type $[a,c_i]$, or only intervals of the type $[d_j,b]$, then the intersection clearly contains a or b. We may thus assume that S contains intervals of both types. We now define d by $d = \sup \{d_j\}$, and we complete the proof by showing that $d \leq c_i$ for every i. Suppose that $c_{i_0} < d$ for some i_0. Then by the definition of d there exists a d_{j_0} such that $c_{i_0} < d_{j_0}$. Since $[a,c_{i_0}] \cap [d_{j_0},b] = \emptyset$, this contradicts the finite intersection property for S and concludes the proof.

The reader should understand that there are elementary proofs of the Heine-Borel theorem which do not use Theorem F or anything like it. Theorem F will render us its major service in connection with the proof of the vital Tychonoff theorem of Sec. 23.

Problems

1. A *countably compact space* is a topological space in which every countable open cover has a finite subcover. Prove that a second countable space is countably compact ⇔ it is compact.

2. Let Y be a subspace of a topological space X. If Z is a non-empty subset of Y, show that Z is compact as a subspace of Y ⇔ it is compact as a subspace of X.

3. Let X be a topological space. If $\{X_i\}$ is a non-empty finite class of compact subspaces of X, show that $\cup_i X_i$ is also a compact subspace of X. If $\{X_j\}$ is a non-empty class of compact subspaces of X each of which is closed, and if $\cap_j X_j$ is non-empty, show that $\cap_j X_j$ is also a compact subspace of X.

4. Let X be a compact space. We know by Theorem A that every closed subspace of X is compact. By considering Example 16-3, show that a compact subspace of X need not be closed.

5. Prove the converse of the Heine-Borel theorem: every compact subspace of the real line is closed and bounded.

6. Generalize the preceding problem by proving that a compact subspace of an arbitrary metric space is closed and bounded. (It should be carefully noted, as Secs. 24 and 25 will show, that a closed and bounded subspace of an arbitrary metric space is not necessarily compact.)

7. Show that a continuous real or complex function defined on a compact space is bounded. More generally, show that a continuous mapping of a compact space into any metric space is bounded.

8. Show that a continuous real function f defined on a compact space X attains its infimum and its supremum in the following sense: if $a = \inf \{f(x) : x \in X\}$ and $b = \sup \{f(x) : x \in X\}$, then there exist points x_1 and x_2 in X such that $f(x_1) = a$ and $f(x_2) = b$.

9. If X is a compact space, and if $\{f_n\}$ is a monotone sequence of continuous real functions defined on X which converges pointwise to a continuous real function f defined on X, show that f_n converges uniformly to f. (The assumption that $\{f_n\}$ is a *monotone sequence* means that either $f_1 \leq f_2 \leq f_3 \leq \cdots$ or $f_1 \geq f_2 \geq f_3 \geq \cdots$.)

22. PRODUCTS OF SPACES

There are two main techniques for making new topological spaces out of old ones. The first of these, and the simplest, is to form subspaces of some given space. The second is to multiply together a number of given spaces. Our purpose in this section is to describe the way in which the latter process is carried out.

In Sec. 4 we defined what is meant by the product P_iX_i of an arbitrary non-empty class of sets. We also defined the projection p_i of this product onto its ith coordinate set X_i. The reader should make certain that these concepts are firmly in mind. If each coordinate set is a topological space, then there is a standard method of defining a topology on the product. It is difficult to exaggerate the importance of this definition, and we examine it with great care in the following discussion.

Let us begin by recalling the discussion in Sec. 18 of open rectangles and open strips in the Euclidean plane R^2. We observed there that the open rectangles form an open base for the topology of R^2, and also that the open strips form an open subbase for this topology whose generated open base consists of all open rectangles, all open strips, the empty set, and the full space. The topology of the Euclidean plane is of course defined in terms of a metric. If we wish, however, we can ignore this fact and regard the topology of R^2 as generated in the sense of Theorem

18-D by the class of all open strips. This situation provides the motivation for the more general ideas we now develop.

Let X_1 and X_2 be topological spaces, and form the product $X = X_1 \times X_2$ of the two sets X_1 and X_2. Consider the class S of all subsets of X of the form $G_1 \times X_2$ and $X_1 \times G_2$, where G_1 and G_2 are open subsets of X_1 and X_2, respectively. The topology on X generated by this class in the sense of Theorem 18-D is called the *product topology* on X. The product topology therefore has S as an open subbase; in fact, it is defined by the requirement that S be an open subbase. The open base generated by S, that is, the class of all finite intersections of its sets, is clearly the class of all sets of the form $G_1 \times G_2$, and the open

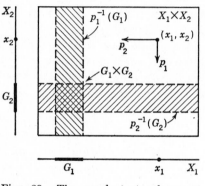

sets in X are all unions of these sets. There are two projections p_1 and p_2 of X onto its coordinate spaces X_1 and X_2, and by definition they carry a typical element (x_1,x_2) of X to x_1 and x_2, respectively. We note that S is precisely the class of all inverse images in X of all open subsets of X_1 and X_2 under these projections: $G_1 \times X_2 = p_1^{-1}(G_1)$ and $X_1 \times G_2 = p_2^{-1}(G_2)$. The product topology is thus a topology on the product with respect to which both projections are continuous mappings, and it is

Fig. 23. The product topology on $X_1 \times X_2$.

evidently the weakest such topology. In terms of the ideas discussed in Sec. 19, the product topology can be regarded as the weak topology generated by the projections. Figure 23 may assist the reader in visualizing some of these notions.

With the above concepts to guide the way, one more step carries us to the product topology in its full generality. Let $\{X_i\}$ be any nonempty class of topological spaces, and consider the product $X = P_iX_i$ of the sets X_i. A typical element x in X is an array $x = \{x_i\}$ of points in the coordinate spaces, where each x_i belongs to the corresponding space X_i; and for each index i, the projection p_i is defined by $p_i(x) = x_i$. We now define the *product topology* on X to be the weak topology generated by the set of all projections. This means the product topology is that generated by the class S of all inverse images in X of open sets in the X_i's, that is, the class S of all subsets of X of the form $S = p_i^{-1}(G_i)$, where i is any index and G_i is any open subset of X_i. It is easy to see that S can also be described as the class of all products of the form $S = P_iG_i$, where G_i is an open subset of X_i which equals X_i for all i's but one. The class S is called the *defining open subbase* for the product topology; and the class of all complements of sets in S—namely, the class of all products of the form P_iF_i, where F_i is a closed subset of X_i which equals X_i for all i's but one—is called the *defining closed subbase*. The open base generated by S, that is, the class of all finite intersections of its sets, is called the *defining open base* for the product topology; and this is evidently the class of all products of the form P_iG_i, where G_i is an open subset of X_i which equals X_i for all but a finite number of i's.   It should be clearly understood that an unrestricted product of open sets in the coordinate spaces need not be open in the product topology.   A convenient way of thinking about this defining open base is that a typical one of its sets consists of all points $x = \{x_i\}$ in the product such that the ith coordinate x_i is required to lie in an open subset G_i of X_i for a finite number of i's, all other coordinates being unrestricted.

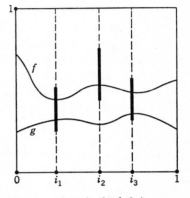

Fig. 24.   A set in the defining open base for a product space.

When the product of a non-empty class of topological spaces is equipped with the product topology defined in the above paragraph, it is called a *product space*, or more simply, the *product* of the spaces involved.[1]   It should be clear from Theorem 18-E and these definitions that all projections of a product space onto its coordinate spaces are automatically both continuous and open.

We conclude this section by analyzing an example which we hope will increase the reader's capacity to "see" the structure of product spaces. Let the index set I consist of all real numbers i in the closed unit interval [0,1]. I is to be considered as a set without any structure.   Now let each index i have attached to it a topological space X_i, and let every X_i be a replica of the closed unit interval [0,1] with its usual topology.   The resulting product space $X = P_iX_i$ is illustrated in Fig. 24.   The base of this figure is the index set I, and each vertical cross section represents the coordinate space X_i attached to the index at its base.   An element of the product space X is an array of points, one of which lies in each X_i. Such an element is essentially a function—if we identify a function with its graph—defined on the set I, with values in the closed unit interval. We can now visualize as follows a typical set in the defining open base for the product topology.   We choose a finite set of indices, say $\{i_1, i_2, i_3\}$, and for each of these we choose an open set on the vertical segment above

---

[1] Since a topological space must first of all be a non-empty set, it is worth remarking here that the axiom of choice guarantees that the product of a non-empty class of non-empty sets is non-empty.

it. Our basic open set then consists of all functions in X whose graphs cross each of these three vertical segments within the given open set on that segment. In the figure, f belongs to our basic open set, but g does not. The product topology on any product space can be visualized in a similar way. All one has to do is imagine the coordinate spaces as fibers, each attached to a specific element of the index set. The resulting mental image of the product space will then look something like a bundle of fibers, or perhaps a bed of reeds growing in a pond.

Problems

1. All projections, being open mappings, send open sets to open sets. Use the Euclidean plane to show that a projection need not send closed sets to closed sets.

2. Show that the relative topology on a subspace of a product space is the weak topology generated by the restrictions of the projections to that subspace.

3. Let f be a mapping of a topological space X into a product space $P_i X_i$, and show that f is continuous $\Leftrightarrow p_i f$ is continuous for each projection p_i.

4. Consider the product space defined and discussed in the last paragraph of the text, and show that this space is not second countable. (*Hint:* recall Theorem 18-B, and observe that the index set is uncountably infinite.)

5. Let X, Y, and Z be topological spaces, and consider a mapping $z = f(x,y)$ of the product set $X \times Y$ into the set Z. We say that f is *continuous in* x if for each fixed y_0 the mapping of X into Z given by $z = f(x,y_0)$ is continuous. The statement that f is *continuous in* y is defined similarly. f is said to be *jointly continuous in* x *and* y if it is continuous as a mapping of the product space $X \times Y$ into the space Z.

    (*a*) If all three spaces are metric spaces, show that f is jointly continuous $\Leftrightarrow x_n \to x$ and $y_n \to y$ implies $f(x_n,y_n) \to f(x,y)$.

    (*b*) Show that if f is jointly continuous, then it is continuous in each variable separately. Show that the converse of this statement is false by considering the real function defined on the Euclidean plane by $f(x,y) = xy/(x^2 + y^2)$ and $f(0,0) = 0$.

23. TYCHONOFF'S THEOREM AND LOCALLY COMPACT SPACES

The main theorem of this section, to the effect that any product of compact spaces is compact, is perhaps the most important single theorem of general topology. We shall use it repeatedly throughout the rest of this book, and the reader will come to see that its commanding position is due largely to the fact that in the higher levels of our subject many spaces constructed for special purposes turn out to be closed subspaces of products of compact spaces. Such a subspace is necessarily compact, and since compact spaces are so pleasant to work with, this makes the resulting theory much cleaner and smoother than would otherwise be the case.

Theorem A (Tychonoff's Theorem). *The product of any non-empty class of compact spaces is compact.*

PROOF. Let $\{X_i\}$ be a non-empty class of compact spaces, and form the product $X = P_i X_i$. Let $\{F_j\}$ be a non-empty subclass of the defining closed subbase for the product topology on X. This means that each F_j is a product of the form $F_j = P_i F_{ij}$, where F_{ij} is a closed subset of X_i which equals X_i for all i's but one. We assume that the class $\{F_j\}$ has the finite intersection property, and by virtue of Theorem 21-F we conclude the proof by showing that $\cap_j F_j$ is non-empty. For a given fixed i, $\{F_{ij}\}$ is a class of closed subsets of X_i with the finite intersection property; and by the assumed compactness of X_i (and Theorem 21-D), there exists a point x_i in X_i which belongs to $\cap_j F_{ij}$. If we do this for each i, we obtain a point $x = \{x_i\}$ in X which is in $\cap_j F_j$.

As our first application of Tychonoff's theorem, we prove an extension of the classical Heine-Borel theorem. We prepare the way for this proof by defining what we mean by open and closed rectangles in the n-dimensional Euclidean space R^n. If (a_i, b_i) is a bounded open interval on the real line for each $i = 1, 2, \ldots, n$, then the subset of R^n defined by

$$P_{i=1}^n (a_i, b_i) = \{(x_1, x_2, \ldots, x_n) : a_i < x_i < b_i \text{ for each } i\}$$

is called an *open rectangle* in R^n. A *closed rectangle* is defined similarly, as a product of n closed intervals.

Theorem B (the Generalized Heine-Borel Theorem). *Every closed and bounded subspace of R^n is compact.*

PROOF. A closed and bounded subspace of R^n is a closed subspace of some closed rectangle, so by Theorem 21-A it suffices to show that any closed rectangle is compact as a subspace of R^n. Let $X = P_{i=1}^n [a_i, b_i]$ be a closed rectangle in R^n. Each coordinate space $[a_i, b_i]$ is compact by the classical Heine-Borel theorem, so by Tychonoff's theorem it suffices to show that the product topology on X is the same as its relative topology as a subspace of R^n. It is easy to see that the open rectangles in R^n form an open base for its usual topology, that is, for its metric topology, and from this it follows that the product topology on R^n is the same as its usual topology.   By Problem 22-2, the relative topology on X is the weak topology generated by its n projections onto the coordinate spaces $[a_i,b_i]$; but this is the product topology on X, so the proof is complete.[1]

The n-dimensional Euclidean space R^n is the most important example of a type of topological space which is of great significance in modern analysis, especially in the theory of integration.   A topological space is said to be *locally compact* if each of its points has a neighborhood with compact closure.   It is easy to see by the above theorem that R^n actually is locally compact, because any open sphere centered on any point is a neighborhood of the point whose closure, being a closed and bounded subspace of R^n, is compact.   It is trivial that any compact space is locally compact, for the full space is a neighborhood with compact closure of every point in the space.   We return to the study of locally compact spaces in Sec. 37, where we give a more detailed analysis of their structure and properties.

Problems

1. Prove in detail that the open rectangles in R^n form an open base.
2. Show that every closed and bounded subspace of the n-dimensional unitary space C^n is compact.
3. Show that a topological space is locally compact $\Leftrightarrow$ there is an open base at each point whose sets all have compact closures.
4. Observe that any discrete space is locally compact.   Assuming that there are topological spaces which are not locally compact (we assure the reader that this is true), show that a continuous image of a locally compact space need not be locally compact.

24. COMPACTNESS FOR METRIC SPACES

In all candor, we must admit that the intuitive meaning of compactness for topological spaces is somewhat elusive.   This concept, however, is so vitally important throughout topology that we consider it worthwhile to devote this and the next section to giving several equivalent forms of compactness for the special case of a metric space.   Some of these are quite useful in applications and are perhaps more directly comprehensible than the open cover definition.   We hope they will help

---

[1] It is worth remarking that the high-powered machinery used in this proof is not really necessary for proving the theorem.   There are other proofs which are more elementary in nature, but we prefer the one given here because it illustrates some of our current concepts and tools.

the reader to achieve a fuller understanding of the geometric significance of compactness.[1]

We begin by recalling the classical *Bolzano-Weierstrass theorem:* if X is a closed and bounded subset of the real line, then every infinite subset of X has a limit point in X. This suggests that we consider the property expressed here as one which a general metric space may or may not possess. A metric space is said to have the *Bolzano-Weierstrass property* if every infinite subset has a limit point. Another property closely allied to this is that of sequential compactness: a metric space is said to be *sequentially compact* if every sequence in it has a convergent subsequence. Our main purpose in this section is to prove that each of these properties is equivalent to compactness in the case of a metric space. The following is an outline of our procedure: we first prove that these two properties are equivalent to one another; next, that compactness implies the Bolzano-Weierstrass property; and finally, that sequential compactness implies compactness. The first two of these steps are relatively simple, but the last involves several stages.

Theorem A. *A metric space is sequentially compact $\Leftrightarrow$ it has the Bolzano-Weierstrass property.*

PROOF. Let X be a metric space, and assume first that X is sequentially compact. We show that an infinite subset A of X has a limit point. Since A is infinite, a sequence $\{x_n\}$ of distinct points can be extracted from A. By our assumption of sequential compactness, this sequence has a subsequence which converges to a point x. Theorem 12-A shows that x is a limit point of the set of points of the subsequence, and since this set is a subset of A, x is also a limit point of A.

We now assume that every infinite subset of X has a limit point, and we prove from this that X is sequentially compact. Let $\{x_n\}$ be an arbitrary sequence in X. If $\{x_n\}$ has a point which is infinitely repeated, then it has a constant subsequence, and this subsequence is clearly convergent. If no point of $\{x_n\}$ is infinitely repeated, then the set A of points of this sequence is infinite. By our assumption, the set A has a limit point x, and it is easy to extract from $\{x_n\}$ a subsequence which converges to x.

Theorem B. *Every compact metric space has the Bolzano-Weierstrass property.*

PROOF. Let X be a compact metric space and A an infinite subset of X. We assume that A has no limit point, and from this we deduce a con-

---

[1] A solid case can be made for the proposition that compact spaces are natural generalizations of spaces with only a finite number of points. For a discussion of this, and of the significance of compactness in analysis, see Hewitt [19].

tradiction.   By our assumption, each point of X is not a limit point of A, so each point of X is the center of an open sphere which contains no point of A different from its center.   The class of all these open spheres is an open cover, and by compactness there exists a finite subcover.   Since A is contained in the set of all centers of spheres in this subcover, A is clearly finite.   This contradicts the fact that A is infinite, and concludes the proof.

Our next task is to prove that compactness is implied by sequential compactness.   We carry this out in several stages, the first of which can be motivated by the following considerations.   Let $\{G_i\}$ be an open cover of a metric space X.   Then each point x in X belongs to at least one G_i, and since the G_i's are open, each point x is the center of some open sphere which is contained in at least one G_i.   If we now move to another point of X, we may be forced to decrease the radius of our open sphere in order to squeeze it into a G_i.   Under special circumstances it may not be necessary to take radii below a certain level as we move from point to point over the entire space.   The following concept is useful for handling this sort of situation.   A real number $a > 0$ is called a *Lebesgue number* for our given open cover $\{G_i\}$ if each subset of X whose diameter is less than a is contained in at least one G_i.

Theorem C (Lebesgue's Covering Lemma).   *In a sequentially compact metric space, every open cover has a Lebesgue number.*

PROOF.   Let X be a sequentially compact metric space, and let $\{G_i\}$ be an open cover.   We say that a subset of X is "big" if it is not contained in any G_i.   If there are no big sets, then any positive real number will serve as our Lebesgue number a.   We may thus assume that big sets do exist, and we define a' to be the greatest lower bound of their diameters. Clearly, $0 \leq a' \leq +\infty$.   It will suffice to show that $a' > 0$; for if $a' = +\infty$, then any positive real number will do for a, and if a' is real, we can take a to be a'.   We therefore assume that $a' = 0$, and we deduce a contradiction from this assumption.   Since every big set must have at least two points, we infer from $a' = 0$ that for each positive integer n there exists a big set B_n such that $0 < d(B_n) < 1/n$.   We now choose a point x_n in each B_n.   Since X is sequentially compact, the sequence $\{x_n\}$ has a subsequence which converges to some point x in X.   The point x belongs to at least one set G_{i_0} in our open cover, and since G_{i_0} is open, x is the center of some open sphere $S_r(x)$ contained in G_{i_0}.   Let $S_{r/2}(x)$ be the concentric open sphere with radius $r/2$.   Since our subsequence of $\{x_n\}$ converges to x, there are infinitely many positive integers n for which x_n is in $S_{r/2}(x)$.   Let n_0 be one of these positive integers which is so large that $1/n_0 < r/2$.   Since $d(B_{n_0}) < 1/n_0 < r/2$, we see by

Problem 10-3 that $B_{n_0} \subseteq S_r(x) \subseteq G_{i_0}$. This contradicts the fact that B_{n_0} is a big set, and completes the proof.

The next stage requires the following concepts. Let X be a metric space. If $\epsilon > 0$ is given, a subset A of X is called an *ϵ-net* if A is finite and $X = \cup_{a \in A} S_\epsilon(a)$, that is, if A is finite and its points are scattered through X in such a way that each point of X is distant by less than ϵ from at least one point of A. The metric space X is said to be *totally bounded* if it has an ϵ-net for each $\epsilon > 0$. It is clear that if X is totally bounded, then it is also bounded; for if A is an ϵ-net, then the diameter of A is finite (since A is a non-empty finite set) and $d(X) \leq d(A) + 2\epsilon$. Total boundedness is actually a much stronger property than boundedness, as we shall see below.

Theorem D. *Every sequentially compact metric space is totally bounded.*
PROOF. Let X be a sequentially compact metric space, and let $\epsilon > 0$ be given. Choose a point a_1 in X and form the open sphere $S_\epsilon(a_1)$. If this open sphere contains every point of X, then the single-element set $\{a_1\}$ is an ϵ-net. If there are points outside of $S_\epsilon(a_1)$, let a_2 be such a point and form the set $S_\epsilon(a_1) \cup S_\epsilon(a_2)$. If this union contains every point of X, then the two-element set $\{a_1, a_2\}$ is an ϵ-net. If we continue in this way, some union of the form $S_\epsilon(a_1) \cup S_\epsilon(a_2) \cup \cdots \cup S_\epsilon(a_n)$ will necessarily contain every point of X; for if this process could be continued indefinitely, then the sequence $\{a_1, a_2, \ldots, a_n, \ldots\}$ would be a sequence with no convergent subsequence, contrary to the assumed sequential compactness of X. We see by this that some finite set of the form $\{a_1, a_2, \ldots, a_n\}$ is an ϵ-net, so X is totally bounded.

We are now in a position to complete this line of thought by proving that compactness is implied by sequential compactness.

Theorem E. *Every sequentially compact metric space is compact.*
PROOF. Let X be a sequentially compact metric space, and let $\{G_i\}$ be an open cover. By Theorem C, this open cover has a Lebesgue number a. We put $\epsilon = a/3$, and use Theorem D to find an ϵ-net

$$A = \{a_1, a_2, \ldots, a_n\}.$$

For each $k = 1, 2, \ldots, n$, we have $d(S_\epsilon(a_k)) \leq 2\epsilon = 2a/3 < a$. By the definition of a Lebesgue number, for each k we can find a G_{i_k} such that $S_\epsilon(a_k) \subseteq G_{i_k}$. Since every point of X belongs to one of the $S_\epsilon(a_k)$'s, the class $\{G_{i_1}, G_{i_2}, \ldots, G_{i_n}\}$ is a finite subcover of $\{G_i\}$. X is therefore compact.

Our results so far can be summarized by the statement that if X is a metric space, then the following three conditions are all equivalent to one another:

(1) X is compact;

(2) X is sequentially compact;

(3) X has the Bolzano-Weierstrass property.

Also, of course, we have as by-products the additional information that a compact metric space is totally bounded and that every open cover of a compact metric space has a Lebesgue number. The latter fact has the following useful consequence.

Theorem F.   *Any continuous mapping of a compact metric space into a metric space is uniformly continuous.*

PROOF.   Let f be a continuous mapping of a compact metric space X into a metric space Y, and let d_1 and d_2 be the metrics on X and Y. Let $\epsilon > 0$ be given. For each point x in X, consider its image $f(x)$ and the open sphere $S_{\epsilon/2}(f(x))$ centered on this image with radius $\epsilon/2$. Since f is continuous, the inverse image of each of these open spheres is an open subset of X, and the class of all such inverse images is an open cover of X. Since X is compact, Theorem C guarantees that this open cover has a Lebesgue number δ. If x and x' are any two points in X for which $d_1(x,x') < \delta$, then the set $\{x,x'\}$ is a set with diameter less than δ, both points belong to the inverse image of some one of the above open spheres, both $f(x)$ and $f(x')$ belong to one of these open spheres, and therefore $d_2(f(x), f(x')) < \epsilon$, which shows that f is indeed uniformly continuous.

We continue our study of compact metric spaces in the next section.

Problems

1.   Let A be a subspace of a metric space X, and show that A is totally bounded $\Leftrightarrow \bar{A}$ is totally bounded.

2.   Show that a subspace of R^n is bounded $\Leftrightarrow$ it is totally bounded.

3.   Prove the Bolzano-Weierstrass theorem for R^n: if X is a closed and bounded subset of R^n, then every infinite subset of X has a limit point in X.

4.   Show that a compact metric space is separable.

25. ASCOLI'S THEOREM

Our previous characterizations of compactness for a metric space strongly suggest that this property is related to completeness and total boundedness in some way yet to be formulated.   We begin by proving a theorem which clarifies this situation.

Theorem A.   *A metric space is compact $\Leftrightarrow$ it is complete and totally bounded.*
PROOF.   Let X be a metric space.   The first half of our proof is easy, for if X is compact, then it is totally bounded by Theorem 24-D, and it is complete by Problem 12-2 and the fact that every sequence (and therefore every Cauchy sequence) has a convergent subsequence.

We now assume that X is complete and totally bounded, and we prove that X is compact by showing that every sequence has a convergent subsequence.   Since X is complete, it suffices to show that every sequence has a Cauchy subsequence.   Consider an arbitrary sequence

$$S_1 = \{x_{11},\, x_{12},\, x_{13},\, \ldots\}.$$

The reason for this notation will soon be clear.   Since X is totally bounded, there exists a finite class of open spheres, each with radius $\frac{1}{2}$, whose union equals X; and from this we see that S_1 has a subsequence $S_2 = \{x_{21},\, x_{22},\, x_{23},\, \ldots\}$ all of whose points lie in some one open sphere of radius $\frac{1}{2}$.   Another application of the total boundedness of X shows similarly that S_2 has a subsequence $S_3 = \{x_{31},\, x_{32},\, x_{33},\, \ldots\}$ all of whose points lie in some one open sphere of radius $\frac{1}{3}$.   We continue forming successive subsequences in this manner, and we let

$$S = \{x_{11},\, x_{22},\, x_{33},\, \ldots\}$$

be the "diagonal" subsequence of S_1.   By the nature of this construction, S is clearly a Cauchy subsequence of S_1, and our proof is complete.

This theorem gives total boundedness a prominent part in determining whether a metric space is compact or not.   As we know, many metric spaces occur as closed subspaces of complete metric spaces, and for these we can make the role of total boundedness even more striking.

Theorem B.   *A closed subspace of a complete metric space is compact $\Leftrightarrow$ it is totally bounded.*
PROOF.   Since a closed subspace of a complete metric space is automatically complete, this is a direct consequence of Theorem A.

What sort of property is total boundedness?   We have seen that it always implies boundedness, and we know by Problem 24-2 that the converse of this is true for subspaces of the finite-dimensional Euclidean space R^n.   It is false, however, that boundedness implies total boundedness for subspaces of the infinite-dimensional Euclidean space R^∞.   In fact, the closed unit sphere in R^∞, defined by $X = \{x : \|x\| \leq 1\}$, is not totally bounded, though it is obviously bounded.   To see this, it suffices to observe that the sequence $\{x_n\}$ in X defined by

$$x_1 = \{1, 0, 0, \ldots, 0, \ldots\},$$
$$x_2 = \{0, 1, 0, \ldots, 0, \ldots\},$$
$$x_3 = \{0, 0, 1, \ldots, 0, \ldots\},$$
$$\cdots \cdots \cdots \cdots \cdots \cdots \cdots$$

has no convergent subsequence, for the distance from any point of the sequence to any other is $2^{\frac{1}{2}}$. This shows that X is not compact, hence not totally bounded. The following fact, which we cannot prove here (see Sec. 47), may add to the reader's intuition about the relation between boundedness and total boundedness: a Banach space is finite-dimensional $\Leftrightarrow$ every bounded subspace is totally bounded.

We now turn to the problem of characterizing compact subspaces of $\mathcal{C}(X,R)$ or $\mathcal{C}(X,C)$. By Theorem B, we know at once that a closed subspace of $\mathcal{C}(X,R)$ or $\mathcal{C}(X,C)$ is compact $\Leftrightarrow$ it is totally bounded. Unfortunately, however, this information is of little value in most applications to analysis. What is needed is a criterion expressed in terms of the individual functions in the subspace. Furthermore, for most of the applications it suffices to consider only the case in which X is a compact metric space. We describe the relevant concept as follows. Let X be a compact metric space with metric d, and let A be a non-empty set of continuous real or complex functions defined on X. If f is a function in A, then by Theorem 24-F this function is uniformly continuous; that is, for each $\epsilon > 0$, there exists $\delta > 0$ such that $d(x,x') < \delta \Rightarrow |f(x) - f(x')| < \epsilon$. In general, δ depends not only on ϵ but also on the function f. A is said to be *equicontinuous* if for each ϵ a δ can be found which serves at once for all functions f in A, that is, if for each $\epsilon > 0$ there exists $\delta > 0$ such that for every f in A $d(x,x') < \delta \Rightarrow |f(x) - f(x')| < \epsilon$.

Theorem C (Ascoli's Theorem). *If X is a compact metric space, then a closed subspace of $\mathcal{C}(X,R)$ or $\mathcal{C}(X,C)$ is compact $\Leftrightarrow$ it is bounded and equicontinuous.*

PROOF. Let d be the metric on X, and let F be a closed subspace of $\mathcal{C}(X,R)$ or $\mathcal{C}(X,C)$.

We first assume that F is compact, and we prove that it is bounded and equicontinuous. Problem 21-6 shows that F is bounded. We prove that F is equicontinuous as follows. Let $\epsilon > 0$ be given. Since F is compact, and therefore totally bounded, we can find an $(\epsilon/3)$-net $\{f_1, f_2, \ldots, f_n\}$ in F. Each f_k is uniformly continuous, so for each $k = 1, 2, \ldots, n$, there exists $\delta_k > 0$ such that $d(x,x') < \delta_k \Rightarrow |f_k(x) - f_k(x')| < \epsilon/3$. We now define δ to be the smallest of the numbers $\delta_1, \delta_2, \ldots, \delta_n$. If f is any function in F and f_k is chosen so that $\|f -$

$f_k\| < \epsilon/3$, then

$$d(x,x') < \delta \Rightarrow |f(x) - f(x')| \leq |f(x) - f_k(x)| + |f_k(x) - f_k(x')|$$
$$+ |f_k(x') - f(x')| < \epsilon/3 + \epsilon/3 + \epsilon/3 = \epsilon.$$

This shows that F is equicontinuous.

We now assume that F is bounded and equicontinuous, and we demonstrate that it is compact by showing that every sequence in it has a convergent subsequence. Since F is closed, and therefore complete, it suffices to show that every sequence in it has a Cauchy subsequence. As we proceed, the reader will see that our proof is similar in structure to the last part of the proof of Theorem A. By Problem 24-4, X has a countable dense subset. Let the points of this subset be arranged in a sequence $\{x_i\} = \{x_2, x_3, \ldots, x_i, \ldots\}$, where we start with the subscript 2 for reasons which will become clear below. Now let

$$S_1 = \{f_{11}, f_{12}, f_{13}, \ldots\}$$

be an arbitrary sequence in F. Our hypothesis that F is bounded means that there exists a real number K such that $\|f\| \leq K$ for every f in F, or equivalently, such that $|f(x)| \leq K$ for every f in F and every x in X. Consider the sequence of numbers $\{f_{1j}(x_2)\}$, $j = 1, 2, 3, \ldots$, and observe that since this sequence is bounded, it has a convergent subsequence. Let $S_2 = \{f_{21}, f_{22}, f_{23}, \ldots\}$ be a subsequence of S_1 such that $\{f_{2j}(x_2)\}$ converges. We next consider the sequence of numbers $\{f_{2j}(x_3)\}$, and in the same way we let $S_3 = \{f_{31}, f_{32}, f_{33}, \ldots\}$ be a subsequence of S_2 such that $\{f_{3j}(x_3)\}$ converges. If we continue this process, we get an array of sequences of the form

$$S_1 = \{f_{11}, f_{12}, f_{13}, \ldots\},$$
$$S_2 = \{f_{21}, f_{22}, f_{23}, \ldots\},$$
$$S_3 = \{f_{31}, f_{32}, f_{33}, \ldots\},$$
$$\cdots \cdots \cdots \cdots \cdots$$

in which each sequence is a subsequence of the one directly above it, and for each i the sequence $S_i = \{f_{i1}, f_{i2}, f_{i3}, \ldots\}$ has the property that $\{f_{ij}(x_i)\}$ is a convergent sequence of numbers. If we define $f_1, f_2, f_3, \ldots$ by $f_1 = f_{11}, f_2 = f_{22}, f_3 = f_{33}, \ldots$, then the sequence $S = \{f_1, f_2, f_3, \ldots\}$ is the "diagonal" subsequence of S_1. It is clear from this construction that for each point x_i in our dense subset of X, the sequence $\{f_n(x_i)\}$ is a convergent sequence of numbers. It remains only to show that S, as a sequence of functions in $\mathcal{C}(X,R)$ or $\mathcal{C}(X,C)$, is a Cauchy sequence. Let $\epsilon > 0$ be given. Since F is equicontinuous, there exists $\delta > 0$ such that $d(x,x') < \delta \Rightarrow |f_n(x) - f_n(x')| < \epsilon/3$ for all functions f_n in S. We now form the open sphere $S_\delta(x_i)$ with radius δ centered on each of the x_i's. Since the x_i's are dense, these open spheres form an open cover of X, and since X is compact, $X = \cup_{i=2}^{i_0} S_\delta(x_i)$ for some i_0. It is easy to see that there exists a positive integer n_0 such that $m, n \geq n_0 \Rightarrow |f_m(x_i) - f_n(x_i)| < \epsilon/3$ for all the points $x_2, x_3, \ldots, x_{i_0}$. Our proof is completed by the remark that if x is an arbitrary point of X, then an i can be found in the set $\{2, 3, \ldots, i_0\}$ such that $d(x,x_i) < \delta$, and that therefore

$$m,n \geq n_0 \Rightarrow |f_m(x) - f_n(x)| \leq |f_m(x) - f_m(x_i)| + |f_m(x_i) - f_n(x_i)|$$
$$+ |f_n(x_i) - f_n(x)| < \epsilon/3 + \epsilon/3 + \epsilon/3 = \epsilon.$$

We observe that the total boundedness in Theorem B is replaced, in Ascoli's theorem, by the weaker condition of boundedness, and that the resulting deficiency is made up by the additional condition of equicontinuity.[1] For several applications of Ascoli's theorem (which is sometimes called *Arzela's theorem*) to problems in analysis, see Goffman [13, pp. 151–156] or Kolmogorov and Fomin [26, vol. 1, secs. 17–20].

Problems

1. Let A be a subspace of a complete metric space, and show that $\bar{A}$ is compact $\Leftrightarrow$ A is totally bounded.

2. Let X be a compact metric space and F a closed subspace of $\mathcal{C}(X,R)$ or $\mathcal{C}(X,C)$. Show that F is compact if it is equicontinuous and $F_x = \{f(x) : f \in F\}$ is a bounded set of numbers for each point x in X.

3. Show that R^∞ is not locally compact.

4. By considering the sequence of functions in $\mathcal{C}[0,1]$ defined by

$$f_n(x) = nx$$

for $0 \leq x \leq 1/n$, $f_n(x) = 1$ for $1/n \leq x \leq 1$, show that $\mathcal{C}[0,1]$ is not locally compact.

---

[1] The following terminology is often used with Ascoli's theorem. Let F be any non-empty set of real or complex functions defined on an arbitrary non-empty set X. The statement that a function f in F is bounded means, of course, that there exists a real number K such that $|f(x)| \leq K$ for every x in X. The functions in F are often said to be *uniformly bounded* (or F is called a *uniformly bounded set of functions*) if there exists a single K which works in this way for all f's in F, i.e., if there is a K such that $|f(x)| \leq K$ for every x in X and every f in F. If we were to use this expression, Ascoli's theorem would take the following form: if X is a compact metric space, then a closed subspace of $\mathcal{C}(X,R)$ or $\mathcal{C}(X,C)$ is compact $\Leftrightarrow$ it is uniformly bounded and equicontinuous. The uniform boundedness here is merely boundedness as a subset of the metric space $\mathcal{C}(X,R)$ or $\mathcal{C}(X,C)$.

CHAPTER FIVE

Separation

A topological space may be very sparsely endowed with open sets. As we know, some spaces have only two, the empty set and the full space. In a discrete space, on the other hand, every set is open. Most of the familiar spaces of geometry and analysis fall somewhere in between these two artificial extremes. The so-called *separation properties* enable us to state with precision that a given topological space has a rich enough supply of open sets to serve whatever purpose we may have in mind.

The separation properties are of concern to us because the supply of open sets possessed by a topological space is intimately linked to its supply of continuous functions; and since continuous functions are of central importance in topology, we naturally wish to guarantee that enough of them are present to make our discussions fruitful. If, for instance, the only open sets in a topological space are the empty set and the full space, then the only continuous functions present are the constants, and very little of interest can be said about these. In general terms, the more open sets there are, the more continuous functions a space has. Discrete spaces have continuous functions in the greatest possible abundance, for all functions are continuous. However, few really important spaces are discrete, so this goes a bit too far. The separation properties make it possible for us to be sure that our spaces have enough continuous functions without committing ourselves to the excesses of discrete spaces.

129

26. T_1-SPACES AND HAUSDORFF SPACES

One of the most natural things to require of a topological space is that each of its points be a closed set.[1] The separation property which relates to this is the following. A T_1-*space* is a topological space in which, given any pair of distinct points, each has a neighborhood which does not contain the other.[2] It is obvious that any subspace of a T_1-space is also a T_1-space. Our first theorem shows that T_1-spaces are precisely those topological spaces in which points are closed.

Theorem A. *A topological space is a T_1-space $\Leftrightarrow$ each point is a closed set.*
PROOF. If X is a topological space, then an arbitrary point x in X is closed $\Leftrightarrow$ its complement is open $\Leftrightarrow$ each point y different from x has a neighborhood which does not contain x $\Leftrightarrow$ X is a T_1-space.

Our next separation property is slightly stronger. A *Hausdorff space* is a topological space in which each pair of distinct points can be separated by open sets, in the sense that they have disjoint neighborhoods. Every Hausdorff space is clearly a T_1-space, and every subspace of a Hausdorff space is also a Hausdorff space.

Theorem B. *The product of any non-empty class of Hausdorff spaces is a Hausdorff space.*
PROOF. Let $X = P_i X_i$ be the product of a non-empty class of Hausdorff spaces X_i. If $x = \{x_i\}$ and $y = \{y_i\}$ are two distinct points in X, then we must have $x_{i_0} \neq y_{i_0}$ for at least one index i_0. Since X_{i_0} is a Hausdorff space, x_{i_0} and y_{i_0} can be separated by open sets in X_{i_0}. These two disjoint open subsets of X_{i_0} give rise to two disjoint sets in the defining open subbase for X, each of which contains one of the points x and y.

Most of the important facts about Hausdorff spaces depend on the following theorem.

Theorem C. *In a Hausdorff space, any point and disjoint compact subspace can be separated by open sets, in the sense that they have disjoint neighborhoods.*
PROOF. Let X be a Hausdorff space, x a point in X, and C a compact subspace of X which does not contain x. We construct a disjoint pair of

---

[1] It is customary here to drop the distinction between a point x in a space and the set $\{x\}$ which contains only that point. This convention often makes it possible to avoid cumbersome modes of expression, and we shall use it freely.

[2] The T_i-space nomenclature, for $i = 0, 1, \ldots, 5$, was introduced by Alexandroff and Hopf in their famous treatise [2]. The T refers to the German word *Trennungsaxiom*, which means "separation axiom." The term T_1-*space* is the only one of these which is still in general use.

open sets G and H such that $x \in G$ and $C \subseteq H$. Let y be a point in C. Since X is a Hausdorff space, x and y have disjoint neighborhoods G_x and H_y. If we allow y to vary over C, we obtain a class of H_y's whose union contains C; and since C is compact, some finite subclass, which we denote by $\{H_1, H_2, \ldots, H_n\}$, is such that $C \subseteq \bigcup_{i=1}^{n} H_i$. If $G_1, G_2, \ldots, G_n$ are the neighborhoods of x which correspond to the H_i's, we put

$$G = \bigcap_{i=1}^{n} G_i$$

and $H = \bigcup_{i=1}^{n} H_i$ and observe that these two sets have the required properties.

In Theorem 21-A we proved that every closed subspace of a compact space is compact, and in Problem 21-4 we saw that a compact subspace of a compact space need not be closed. We now use the preceding theorem to show that compact subspaces of Hausdorff spaces are always closed.

Theorem D. *Every compact subspace of a Hausdorff space is closed.*

PROOF. Let C be a compact subspace of a Hausdorff space X. We prove that C is closed by showing that its complement C' is open. C' is open if it is empty, so we may assume that it is non-empty. Let x be any point in C'. By Theorem C, x has a neighborhood G such that $x \in G \subseteq C'$. This shows that C' is a union of open sets and is therefore open itself.

One of the most useful consequences of this result is

Theorem E. *A one-to-one continuous mapping of a compact space onto a Hausdorff space is a homeomorphism.*

PROOF. Let $f : X \to Y$ be a one-to-one continuous mapping of a compact space X onto a Hausdorff space Y. We must show that $f(G)$ is open in Y whenever G is open in X, and for this it suffices to show that $f(F)$ is closed in Y whenever F is closed in X. If F is empty, $f(F)$ is also empty and therefore closed, so we may assume that F is non-empty. By Theorem 21-A, F is a compact subspace of X; by Theorem 21-B, $f(F)$ is a compact subspace of Y; and we complete the proof by using the preceding theorem to conclude that $f(F)$ is a closed subspace of Y.

Compact Hausdorff spaces are among the most important of all topological spaces, and in the following sections and chapters we shall become thoroughly acquainted with their major properties.

Problems

1. Show that the topological space defined in Example 16-5 is not a T_1-space.

2. Show that the topological space defined in Example 16-4 is a T_1-space but not a Hausdorff space.
3. Show that any finite T_1-space is discrete.
4. If X is a T_1-space with at least two points, show that an open base which contains X as a member remains an open base if X is dropped.
5. Let X be a topological space, Y a Hausdorff space, and A a subspace of X. Show that a continuous mapping of A into Y has at most one continuous extension to a mapping of $\bar{A}$ into Y. Problem 17-2 is a special case of this statement.
6. If f is a continuous mapping of a topological space X into a Hausdorff space Y, prove that the graph of f is a closed subset of the product $X \times Y$.
7. Let X be any non-empty set, and prove that in the lattice of all topologies on X each chain has at most one compact Hausdorff topology as a member. (It is interesting to speculate about whether a compact Hausdorff topology can be defined on an arbitrary non-empty set.)
8. Let X be an arbitrary topological space and $\{x_n\}$ a sequence of points in X. This sequence is said to be *convergent* if there exists a point x in X such that for each neighborhood G of x a positive integer n_0 can be found with the property that x_n is in G for all $n \geq n_0$. The point x is called a *limit* of the sequence, and we say that x_n *converges to* x (and symbolize this by $x_n \to x$).
    (a) Show that in Example 16-3 any sequence converges to every point of the space. This is the reason why the above point x is called *a* limit instead of *the* limit.
    (b) If X is a Hausdorff space, show that every convergent sequence in X has a unique limit.
    (c) Show that if $f : X \to Y$ is a continuous mapping of one topological space into another, then $x_n \to x$ in $X \Rightarrow f(x_n) \to f(x)$ in Y. Prove that the converse of this is true if each point in X has a countable open base.[1]

27. COMPLETELY REGULAR SPACES AND NORMAL SPACES

Let X be an arbitrary topological space, and consider the set $\mathfrak{C}(X,R)$ of all bounded continuous real functions defined on X. If for each pair of distinct points x and y in X there exists a function f in $\mathfrak{C}(X,R)$ such that $f(x) \neq f(y)$, we say that $\mathfrak{C}(X,R)$ *separates points*. It is easy to see

---

[1] The facts brought out in this problem are the main reasons why the concept of a convergent sequence is not very important in the general theory of topological spaces.

that if $\mathcal{C}(X,R)$ does separate points, then X is necessarily a Hausdorff space; for assuming that $f(x) < f(y)$ and that r is a real number such that $f(x) < r < f(y)$, then the sets $\{z:f(z) < r\}$ and $\{z:f(z) > r\}$ are disjoint neighborhoods of x and y.

It is convenient to strengthen this separation property slightly by allowing one of the points to be an arbitrary closed subspace of X. A *completely regular space* is a T_1-space X with the property that if x is any point in X and F any closed subspace of X which does not contain x, then there exists a function f in $\mathcal{C}(X,R)$, all of whose values lie in the closed unit interval $[0,1]$, such that $f(x) = 0$ and $f(F) = 1$. It is worth noticing that since constants are continuous, we could just as well have required here that f be 1 at x and 0 on F, for the function $g = 1 - f$ has these properties. We may think of completely regular spaces as T_1-spaces in which continuous functions separate points and disjoint closed subspaces. Since points are closed in a completely regular space, it is permissible to take the closed subspace F to be a point, and it is clear by the above paragraph that every completely regular space is a Hausdorff space. It is also easy to see that every subspace of a completely regular space is completely regular. In Sec. 30 we give an explicit characterization of all completely regular spaces in terms of product spaces.

Our next (and last) separation property is similar to that which defines a Hausdorff space, except that it applies to disjoint closed sets instead of merely to distinct points. A *normal space* is a T_1-space in which each pair of disjoint closed sets can be separated by open sets, in the sense that they have disjoint neighborhoods. We shall see in the next section (as a consequence of Urysohn's lemma) that every normal space is completely regular.

Figure 25 is intended to illustrate and clarify the relations among our various separation properties: a topological space which possesses any one property, in the order of their definition, also possesses all properties which precede it; in other words, they have been defined in order of increasing strength. The figure also indicates that metric spaces and compact Hausdorff spaces are normal. We established the first of these facts in Problem 11-1b, and we now prove the second.

Theorem A. *Every compact Hausdorff space is normal.*

PROOF. Let X be a compact Hausdorff space, and A and B disjoint closed subsets of X. We must produce a disjoint pair of open sets G and H such that $A \subseteq G$ and $B \subseteq H$. If either closed set is empty, we can take the empty set as a neighborhood of it and the full space as a neighborhood of the other. We may therefore assume that both A and B are non-empty. Since X is compact, A and B are disjoint compact subspaces of X. Let x be a point of A. By Theorem 26-C and our hypothesis that X is Hausdorff, x and B have disjoint neighborhoods G_x and H_B. If we allow x to vary over A, we obtain a class of G_x's whose union contains A; and since A is compact, some finite subclass, which we denote by $\{G_1, G_2, \ldots, G_n\}$, is such that $A \subseteq \bigcup_{i=1}^{n} G_i$. If $H_1, H_2, \ldots, H_n$ are the neighborhoods of B which correspond to the G_i's, it is clear that $G = \bigcup_{i=1}^{n} G_i$ and $H = \bigcap_{i=1}^{n} H_i$ are disjoint neighborhoods of A and B.

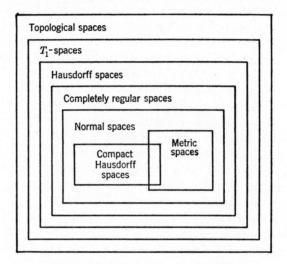

Fig. 25.   The separation properties.

In Sec. 29 we investigate the manner in which normal spaces, compact Hausdorff spaces, and metric spaces are related to one another.

Problems

1. Show that a closed subspace of a normal space is normal.
2. Let X be a T_1-space, and show that X is normal $\Leftrightarrow$ each neighborhood of a closed set F contains the closure of some neighborhood of F.
3. Assume, as Fig. 25 suggests, that a compact Hausdorff space X is completely regular and that therefore $\mathcal{C}(X,R)$ separates points. Use this to prove that the weak topology generated by $\mathcal{C}(X,R)$ equals the given topology. Show further that if S is any subset of $\mathcal{C}(X,R)$ which also separates points, then the weak topology generated by S also equals the given topology.
4. Let X be a completely regular space, and show from the definition that the weak topology generated by $\mathcal{C}(X,R)$ equals the given topology.

28. URYSOHN'S LEMMA AND THE TIETZE EXTENSION THEOREM

As we suggested in the introduction to this chapter, one of the main purposes served by assuming that a topological space is rich in open sets is to guarantee that it is also rich in continuous functions. The following is the fundamental theorem in this direction.

Theorem A (Urysohn's Lemma). *Let X be a normal space, and let A and B be disjoint closed subspaces of X. Then there exists a continuous real function f defined on X, all of whose values lie in the closed unit interval $[0,1]$, such that $f(A) = 0$ and $f(B) = 1$.*

PROOF. B' is a neighborhood of the closed set A, so by the normality of X and Problem 27-2, A has a neighborhood $U_{\frac{1}{2}}$ such that

$$A \subseteq U_{\frac{1}{2}} \subseteq \overline{U_{\frac{1}{2}}} \subseteq B'.$$

$U_{\frac{1}{2}}$ and B' are neighborhoods of the closed sets A and $\overline{U_{\frac{1}{2}}}$, so in the same way there exist open sets $U_{\frac{1}{4}}$ and $U_{\frac{3}{4}}$ such that

$$A \subseteq U_{\frac{1}{4}} \subseteq \overline{U_{\frac{1}{4}}} \subseteq U_{\frac{1}{2}} \subseteq \overline{U_{\frac{1}{2}}} \subseteq U_{\frac{3}{4}} \subseteq \overline{U_{\frac{3}{4}}} \subseteq B'.$$

If we continue this process, for each dyadic rational number of the form $t = m/2^n$ (where $n = 1, 2, 3, \ldots$ and $m = 1, 2, \ldots, 2^n - 1$) we have an open set of the form U_t, such that

$$t_1 < t_2 \Rightarrow A \subseteq U_{t_1} \subseteq \overline{U_{t_1}} \subseteq U_{t_2} \subseteq \overline{U_{t_2}} \subseteq B'.$$

We now define our function f by $f(x) = 0$ if x is in every U_t and

$$f(x) = \sup \{t : x \notin U_t\}$$

otherwise. It is clear that the values of f lie in $[0,1]$, and that $f(A) = 0$ and $f(B) = 1$. All that remains to be proved is that f is continuous. All intervals of the form $[0,a)$ and $(a,1]$, where $0 < a < 1$, constitute an open subbase for $[0,1]$. It therefore suffices to show that $f^{-1}([0,a))$ and $f^{-1}((a,1])$ are open. It is easy to see that $f(x) < a \Leftrightarrow x$ is in some U_t for $t < a$; and from this it follows that $f^{-1}([0,a)) = \{x : f(x) < a\} = \bigcup_{t<a} U_t$, which is an open set. Similarly, $f(x) > a \Leftrightarrow x$ is outside of $\overline{U_t}$ for some $t > a$; and therefore $f^{-1}((a,1]) = \{x : f(x) > a\} = \bigcup_{t>a} \overline{U_t}'$, which is an open set.

It is clear from this theorem that every normal space is completely regular: all that is necessary is to take the closed subspace A to be a single point and to observe that the function f is exactly what is required in the definition of complete regularity.

The following slightly more flexible form of Urysohn's lemma will be useful in applications.

Theorem B. *Let X be a normal space, and let A and B be disjoint closed subspaces of X. If $[a,b]$ is any closed interval on the real line, then there exists a continuous real function f defined on X, all of whose values lie in $[a,b]$, such that $f(A) = a$ and $f(B) = b$.*

PROOF. If $a = b$, we have only to define f by $f(x) = a$ for every x, so we may assume that $a < b$. If g is a function with the properties stated in Urysohn's lemma, then $f = (b - a)g + a$ has the properties required by our theorem.

If there is given a continuous function defined on a subspace of a topological space, Urysohn's lemma has an important bearing on the interesting question of whether this function can be extended continuously to the full space. The following is the classic theorem along these lines.

Theorem C (the **Tietze Extension Theorem**). *Let X be a normal space, F a closed subspace, and f a continuous real function defined on F whose values lie in the closed interval $[a,b]$. Then f has a continuous extension f' defined on all of X whose values also lie in $[a,b]$.*

PROOF. If $a = b$, the conclusion of our theorem is obvious, so we may assume that $a < b$. We may clearly assume that $[a,b]$ is the smallest closed interval which contains the range of f. Furthermore, the device used in the proof of Theorem B enables us to assume that $a = -1$ and $b = 1$. We begin by defining f_0 to be f. The domain of f_0 is our closed subspace F, and we define two subsets A_0 and B_0 of F by

$$A_0 = \{x : f_0(x) \leq -\tfrac{1}{3}\}$$

and $B_0 = \{x : f_0(x) \geq \tfrac{1}{3}\}$. A_0 and B_0 are disjoint, non-empty, and closed in F; and since F is closed, they are closed in X. A_0 and B_0 are thus a disjoint pair of closed subspaces of X, and by Theorem B there exists a continuous function $g_0 \colon X \to [-\tfrac{1}{3}, \tfrac{1}{3}]$ such that $g_0(A_0) = -\tfrac{1}{3}$ and $g_0(B_0) = \tfrac{1}{3}$. We next define f_1 on F by $f_1 = f_0 - g_0$, and we observe that $|f_1(x)| \leq \tfrac{2}{3}$. If $A_1 = \{x : f_1(x) \leq (-\tfrac{1}{3})(\tfrac{2}{3})\}$ and

$$B_1 = \{x : f_1(x) \geq (\tfrac{1}{3})(\tfrac{2}{3})\},$$

then in the same way as above there exists a continuous function $g_1 \colon X \to [(-\tfrac{1}{3})(\tfrac{2}{3}), (\tfrac{1}{3})(\tfrac{2}{3})]$ such that $g_1(A_1) = (-\tfrac{1}{3})(\tfrac{2}{3})$ and

$$g_1(B_1) = (\tfrac{1}{3})(\tfrac{2}{3}).$$

We next define f_2 on F by $f_2 = f_1 - g_1 = f_0 - (g_0 + g_1)$, and we observe that $|f_2(x)| \leq (\tfrac{2}{3})^2$. By continuing in this manner, we get a sequence $\{f_0, f_1, f_2, \ldots\}$ defined on F such that $|f_n(x)| \leq (\tfrac{2}{3})^n$, and a sequence $\{g_0, g_1, g_2, \ldots\}$ defined on X such that $|g_n(x)| \leq (\tfrac{1}{3})(\tfrac{2}{3})^n$, with the property that on F we have $f_n = f_0 - (g_0 + g_1 + \cdots + g_{n-1})$. We now define s_n by $s_n = g_0 + g_1 + \cdots + g_{n-1}$, and we regard the s_n's as the partial sums of an infinite series of functions in $\mathfrak{C}(X,R)$. We know that $\mathfrak{C}(X,R)$ is complete, so by $|g_n(x)| \leq (\frac{1}{3})(\frac{2}{3})^n$ and the fact that $\Sigma_{n=0}^{\infty} (\frac{1}{3})(\frac{2}{3})^n = 1$, we see that s_n converges uniformly on X to a bounded continuous real function f' such that $|f'(x)| \leq 1$. We conclude our proof by noting that since $|f_n(x)| \leq (\frac{2}{3})^n$, s_n converges uniformly on F to $f_0 = f$, and that therefore f' equals f on F and is a continuous extension of f to the full space X which has the desired property.

It is of some interest to observe that this theorem becomes false if we omit the assumption that the subspace F is closed. This is easily seen by means of the following example. Let X be the closed unit interval $[0,1]$, F the subspace $(0,1]$, and f the function defined on F by $f(x) = \sin (1/x)$. X is clearly normal, F is not closed as a subspace of X, and f cannot be extended continuously to X in any manner whatsoever.

Problems

1. In the text we used Urysohn's lemma as a tool to prove Tietze's theorem. Reverse this process, and deduce Urysohn's lemma from Tietze's theorem.
2. State and prove a generalization of Tietze's theorem which relates to functions whose values lie in R^n.
3. Justify the assertion in the last paragraph of the text that the function defined there cannot be extended continuously to X.

29. THE URYSOHN IMBEDDING THEOREM

In Chap. 3, we generalized metric spaces to topological spaces. We now reverse this procedure and seek out simple conditions which guarantee that a topological space is essentially a metric space, that is, which imply that it is metrizable. Problem 16-12 shows that we must look for properties of a topological space X which enable us to construct a homeomorphism f of X onto a subspace of some metric space; for the metric on this subspace can then be carried back by f to X, and we can infer that X is metrizable. The simplest property of this kind is discreteness; for if X is a discrete space, then its underlying set of points, equipped with the metric defined in Example 9-1, is a homeomorphic image of X under the identity mapping. We can lift our discussion to a more meaningful level by observing that since every metric space is normal, normality must be among the properties assumed of X, or it must be implied by them.

As motivation for our main theorem, we note that since the metric space R^∞ is second countable, every subspace of it is also second countable. It turns out that second countability, in addition to normality, suffices to guarantee that a topological space is homeomorphic to a subspace of R^∞. In effect, we imbed such a space homeomorphically in R^∞.

Theorem A (the Urysohn Imbedding Theorem). *If X is a second countable normal space, then there exists a homeomorphism f of X onto a subspace of R^∞, and X is therefore metrizable.*

PROOF. We may assume that X has infinitely many points, for otherwise it would be finite and discrete, and clearly homeomorphic to any subspace of R^∞ with the same number of points. Since X is second countable, it has a countably infinite open base $\{G_1, G_2, G_3, \ldots\}$ each of whose sets is different from the empty set and the full space. If G_j and $x \in G_j$ are given, then by normality there exists a G_i such that $x \in G_i \subseteq \overline{G_i} \subseteq G_j$. The set of all ordered pairs (G_i, G_j) such that $\overline{G_i} \subseteq G_j$ is countably infinite, and we can arrange them in a sequence P_1, $P_2, \ldots, P_n \ldots$. By Urysohn's lemma, for each ordered pair $P_n = (G_i, G_j)$ there exists a continuous real function $f_n : X \to [0,1]$ such that $f_n(\overline{G_i}) = 0$ and $f_n(G_j') = 1$. For each x in X we define $f(x)$ to be the sequence given by $f(x) = \{f_1(x), f_2(x)/2, \ldots, f_n(x)/n, \ldots\}$. If we recall that the infinite series $\Sigma_{n=1}^\infty 1/n^2$ converges, it is easy to see that f is a one-to-one mapping of X into R^∞. It remains to be proved that f and f^{-1} are continuous.

To prove that f is continuous, it suffices to show that given x_0 in X and $\epsilon > 0$, there exists a neighborhood H of x_0 such that $y \in H \Rightarrow \|f(y) - f(x_0)\| < \epsilon$. Since an infinite series of functions converges uniformly if its terms are bounded by the terms of a convergent infinite series of constants, it is easy to see that there exists a positive integer n_0 such that for every y in X we have

$$\|f(y) - f(x_0)\|^2 = \Sigma_{n=1}^\infty |[f_n(y) - f_n(x_0)]/n|^2$$
$$< \Sigma_{n=1}^{n_0} |[f_n(y) - f_n(x_0)]/n|^2 + \epsilon^2/2.$$

By the continuity of the f_n's, for each $n = 1, 2, \ldots, n_0$ there exists a neighborhood H_n of x_0 such that $y \in H_n \Rightarrow |[f_n(y) - f_n(x_0)]/n|^2 < \epsilon^2/2n_0$. If we define H by $H = \cap_{n=1}^{n_0} H_n$, it is clear that H is a neighborhood of x_0 such that $y \in H \Rightarrow \|f(y) - f(x_0)\|^2 < \epsilon^2 \Rightarrow \|f(y) - f(x_0)\| < \epsilon$.

We conclude our proof by showing that f^{-1} is continuous as a mapping of $f(X)$ onto X. It suffices to show that given x_0 in X and a basic neighborhood G_j of x_0, there exists $\epsilon > 0$ such that $\|f(y) - f(x_0)\| < \epsilon \Rightarrow y \in G_j$. G_j is the second member of some ordered pair $P_{n_0} = (G_i, G_j)$ such that $x_0 \in G_i \subseteq \overline{G_i} \subseteq G_j$. If we choose $\epsilon < 1/2n_0$, then we see that $\|f(y) - f(x_0)\| < \epsilon \Rightarrow \Sigma_{n=1}^\infty |[f_n(y) - f_n(x_0)]/n|^2 < (1/2n_0)^2 \Rightarrow |f_{n_0}(y)$

$- f_{n_0}(x_0)| < \frac{1}{2}$. Since x_0 is in G_i, $f_{n_0}(x_0) = 0$, and therefore $|f_{n_0}(y)| < \frac{1}{2}$. Since $f_{n_0}(G_j') = 1$, we see that y is in G_j.

This theorem puts us in a position to answer several natural questions which arise in connection with the inner portions of Fig. 25. We ask the reader to deal with these matters in the following problems.

Problems

1. We know that every metric space is normal, and also that a normal space, if second countable, is metrizable. Give an example of a normal space which is not metrizable (*hint:* see Problem 22-4). This shows that metrizable spaces cannot be characterized among topological spaces by the property of normality.
2. Among normal spaces, second countability implies metrizability. Give an example of a metric space which is not second countable. This shows that metrizable spaces cannot be characterized among normal spaces by the property of second countability.
3. Show that a compact Hausdorff space is metrizable $\Leftrightarrow$ it is second countable.[1]

30. THE STONE-ČECH COMPACTIFICATION

In the preceding section we showed that if a normal space is second countable, then it can be imbedded as a subspace in the familiar metric space R^∞. We now develop a similar imbedding theorem for completely regular spaces.

In order to motivate this theorem properly, we remark that if X is a topological space which occurs as a subspace of a compact Hausdorff space Y, then since Y is completely regular, X is also completely regular, and is a dense subspace of the compact Hausdorff space $\bar{X}$. We see in this way that many completely regular spaces are dense subspaces of compact Hausdorff spaces. Our purpose in this section is to show that any completely regular space X can be imbedded as a dense subspace in a special compact Hausdorff space denoted by $\beta(X)$, and that $\beta(X)$ has the remarkable property that every bounded continuous real function defined on X has a unique extension to a bounded continuous real function defined on $\beta(X)$.

---

[1] The results of this section have characterized metrizable spaces among second countable spaces (by normality) and among compact Hausdorff spaces (by second countability), but not among topological spaces in general. This more difficult problem was solved by Smirnov [38]. For an exposition of his solution, see Kelley [25, pp. 126–130].

How truly remarkable this extension property is can be seen by considering the example given at the end of Sec. 28. Here the completely regular space X is the interval $(0,1]$. This space is clearly a dense subspace of the compact Hausdorff space $[0,1]$. The function f defined on $(0,1]$ by $f(x) = \sin(1/x)$ is a bounded continuous real function defined on X, but it cannot be extended continuously to $[0,1]$. The space $[0,1]$, though it is *a* compact Hausdorff space which contains $(0,1]$ as a dense subspace, is evidently not the space $\beta(X)$. The latter is much too complicated for any simple description of it to be possible.

Before we start our discussion, we recall two items from the previous sections:

    (1)   if X is a completely regular space, then the weak topology generated by $\mathcal{C}(X,R)$ equals the given topology;

    (2)   the relative topology on a subspace of a product space equals the weak topology generated by the restrictions of the projections to that subspace.

These facts (they are Problems 27-4 and 22-2) are the basic principles on which the following analysis rests.

We begin with an arbitrary topological space X and the set $\mathcal{C}(X,R)$ of all bounded continuous real functions defined on X. Let the functions in $\mathcal{C}(X,R)$ be indexed by a set of indices i, so that $\mathcal{C}(X,R) = \{f_i\}$. For each index i, let I_i be the smallest closed interval which contains the range of the function f_i. Each I_i is a compact Hausdorff space, so their product $P = P_i I_i$ is also a compact Hausdorff space, and every subspace of P is completely regular. We next define a mapping f of X into this product space by means of $f(x) = \{f_i(x)\}$, that is, in such a way that $f(x)$ is that point in the product space P whose ith coordinate is the real number $f_i(x)$. By Problem 22-3 and the fact that $p_i f = f_i$ for each projection p_i, it is clear that f is a continuous mapping of X into P.

We now assume that $\mathcal{C}(X,R)$ separates the points of X. This is a weaker assumption than complete regularity and is exactly the requirement that f be a one-to-one mapping. At this stage, we use f to replace $f(X)$ as a set by X; that is, *we imbed X in P as a set.* X is thus a subset of P which has two topologies: its own, and the relative topology which it has as a subspace of P. We observe two features of this situation. First, since f is continuous, the given topology on X is stronger than its relative topology. Second, $\mathcal{C}(X,R)$ is precisely the set of all restrictions to X of the projections p_i of P onto its coordinate spaces I_i. It is now clear that if X is completely regular, so that by statement (1) its given topology equals the weak topology generated by $\mathcal{C}(X,R)$, then by statement (2) its given topology equals its relative topology, and X can be regarded as a subspace of P.

In accordance with these ideas, we now assume that X is completely regular, and we fully identify it, both as a set and as a topological space, with the subspace $f(X)$ of P. It is easy to see that the closure $\bar{X}$ of X in P is a compact Hausdorff space in which X is imbedded as a dense subspace. Furthermore, each f_i in $\mathcal{C}(X,R)$—that is, each projection p_i restricted to X—has an extension to a bounded continuous real function defined on $\bar{X}$; this extension is p_i restricted only to $\bar{X}$, and it is unique by Problem 26-5. The space $\bar{X}$ is commonly denoted by $\beta(X)$.

We summarize these results in the following theorem.

Theorem A. *Let X be an arbitrary completely regular space. Then there exists a compact Hausdorff space $\beta(X)$ with the following properties: (1) X is a dense subspace of $\beta(X)$; (2) every bounded continuous real function defined on X has a unique extension to a bounded continuous real function defined on $\beta(X)$.*

We shall prove in Chap. 14 that the space $\beta(X)$ is essentially unique, in the sense that any other compact Hausdorff space with properties (1) and (2) is homeomorphic to $\beta(X)$. It is called the *Stone-Čech compactification* of the given completely regular space.[1]

Even before our work in this section, it was clear that every subspace of a product of closed intervals is completely regular. It is worthy of special emphasis that the above discussion shows, conversely, that *every* completely regular space is homeomorphic to a subspace of such a product.

Problems

1. If X is completely regular, show that every bounded continuous complex function defined on X has a unique extension to a bounded continuous complex function defined on $\beta(X)$.

2. Every closed subspace of a product of closed intervals is a compact Hausdorff space. Show, conversely, that every compact Hausdorff space is homeomorphic to a closed subspace of such a product.

3. Prove the following generalization of the Tietze extension theorem. If X is a normal space, F a closed subspace of X, and f a continuous mapping of F into a completely regular space Y, then f can be extended continuously to a mapping f' of X into a compact Hausdorff space Z which contains Y as a subspace.

[1] For Stone's own version of these ideas, see his paper [39].

CHAPTER SIX

Connectedness

From the intuitive point of view, a *connected space* is a topological space which consists of a single piece. This property is perhaps the simplest which a topological space may have, and yet it is one of the most important for the applications of topology to analysis and geometry.

On the real line, for instance, intervals are connected subspaces, and we shall see that they are the only connected subspaces. Continuous real functions are often defined on intervals, and functions of this kind have many pleasant properties. For example, such a function assumes as a value every number between any two of its values (the *Weierstrass intermediate value theorem*); furthermore, its graph is a connected subspace of the Euclidean plane. Connectedness is also a basic notion in complex analysis, for the regions on which analytic functions are studied are generally taken to be connected open subspaces of the complex plane.

In the portion of topology which deals with continuous curves and their properties, connectedness is of great significance, for whatever else a continuous curve may be, it is certainly a connected topological space. We describe some of the central ideas of this field in Appendix 2.

Spaces which are not connected are also interesting. One of the outstanding characteristics of the Cantor set is the extreme degree in which it fails to be connected. Much the same is true of the subspace of the real line which consists of all rational numbers. These spaces are so badly disconnected that they are almost granular in texture.

Our purpose in this chapter is to convert these rather vague notions into precise mathematical ideas, and also to establish the fundamental facts in the theory of connectedness which rests upon them.

31. CONNECTED SPACES

A *connected space* is a topological space X which cannot be repre-sented as the union of two disjoint non-empty open sets. If $X = A \cup B$, where A and B are disjoint and open, then A and B are also closed, so that X is the union of two disjoint closed sets, and conversely. We see by this that X is connected $\Leftrightarrow$ it cannot be represented as the union of two disjoint non-empty closed sets. It is also clear that the connectedness of X amounts to the condition that $\emptyset$ and X are its only subsets which are both open and closed. A *connected subspace* of X is a subspace Y which is connected as a topological space in its own right. By the defini-tion of the relative topology on Y, this is equivalent to the condition that Y is not contained in the union of two open subsets of X whose inter-sections with Y are disjoint and non-empty.

Our space X is said to be *disconnected* if it is not connected, that is, if it can be represented in the form $X = A \cup B$, where A and B are disjoint, non-empty, and open; and if X is disconnected, a representation of it in this form (there may be many) is called a *disconnection* of X.

We begin by proving a theorem which supports a considerable part of the theory of connectedness.

Theorem A.   *A subspace of the real line R is connected $\Leftrightarrow$ it is an interval. In particular, R is connected.*

PROOF.   Let X be a subspace of R. We first prove that if X is connected, then it is an interval. We do this by assuming that X is not an interval and by using this assumption to show that X is not connected. To say that X is not an interval is to say that there exist real numbers x, y, z such that $x < y < z$, x and z are in X, and y is not in X. It is easy to see from this that $X = [X \cap (-\infty, y)] \cup [X \cap (y, +\infty)]$ is a disconnection of X, so X is disconnected.

We complete the proof by showing that if X is an interval, then it is necessarily connected. Our strategy here is to assume that X is dis-connected and to deduce a contradiction from this assumption. Let $X = A \cup B$ be a disconnection of X. Since A and B are non-empty, we can choose a point x in A and a point z in B. A and B are disjoint, so $x \neq z$, and by altering our notation if necessary, we may assume that $x < z$. Since X is an interval, $[x, z] \subseteq X$, and each point in $[x, z]$ is in either A or B. We now define y by $y = \sup ([x, z] \cap A)$. It is clear that $x \leq y \leq z$, so y is in X. Since A is closed in X, the definition of y shows that y is in A. From this we conclude that $y < z$. Again by the definition of y, $y + \epsilon$ is in B for every $\epsilon > 0$ such that $y + \epsilon \leq z$, and since B is closed in X, y is in B. We have proved that y is in both A and B, which contradicts our assumption that these sets are disjoint.

Our next theorem asserts that the property of connectedness is preserved by continuous mappings.

Theorem B. *Any continuous image of a connected space is connected.*

PROOF. Let $f: X \to Y$ be a continuous mapping of a connected space X into an arbitrary topological space Y. We must show that $f(X)$ is connected as a subspace of Y. Assume that $f(X)$ is disconnected. As we have seen, this means that there exist two open subsets G and H of Y whose union contains $f(X)$ and whose intersections with $f(X)$ are disjoint and non-empty. This implies, however, that $X = f^{-1}(G) \cup f^{-1}(H)$ is a disconnection of X, which contradicts the connectedness of X.

As a direct consequence of the two theorems just proved, we have the following generalization of the Weierstrass intermediate value theorem.

Theorem C. *The range of a continuous real function defined on a connected space is an interval.*

It is a trivial observation that any two discrete spaces with the same number of points are essentially identical; for any one-to-one mapping of one onto the other (there is at least one) is a homeomorphism, and we may think of them as differing only in the symbols used to designate their points. It is in this sense that there is only one discrete space with any given number of points. The *discrete two-point space*, which is obviously disconnected, is a useful tool in the theory of connectedness. We denote its points by the symbols 0 and 1, and we think of them as real numbers.

Theorem D. *A topological space X is disconnected $\Leftrightarrow$ there exists a continuous mapping of X onto the discrete two-point space $\{0,1\}$.*

PROOF. If X is disconnected and $X = A \cup B$ is a disconnection, then we define a continuous mapping f of X onto $\{0,1\}$ by the requirement that $f(x) = 0$ if x is in A and $f(x) = 1$ if x is in B. This is a valid definition by the fact that A and B are disjoint and their union is X. Since A and B are non-empty and open, f is clearly onto and continuous.

On the other hand, if there exists such a mapping, then X is disconnected; for if X were connected, Theorem B would imply that $\{0,1\}$ is connected, and this would be a contradiction.

This result is a useful tool for the proof of our next theorem.

Theorem E. *The product of any non-empty class of connected spaces is connected.*

PROOF. Let $\{X_i\}$ be a non-empty class of connected spaces, and form their product $X = P_i X_i$. We assume that X is disconnected, and we deduce a contradiction from this assumption. By Theorem D, there exists a continuous mapping f of X onto the discrete two-point space

$\{0,1\}$. Let $a = \{a_i\}$ be a fixed point in X, and consider a particular index i_1. We define a mapping f_{i_1} of X_{i_1} into X by means of $f_{i_1}(x_{i_1}) = \{y_i\}$, where $y_i = a_i$ for $i \neq i_1$ and $y_{i_1} = x_{i_1}$. This is clearly a continuous mapping, so ff_{i_1} is a continuous mapping of X_{i_1} into $\{0,1\}$. Since X_{i_1} is connected, we see by Theorem D that ff_{i_1} is constant and that

$$(ff_{i_1})(x_{i_1}) = f(a)$$

for every point x_{i_1} in X_{i_1}. This shows that $f(x) = f(a)$ for all x's in X which equal a in all coordinate spaces except X_{i_1}. By repeating this process with another index i_2, etc., we see that $f(x) = f(a)$ for all x's in X which equal a in all but a finite number of coordinate spaces. The set of all x's of this kind is a dense subset of X, so by Problem 26-5, f is a constant mapping. This contradicts the assumption that f maps X onto $\{0,1\}$, and completes the proof.

As an application of this result, we show that all finite-dimensional Euclidean and unitary spaces are connected.

Theorem F.   *The spaces R^n and C^n are connected.*

PROOF. We showed in the proof of Theorem 23-B that R^n, as a topological space, can be regarded as the product of n replicas of the real line R. We have seen in Theorem A that R is connected, so R^n is connected by Theorem E. We next prove that C^n and R^{2n} are essentially the same as topological spaces by exhibiting a homeomorphism f of C^n onto R^{2n}. Let $z = (z_1, z_2, \ldots , z_n)$ be an arbitrary element in C^n, and let each coordinate z_k be written out in the form $z_k = a_k + ib_k$, where a_k and b_k are its real and imaginary parts. We define f by

$$f(z) = (a_1, b_1, a_2, b_2, \ldots , a_n, b_n).$$

f is clearly a one-to-one mapping of C^n onto R^{2n}, and if we observe that $\|f(z)\| = \|z\|$, it is easy to see that f is a homeomorphism. The fact that R^{2n} is connected now shows that C^n is also connected.

The techniques developed in the next section will make it possible to give an easy proof of a much more general theorem than this, to the effect that any Banach space is connected.

Problems

1.   Show that a topological space is connected $\Leftrightarrow$ every non-empty proper subset has a non-empty boundary.
2.   Show that a topological space X is connected $\Leftrightarrow$ for every two points in X there is some connected subspace of X which contains both.

3. Prove that a subspace of a topological space X is disconnected $\Leftrightarrow$ it can be represented as the union of two non-empty sets each of which is disjoint from the closure (in X) of the other.

4. Show that the graph of a continuous real function defined on an interval is a connected subspace of the Euclidean plane.

5. Show that if a connected space has a non-constant continuous real function defined on it, then its set of points is uncountably infinite.

6. If X is a completely regular space, use Theorem D to prove that X is connected $\Leftrightarrow \beta(X)$ is connected.

32. THE COMPONENTS OF A SPACE

If a space is not itself connected, then the next best thing is to be able to decompose it into a disjoint class of maximal connected subspaces. Our present objective is to show that this can always be done.

A maximal connected subspace of a topological space, that is, a connected subspace which is not properly contained in any larger connected subspace, is called a *component* of the space. A connected space clearly has only one component, namely, the space itself. In a discrete space, it is easy to see that each point is a component.

The following two theorems will be useful in obtaining the desired decomposition for a general space.

Theorem A. *Let X be a topological space. If $\{A_i\}$ is a non-empty class of connected subspaces of X such that $\cap_i A_i$ is non-empty, then $A = \cup_i A_i$ is also a connected subspace of X.*

PROOF. Assume that A is disconnected. This means that there exist two open subsets G and H of X whose union contains A and whose intersections with A are disjoint and non-empty. All the A_i's are connected, and each lies in $G \cup H$, so each A_i lies entirely in G or entirely in H and is disjoint from the other. Since $\cap_i A_i$ is non-empty, either all the A_i's lie in G and are disjoint from H, or all lie in H and are disjoint from G. We see by this that A itself is disjoint from either G or H, and this contradiction shows that our assumption that A is disconnected is untenable.

Theorem B. *Let X be a topological space and A a connected subspace of X. If B is a subspace of X such that $A \subseteq B \subseteq \bar{A}$, then B is connected; in particular, $\bar{A}$ is connected.*

PROOF. Assume that B is disconnected, that is, that there exist two open subsets G and H of X whose union contains B and whose intersections with B are disjoint and non-empty. Since A is connected and contained in $G \cup H$, A is contained in either G or H and is disjoint from the other. Let us say, just to be specific, that A is disjoint from H. This implies that $\bar{A}$ is also disjoint from H, and since $B \subseteq \bar{A}$, B is disjoint from H. This contradiction shows that B cannot be disconnected, and proves our theorem.

We are now in a position to state and prove the main facts about components.

Theorem C. *If X is an arbitrary topological space, then we have the following: (1) each point in X is contained in exactly one component of X; (2) each connected subspace of X is contained in a component of X; (3) a connected subspace of X which is both open and closed is a component of X; and (4) each component of X is closed.*

PROOF. To prove (1), let x be a point in X. Consider the class $\{C_i\}$ of all connected subspaces of X which contain x. This class is non-empty, since x itself is connected. By Theorem A, $C = \cup_i C_i$ is a connected subspace of X which contains x. C is clearly maximal, and therefore a component of X, because any connected subspace of X which contains C is one of the C_i's and is thus contained in C. Finally, C is the only component of X which contains x. For if C^* is another, it is clearly among the C_i's and is therefore contained in C, and since C^* is maximal as a connected subspace of X, we must have $C^* = C$.

Part (2) is a direct consequence of the construction in the above paragraph, for by this construction, a connected subspace of X is contained in the component which contains any one of its points.

We prove (3) as follows. Let A be a connected subspace of X which is both open and closed. By (2), A is contained in some component C. If A is a proper subset of C, then it is easy to see that

$$C = (C \cap A) \cup (C \cap A')$$

is a disconnection of C. This contradicts the fact that C, being a component, is connected, and we conclude that $A = C$.

Part (4) follows immediately from Theorem B; for if a component C is not closed, then its closure $\bar{C}$ is a connected subspace of X which properly contains C, and this contradicts the maximality of C as a connected subspace of X.

In view of parts (3) and (4) of this theorem, it is natural to ask if a component of a space is necessarily open. The answer is no, as the following example shows. Let X be the subspace of the real line which consists of all rational numbers. We observe two facts about X. First, if x and z are any two distinct rationals, and if $x < z$, then there exists an irrational y such that $x < y < z$, and therefore

$$X = [X \cap (-\infty, y)] \cup [X \cap (y, +\infty)]$$

is a disconnection of X which separates x and z. It is easy to see from this that any subspace of X with more than one point is disconnected, so the components of X are its points. Second, the points of X are not open, for any open subset of R which contains a given rational number also contains others different from it. Here, then, is a space whose components are its points and whose points are not open. This example also shows that a space need not be discrete in order that each of its points be a component.

Problems

1. If $A_1, A_2, \ldots, A_n, \ldots$ is a sequence of connected subspaces of a topological space each of which intersects its successor, show that $\bigcup_{n=1}^{\infty} A_n$ is connected.

2. Show that the union of any non-empty class of connected subspaces of a topological space each pair of which intersects is connected.

3. In Theorem 31-E we proved that a product space is connected if its coordinate spaces are. Devise a different proof of this fact, based on Theorem A, for the case in which there are only two coordinate spaces.

4. Use Theorem A to prove that an arbitrary Banach space B is connected. (*Hint:* if x is a vector, show that the set of all scalar multiples of x is a connected subspace of B.)

5. Let B be an arbitrary Banach space. A *convex set* in B is a non-empty subset S with the property that if x and y are in S, then

$$z = x + t(y - x) = (1 - t)x + ty$$

is also in S for every real number t such that $0 \leq t \leq 1$. Intuitively, a convex set is a non-empty set which contains the segment joining any pair of its points. Prove that every convex subspace of B is connected. Prove also that every sphere (open or closed) in B is convex, and is therefore connected.

6. Show that an open subspace of the complex plane is connected $\Leftrightarrow$ every two points in it can be joined by a polygonal line.

7. Consider the union of two open discs in the complex plane which are externally tangent to each other. State whether this subspace of the plane is connected or disconnected, and justify your answer. Do the same when one disc is open and the other closed, and when both are closed.

8. Consider the following subspace of the Euclidean plane: $\{(x,y) : x \neq 0$ and $y = \sin (1/x)\}$. Is this connected or disconnected? Why? Answer the same questions for the subspace $\{(x,y) : x \neq 0$ and $y = \sin (1/x)\} \cup \{(x,y) : x = 0$ and $-1 \leq y \leq 1\}$.

33. TOTALLY DISCONNECTED SPACES

We have seen that a connected space is one for which no disconnection is possible. We now consider spaces which have a great many disconnections, and which therefore lie, in a manner of speaking, at the opposite end of the connectivity spectrum.

A *totally disconnected space* is a topological space X in which every pair of distinct points can be separated by a disconnection of X. This means that for every pair of points x and y in X such that $x \neq y$, there exists a disconnection $X = A \cup B$ with $x \in A$ and $y \in B$. Such a space is evidently a Hausdorff space, and if it has more than one point, it is disconnected. Oddly enough, a one-point space is both connected and totally disconnected.

The discrete spaces are the simplest totally disconnected spaces. A more interesting example is the space discussed at the end of the previous section, that is, the set of all rational numbers considered as a subspace of the real line. The set of all irrational numbers is also a totally disconnected subspace of the real line, and this is proved in much the same way, from the fact that there exists a rational number between any two irrationals. The Cantor set is yet another totally disconnected subspace of the real line, this time one which is compact.

Our first theorem should not come as a surprise to anyone.

Theorem A. *The components of a totally disconnected space are its points.*
PROOF. If X is a totally disconnected space, it suffices to show that every subspace Y of X which contains more than one point is disconnected. Let x and y be distinct points in Y, and let $X = A \cup B$ be a disconnection of X with $x \in A$ and $y \in B$. It is obvious that

$$Y = (Y \cap A) \cup (Y \cap B)$$

is a disconnection of Y.

Total disconnectedness is closely related to another interesting property.

Theorem B. *Let X be a Hausdorff space. If X has an open base whose sets are also closed, then X is totally disconnected.*
PROOF. Let x and y be distinct points in X. Since X is Hausdorff, x has a neighborhood G which does not contain y. By our assumption, there exists a basic open set B which is also closed such that $x \in B \subseteq G$. $X = B \cup B'$ is clearly a disconnection of X which separates x and y.

If the space X in this theorem is also compact, then the implication can be reversed, and the two conditions are equivalent.

Theorem C. *Let X be a compact Hausdorff space.   Then X is totally disconnected $\Leftrightarrow$ it has an open base whose sets are also closed.*

PROOF.   In view of Theorem B, it suffices to assume that X is totally disconnected and to prove that the class of all subsets of X which are both open and closed forms an open base.   Let x be a point and G an open set which contains it.   We must produce a set B which is both open and closed such that $x \in B \subseteq G$.   We may assume that G is not the full space, for if $G = X$, then we can satisfy our requirement by taking $B = X$. G' is thus a closed subspace of X, and since X is compact, G' is also compact.   By the assumption that X is totally disconnected, for each point y in G' there exists a set H_y which is both open and closed and contains y but not x. G' is compact, so there exists some finite class of H_y's, which we denote by $\{H_1, H_2, \ldots , H_n\}$, with the property that its union contains G' but not x.   We define H by $H = \bigcup_{i=1}^{n} H_i$, and we observe that since this is a finite union and all the H_i's are closed as well as open, H is both open and closed, it contains G', and it does not contain x.   If we now define B to be H', then B clearly has the properties required of it.

Totally disconnected spaces are of considerable significance in several parts of topology, notably in dimension theory (see Hurewicz and Wallman [21]) and in the classic representation theory for Boolean algebras given in Appendix 3.[1]

Problems

1.   Prove that the product of any non-empty class of totally disconnected spaces is totally disconnected.
2.   Prove that a totally disconnected compact Hausdorff space is homeomorphic to a closed subspace of a product of discrete two-point spaces.

34. LOCALLY CONNECTED SPACES

In Sec. 23 we encountered the concept of a locally compact space, that is, of a space which is compact around each point but need not be compact as a whole.   We now study another "local" property which a

---

[1] The reader should be made aware of the fact that several different definitions of total disconnectedness are commonly found in the literature.   The definition given above seems to the present writer to have the logic of language behind it; and Theorem C shows that this definition is equivalent (in the case of a compact Hausdorff space) to the most important of these alternative definitions.

topological space may have, that of being connected in the vicinity of each of its points.

A *locally connected space* is a topological space with the property that if x is any point in it and G any neighborhood of x, then G contains a connected neighborhood of x. This is evidently equivalent to the condition that each point of the space have an open base whose sets are all

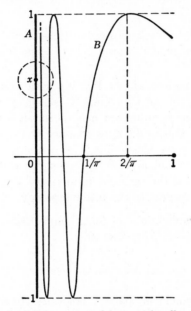

Fig. 26. $A \cup B$ is connected but not locally connected.

connected subspaces.   Locally connected spaces are quite abundant, for, as we have seen in Problem 32-5, every Banach space is locally connected.

We know that local compactness is implied by compactness.   Local connectedness, however, neither implies, nor is implied by, connectedness. The union of two disjoint open intervals on the real line is a simple example of a space which is locally connected but not connected.   A space can also be connected without being locally connected, as the following example shows.   Let X be the subspace of the Euclidean plane defined by $X = A \cup B$, where $A = \{(x,y): x = 0$ and $-1 \leq y \leq 1\}$ and $B = \{(x,y): 0 < x \leq 1$ and $y = \sin(1/x)\}$ (see Fig. 26). B is the image of the interval $(0,1]$ under the continuous mapping f defined by

$$f(x) = (x, \sin(1/x)),$$

so B is connected by Theorem 31-B; and since $X = \bar{B}$, X is connected by Theorem 32-B. X is not locally connected, however, for it is reasonably easy to see that each point x in A has a neighborhood which does not contain any connected neighborhood of x.

We know by Theorem 32-C that the components of an arbitrary topological space X are always closed sets, and from this we see at once that the components of any closed subspace of X are also closed in X. The reader may feel, with some justification, that the components of a well-behaved space ought to be open sets. This is true for locally connected spaces.

Theorem A. *Let X be a locally connected space. If Y is an open subspace of X, then each component of Y is open in X. In particular, each component of X is open.*

PROOF. Let C be a component of Y. We wish to show that C is open in X. Let x be a point in C. Since X is locally connected and Y is open in X, Y contains a connected neighborhood G of x. It suffices to show that $G \subseteq C$. This will follow at once from the fact that C is a component of Y if we can show that G is connected as a subspace of Y. But this is clear by Problem 16-6, according to which the topology of G as a subspace of Y is the same as its topology as a subspace of X; for G is connected with respect to the latter topology.

The principal applications of local connectedness lie in the theory of continuous curves (see Appendix 2).

Problems

1. Prove that a topological space X is locally connected if the components of every open subspace of X are open in X.

2. A connected subspace of a locally connected space X is locally connected if X is the real line. Why? Is this true if X is an arbitrary locally connected space?

3. Show that a compact locally connected space has a finite number of components.

4. Show that the image of a locally connected space under a mapping which is both continuous and open is locally connected.

5. Prove that the product of any non-empty finite class of locally connected spaces is locally connected.

6. Show that the product of an arbitrary non-empty class of locally connected spaces can fail to be locally connected. (*Hint:* consider a product of discrete two-point spaces.)

7. Prove that the product of any non-empty class of connected locally connected spaces is locally connected.

CHAPTER SEVEN

Approximation

Our work in the present chapter centers around the famous theorem of Weierstrass on the approximation by polynomials of continuous real functions defined on closed intervals. This theorem, important as it is in classical analysis, has been overshadowed in recent years by a generalized form of it discovered by Stone. The latter relates to continuous functions defined on compact Hausdorff spaces, and has become an indispensable tool in topology and modern analysis.

We prove the Weierstrass theorem and then the two forms of the Stone-Weierstrass theorem which deal separately with real and complex functions. Finally, after an excursion into the theory of locally compact Hausdorff spaces, we extend the Stone-Weierstrass theorems to this context.

35. THE WEIERSTRASS APPROXIMATION THEOREM

Let us consider a closed interval $[a,b]$ on the real line and a polynomial

$$p(x) = a_0 + a_1x + \cdots + a_nx^n,$$

with real coefficients, defined on $[a,b]$.[1] Every such polynomial is obviously a continuous real function, and as a consequence of the second lemma in Sec. 20, we know that the limit of any uniformly convergent

---

[1] This polynomial can of course be regarded as a function defined on the entire real line. We ignore this fact and consider only x's which lie in $[a,b]$.

sequence of such polynomials is also a continuous real function.   The Weierstrass theorem states that the converse of this is also true, that is, that any continuous real function defined on $[a,b]$ is the limit of some uniformly convergent sequence of polynomials.   This is clearly equivalent to the statement that such a function can be uniformly approximated by polynomials to within any given degree of accuracy.   Many proofs of this classic theorem are known, and the one we give is perhaps as concise and elementary as most.

Theorem A (the Weierstrass Approximation Theorem). *Let f be a continuous real function defined on a closed interval* $[a,b]$, *and let* $\epsilon > 0$ *be given.   Then there exists a polynomial p with real coefficients such that* $|f(x) - p(x)| < \epsilon$ *for all x in* $[a,b]$.

PROOF.   As a first step, we show that it suffices to prove the theorem for the special case in which $a = 0$ and $b = 1$.   If $a = b$, the conclusion follows at once on taking p to be the constant polynomial defined by $p(x) = f(a)$.   We may thus assume that $a < b$.   We next observe that $x = [b - a]x' + a$ gives a continuous mapping of $[0,1]$ onto $[a,b]$, so that the function g defined by $g(x') = f([b - a]x' + a)$ is a continuous real function defined on $[0,1]$.   If our theorem is proved for the case in which $a = 0$ and $b = 1$, then there exists a polynomial p' defined on $[0,1]$ such that $|g(x') - p'(x')| < \epsilon$ for all x' in $[0,1]$.   If we now express this inequality in terms of x, we obtain $|f(x) - p'([x - a]/[b - a])| < \epsilon$ for all x in $[a,b]$; and defining a polynomial p by $p(x) = p'([x - a]/[b - a])$ yields our theorem in the general case.   Accordingly, we may assume that $a = 0$ and $b = 1$.

We next recall that if n is a positive integer and k an integer such that $0 \leq k \leq n$, then the binomial coefficient $\binom{n}{k}$ is defined by

$$\binom{n}{k} = n!/k!(n - k)!.$$

The polynomials B_n—one for each n—defined by

$$B_n(x) = \sum_{k=0}^{n} \binom{n}{k} x^k (1 - x)^{n-k} f\left(\frac{k}{n}\right)$$

are called the *Bernstein polynomials* associated with f.   We prove our theorem by finding a Bernstein polynomial with the required property.

Several identities will be needed for this.   The first is a special case of the binomial theorem:

$$\sum_{k=0}^{n} \binom{n}{k} x^k (1 - x)^{n-k} = [x + (1 - x)]^n = 1. \tag{1}$$

If we differentiate (1) with respect to x, we get

$$\sum_{k=0}^{n} \binom{n}{k} [kx^{k-1}(1-x)^{n-k} - (n-k)x^k(1-x)^{n-k-1}]$$

$$= \sum_{k=0}^{n} \binom{n}{k} x^{k-1}(1-x)^{n-k-1}(k-nx) = 0,$$

and multiplying through by $x(1-x)$ gives

$$\sum_{k=0}^{n} \binom{n}{k} x^k(1-x)^{n-k}(k-nx) = 0. \tag{2}$$

On differentiating (2) with respect to x and considering $x^k(1-x)^{n-k}$ as one of the two factors in applying the product rule, we get

$$\sum_{k=0}^{n} \binom{n}{k} [-nx^k(1-x)^{n-k} + x^{k-1}(1-x)^{n-k-1}(k-nx)^2] = 0. \tag{3}$$

Applying (1) to (3) gives

$$\sum_{k=0}^{n} \binom{n}{k} x^{k-1}(1-x)^{n-k-1}(k-nx)^2 = n;$$

and on multiplying this through by $x(1-x)$, we find that

$$\sum_{k=0}^{n} \binom{n}{k} x^k(1-x)^{n-k}(k-nx)^2 = nx(1-x),$$

or, on dividing both sides by n^2,

$$\sum_{k=0}^{n} \binom{n}{k} x^k(1-x)^{n-k}\left(x - \frac{k}{n}\right)^2 = \frac{x(1-x)}{n}. \tag{4}$$

Identities (1) and (4) will be our main tools in showing that $B_n(x)$ is uniformly close to $f(x)$ for all sufficiently large n.

Now for the proof of the fact just stated. By using (1), we see that

$$f(x) - B_n(x) = \sum_{k=0}^{n} \binom{n}{k} x^k(1-x)^{n-k}\left[f(x) - f\left(\frac{k}{n}\right)\right],$$

so that

$$|f(x) - B_n(x)| \le \sum_{k=0}^{n} \binom{n}{k} x^k(1-x)^{n-k}\left|f(x) - f\left(\frac{k}{n}\right)\right|. \tag{5}$$

Since f is uniformly continuous on $[0,1]$, we can find a $\delta > 0$ such that $|x - k/n| < \delta \Rightarrow |f(x) - f(k/n)| < \epsilon/2$. We now split the sum on the right of (5) into two parts, denoted by Σ and Σ', where Σ is the sum of those terms for which $|x - k/n| < \delta$ (we think of x as fixed but arbitrary) and where Σ' is the sum of the remaining terms.   It is easy to see that $\Sigma < \epsilon/2$. We complete the proof by showing that if n is taken sufficiently large, then Σ' can be made less than $\epsilon/2$ independently of x. Since f is bounded, there exists a positive real number K such that $|f(x)| \leq K$ for all x in $[0,1]$.   From this it follows that

$$\Sigma' \leq 2K \sum \binom{n}{k} x^k (1 - x)^{n-k},$$

where the sum on the right—denote it by Σ''—is taken over all k such that $|x - k/n| \geq \delta$.   It now suffices to show that if n is taken sufficiently large, then Σ'' can be made less than $\epsilon/4K$ independently of x.   Identity (4) shows that

$$\delta^2 \sum{}'' \leq \frac{x(1 - x)}{n},$$

so

$$\sum{}'' \leq \frac{x(1 - x)}{\delta^2 n}.$$

The maximal value of $x(1 - x)$ on $[0,1]$ is $\frac{1}{4}$, so

$$\sum{}'' \leq \frac{1}{4\delta^2 n}.$$

If we take n to be any integer greater than $K/\delta^2\epsilon$, then $\Sigma'' < \epsilon/4K$, $\Sigma' < \epsilon/2$, $|f(x) - B_n(x)| < \epsilon$ for all x in $[0,1]$, and our theorem is fully proved.

The Weierstrass theorem clearly amounts to the assertion that for any closed interval $[a,b]$ on the real line, the polynomials are dense in the metric space $\mathcal{C}[a,b]$.   This is the form of the theorem which we shall generalize in the next section to $\mathcal{C}(X,R)$, where X is an arbitrary compact Hausdorff space.

The slightly restricted statement that the polynomials are dense in $\mathcal{C}[0,1]$ has another generalization, in a different direction. This result is so remarkable that we state it because of its intrinsic interest, though we give no proof.   The Weierstrass theorem for $\mathcal{C}[0,1]$ says, in effect, that all real linear combinations of the functions

$$1, x, x^2, \ldots, x^n, \ldots$$

are dense in $\mathcal{C}[0,1]$, where by a real linear combination of these functions we mean the result of choosing any finite set of them, multiplying each by a real number, and adding.   Instead of working with all positive powers of x, let us permit gaps to occur, and consider the infinite set of functions

$$1, x^{n_1}, x^{n_2}, \ldots, x^{n_k}, \ldots,$$

the n_k's being positive integers for which $n_1 < n_2 < \cdots < n_k < \cdots$. The result we have in mind is called *Müntz's theorem*, and asserts that all real linear combinations of these functions are dense in $\mathbb{C}[0,1]$ $\Leftrightarrow$ the series $\Sigma_{k=1}^{\infty} 1/n_k$ diverges. For a proof, we refer the interested reader to Lorentz [29, pp. 46–48] or Achieser [1, pp. 43–46].

Problems

1. Prove that $\mathbb{C}[a,b]$ is separable.
2. Let f be a continuous real function defined on $[0,1]$. The *moments* of f are the numbers $\int_0^1 f(x)x^n \, dx$, where $n = 0, 1, 2, \ldots$. Prove that two continuous real functions defined on $[0,1]$ are identical if they have the same sequence of moments.
3. Use the Weierstrass theorem to prove that the polynomials are dense in $\mathbb{C}(X,R)$ for any closed and bounded subspace X of the real line.

36. THE STONE-WEIERSTRASS THEOREMS

Our purpose in this section is to lay bare the true nature of the Weierstrass approximation theorem. We achieve this end by generalizing the theorem in such a manner that its inessential features are stripped away.

Our starting point is the fact that the polynomials are dense in $\mathbb{C}[a,b]$ for any closed interval $[a,b]$. We wish to replace $[a,b]$ by an arbitrary compact Hausdorff space X and to make a similar statement about $\mathbb{C}(X,R)$. The most obvious difficulty in this program is that it is meaningless to speak of polynomials on X. This obstacle will disappear when we take a closer look at what polynomials are.

Consider the two functions 1 and x defined on $[a,b]$. The set P of all polynomials on $[a,b]$ is identical with the set of all functions which can be built from these two by applying the following three operations: multiplication, multiplication by real numbers, and addition. P is an algebra of real functions on $[a,b]$, for it is closed with respect to these three operations. Even more, it is a subalgebra of $\mathbb{C}[a,b]$. We say that P is *the subalgebra of* $\mathbb{C}[a,b]$ *generated by* $\{1,x\}$, for it is a subalgebra containing $\{1,x\}$ which is contained in every subalgebra with this property. We know by Problem 20-3 that the closure of a subalgebra of $\mathbb{C}(X,R)$—for any topological space X—is also a subalgebra of $\mathbb{C}(X,R)$. We may therefore speak of the closure $\bar{P}$ of P as *the closed subalgebra of* $\mathbb{C}[a,b]$ *generated by* $\{1,x\}$. As above, this means that $\bar{P}$ is a closed subalgebra containing $\{1,x\}$ which is contained in every closed subalgebra with this property. These ideas make it possible for us to state the Weierstrass theorem in the following equivalent forms:

(1)   the closed subalgebra of $\mathcal{C}[a,b]$ generated by $\{1,x\}$ equals $\mathcal{C}[a,b]$;

(2)   any closed subalgebra of $\mathcal{C}[a,b]$ which contains $\{1,x\}$ equals $\mathcal{C}[a,b]$.

These are potent statements, saying, as they do, that the very small set of functions $\{1,x\}$ suffices to generate the much more extensive set $\mathcal{C}[a,b]$. As our theorems below will show, these statements depend only on the fact that a closed subalgebra of $\mathcal{C}[a,b]$ which contains the set $\{1,x\}$ separates points in the sense of Sec. 27 (for it contains the function x) and contains all constant functions (for it contains the non-zero constant function 1).

Before we go further, it is worth observing that statement (1) is not true in general if either 1 or x is omitted from the generating set. If x is omitted, then the closed subalgebra generated by $\{1\}$ consists of the constant functions, and this is not equal to $\mathcal{C}[a,b]$ unless $a = b$. On the other hand, if 1 is omitted, then the closed subalgebra generated by $\{x\}$ contains only functions which vanish at 0, and if 0 is in $[a,b]$, then the non-zero constant functions, among others, are not in this closed subalgebra.

Our theorems rest on two lemmas, both of which have to do with the fact that $\mathcal{C}(X,R)$ is a lattice for any topological space X. If f and g are functions in $\mathcal{C}(X,R)$, we recall that their join and meet are defined by

$$(f \vee g)(x) = \max \{f(x),g(x)\}$$
and $\qquad (f \wedge g)(x) = \min \{f(x),g(x)\}.$

Our first lemma states conditions which guarantee that a closed sublattice of $\mathcal{C}(X,R)$ equals $\mathcal{C}(X,R)$.

Lemma. *Let X be a compact Hausdorff space with more than one point, and let L be a closed sublattice of $\mathcal{C}(X,R)$ with the following property: if x and y are distinct points of X and a and b any two real numbers, then there exists a function f in L such that $f(x) = a$ and $f(y) = b$. Then L equals $\mathcal{C}(X,R)$.*[1]

PROOF.   Let f be an arbitrary function in $\mathcal{C}(X,R)$. We must show that f is in L. Let $\epsilon > 0$ be given. Since L is closed, it suffices to construct a function g in L such that $f(z) - \epsilon < g(z) < f(z) + \epsilon$ for all points z in

---

[1] If X has only one point, then a single constant function constitutes a closed sublattice of $\mathcal{C}(X,R)$ with the stated property which does not equal $\mathcal{C}(X,R)$. It is therefore necessary to assume that X has more than one point. Further, the reader will notice that the proof given below makes no use of the assumption that X is Hausdorff. However, if there exists a closed sublattice of $\mathcal{C}(X,R)$ with the stated property, then X is necessarily Hausdorff, so there is nothing to be gained by omitting this assumption.

X, for it will follow from this that $\|f - g\| < \epsilon$. We now construct such a function.

Let x be a point in X which is fixed throughout this paragraph, and let y be a point different from x. By our assumption about L, there exists a function f_y in L such that $f_y(x) = f(x)$ and $f_y(y) = f(y)$. Now consider the open set $G_y = \{z : f_y(z) < f(z) + \epsilon\}$. It is clear that both x and y belong to G_y, so the class of G_y's for all points y different from x is an open cover of X. Since X is compact, this open cover has a finite subcover, which we denote by $\{G_1, G_2, \ldots, G_n\}$. If the corresponding functions in L are denoted by $f_1, f_2, \ldots, f_n$, then

$$g_x = f_1 \wedge f_2 \wedge \cdots \wedge f_n$$

is evidently a function in L such that $g_x(x) = f(x)$ and $g_x(z) < f(z) + \epsilon$ for all points z in X.

We next consider the open set $H_x = \{z : g_x(z) > f(z) - \epsilon\}$. Since x belongs to H_x, the class of H_x's for all points x in X is an open cover of X. The compactness of X implies that this open cover has a finite subcover, which we denote by $\{H_1, H_2, \ldots, H_m\}$. We denote the corresponding functions in L by $g_1, g_2, \ldots, g_m$, and we define g by $g = g_1 \vee g_2 \vee \cdots \vee g_m$. It is clear that g is a function in L with the property that $f(z) - \epsilon < g(z) < f(z) + \epsilon$ for all points z in X, so our proof is complete.

In our next lemma we make use of the concept of the absolute value of a function. If f is a real or complex function defined on a topological space X, then the function $|f|$—called the *absolute value* of f—is defined by $|f|(x) = |f(x)|$. If f is continuous, then $|f|$ is also continuous. We observe that the lattice operations in $\mathcal{C}(X,R)$ are expressible in terms of addition, scalar multiplication, and the formation of absolute values:

$$f \vee g = \frac{f + g + |f - g|}{2}$$

and

$$f \wedge g = \frac{f + g - |f - g|}{2}.$$

These identities show that any linear subspace of $\mathcal{C}(X,R)$ which contains the absolute value of each of its functions is a sublattice of $\mathcal{C}(X,R)$.

Lemma. *Let X be an arbitrary topological space. Then every closed subalgebra of $\mathcal{C}(X,R)$ is also a closed sublattice of $\mathcal{C}(X,R)$.*

PROOF. Let A be a closed subalgebra of $\mathcal{C}(X,R)$. By the above remarks, it suffices to show that if f is in A, then $|f|$ is also in A. Let $\epsilon > 0$ be given. Since $|t|$ is a continuous function of the real variable t, by the Weierstrass approximation theorem there exists a polynomial p' with the property that $|\,|t| - p'(t)| < \epsilon/2$ for every t on the closed interval

$[-\|f\|,\|f\|]$. If p is the polynomial which results when the constant term of p' is replaced by 0, then p is a polynomial with 0 as its constant term which has the property that $\mid |t| - p(t)| < \epsilon$ for every t in $[-\|f\|,\|f\|]$. Since A is an algebra, the function $p(f)$ in $\mathcal{C}(X,R)$ is in A. By the stated property of p, it is easy to see that $\mid |f(x)| - p(f(x))| < \epsilon$ for every point x in X, and from this it follows that $\| |f| - p(f)\| < \epsilon$. We conclude the proof by remarking that since A is closed, the fact that $|f|$ can be approximated by the function $p(f)$ in A shows that $|f|$ is in A.

We are now in a position to prove the Stone-Weierstrass theorems.

Theorem A (the Real Stone-Weierstrass Theorem). *Let X be a compact Hausdorff space, and let A be a closed subalgebra of $\mathcal{C}(X,R)$ which separates points and contains a non-zero constant function. Then A equals $\mathcal{C}(X,R)$.*

PROOF. If X has only one point, then $\mathcal{C}(X,R)$ contains only constant functions; and since A contains a non-zero constant function and is an algebra, it contains all constant functions and equals $\mathcal{C}(X,R)$. We may thus assume that X has more than one point. By the above lemmas, it suffices to show that if x and y are distinct points of X, and if a and b are any two real numbers, then there exists a function f in A such that $f(x) = a$ and $f(y) = b$. Since A separates points, there exists a function g in A such that $g(x) \neq g(y)$. If we now define f by

$$f(z) = a\,\frac{g(z) - g(y)}{g(x) - g(y)} + b\,\frac{g(z) - g(x)}{g(y) - g(x)},$$

then f clearly has the required properties.

We next turn our attention to the complex case, that is, to conditions which guarantee that a closed subalgebra of $\mathcal{C}(X,C)$ equals $\mathcal{C}(X,C)$. It is first of all necessary to understand that the conditions of Theorem **A** will not suffice. The simplest example which shows this requires a little knowledge of the theory of analytic functions, and the reader without such knowledge may skip at once to the next paragraph. Let X be the closed unit disc $\{z:|z| \leq 1\}$ in the complex plane. X is clearly a compact Hausdorff space. Consider the set A of all functions in $\mathcal{C}(X,C)$ which are analytic in the interior of X. A is evidently a subalgebra of $\mathcal{C}(X,C)$, and one sees that it is closed by using Morera's theorem. A separates points, for it contains the function f defined by $f(z) = z$. It also contains all constant functions. In spite of this, A does not equal $\mathcal{C}(X,C)$; for the function g defined by $g(z) = \bar{z}$ is in $\mathcal{C}(X,C)$ but is not in A, since it is not differentiable at any point.

What can be done to salvage Theorem A in the complex case? The answer lies in the operation of conjugation discussed at the end of Sec. 20. If f is a complex function defined on a topological space X, we remind the reader that its *conjugate* $\bar{f}$ is defined by $\bar{f}(x) = \overline{f(x)}$.   It will also be convenient for us to define the *real part* and the *imaginary part* of f:

$$R(f) = \frac{f + \bar{f}}{2} \quad \text{and} \quad I(f) = \frac{f - \bar{f}}{2i}. \tag{1}$$

We observe that if a complex function f has different values at two distinct points of X, then at least one of the functions $R(f)$ and $I(f)$ also has different values at these points.

Theorem B (the Complex Stone-Weierstrass Theorem). *Let X be a compact Hausdorff space, and let A be a closed subalgebra of $\mathcal{C}(X,C)$ which separates points, contains a non-zero constant function, and contains the conjugate of each of its functions.   Then A equals $\mathcal{C}(X,C)$.*

PROOF.   The real functions in A clearly form a closed subalgebra B of $\mathcal{C}(X,R)$.   Let us assume for a moment that B equals $\mathcal{C}(X,R)$.   If f is an arbitrary function in $\mathcal{C}(X,C)$, then $R(f)$ and $I(f)$ are in $\mathcal{C}(X,R)$, and are thus in B.   But since $f = R(f) + iI(f)$ and A is an algebra, f is in A and A equals $\mathcal{C}(X,C)$.   It therefore suffices to show that B equals $\mathcal{C}(X,R)$. We prove this by applying Theorem A.

We begin by showing that B separates points.   Let x and y be distinct points in X.   Since A separates points, there exists a function f in A which has different values at x and y.   As we saw in the above remarks, $R(f)$ or $I(f)$ also has different values at x and y.   Since A is an algebra which contains the conjugate of each of its functions, formulas (1) show that both $R(f)$ and $I(f)$ are in B, so B separates points.   We next show that B contains a non-zero constant function.   By our hypothesis, A contains some non-zero constant function g. A is an algebra which contains the conjugate of each of its functions, so $g\bar{g} = |g|^2$ is a non-zero constant function in B.   Theorem A now implies directly that B equals $\mathcal{C}(X,R)$, and our proof is complete.

The two Stone-Weierstrass theorems are among the most important facts in modern analysis.   The theory developed in the last three chapters of this book could hardly exist without them, and they have many other applications as well.[1]

Problems

1.   Prove the two-variable Weierstrass approximation theorem: if $f(x,y)$ is a real function defined and continuous on the closed rectangle $X = [a,b] \times [c,d]$ in the Euclidean plane R^2, then f can be uniformly approximated on X by polynomials in x and y with real coefficients.

[1] See Stone [40].

2. Let X be the closed unit disc in the complex plane, and show that any function f in $\mathcal{C}(X,C)$ can be uniformly approximated on X by polynomials in z and $\bar{z}$ with complex coefficients.

3. Let X and Y be compact Hausdorff spaces, and f a function in $\mathcal{C}(X \times Y,C)$. Show that f can be uniformly approximated by functions of the form $\Sigma_{i=1}^{n} f_i g_i$, where the f_i's are in $\mathcal{C}(X,C)$ and the g_i's are in $\mathcal{C}(Y,C)$.

37. LOCALLY COMPACT HAUSDORFF SPACES

In Sec. 23 we defined a *locally compact space* to be a topological space in which each point has a neighborhood with compact closure. Locally compact spaces often arise in the applications of topology to geometry and analysis, and since those which do are almost always Hausdorff spaces, we restrict our attention in this section to locally compact Hausdorff spaces.

The main fact about such a space is that it can be converted into a compact Hausdorff space by suitably adjoining a single point. The reader is perhaps familiar from analysis with the prototype of this process, in which the complex plane C is enlarged by adjoining to it an "ideal point" called the *point at infinity* and denoted by ∞. This ideal point can be thought of as any object not in C, and we denote by C_∞ the larger set $C \cup \{\infty\}$. C_∞ is called the *extended complex plane* when the neighborhoods of ∞ (other than C_∞ itself) are taken to be the complements in C_∞ of the closed and bounded subsets (i.e., the compact subspaces) of C. These ideas add nothing to our understanding of the complex plane, but they do clarify many proofs and simplify the statements of many theorems, and they are valuable for this reason. Figure 27 gives an easy way of visualizing the extended complex plane. In this figure, the surface S of a sphere of radius $\frac{1}{2}$ is rested tangentially on C at the origin. It is customary to call the point of contact the *south pole* and the opposite point the *north pole*. The indicated projection from the north pole establishes a homeomorphism between S minus its north pole and C, so

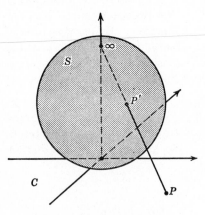

Fig. 27.   The Riemann sphere.

from the topological point of view, S minus its north pole can be regarded as essentially identical with the complex plane C. The north pole of S can be considered to be the point at infinity, and passing from C to C_∞ amounts to using the point ∞ to plug up the hole in C at the north pole. When S is identified in this manner with the extended complex plane, it is usually called the *Riemann sphere*. In summary, the locally compact Hausdorff space C has been made into the compact Hausdorff space S by adding the single point ∞.

We now generalize the construction outlined above to the case of an arbitrary locally compact Hausdorff space X. Let ∞ be an object not in X, and form the set $X_\infty = X \cup \{\infty\}$. We define a topology on X_∞ by specifying the following as open sets: (i) the open subsets of X, regarded as subsets of X_∞; (ii) the complements in X_∞ of the compact subspaces of X; and (iii) the full space X_∞. If we keep in mind the fact that a compact subspace of a Hausdorff space is closed, then it is easy to show that this class of sets actually is a topology on X_∞, and also that the given topology on X equals its relative topology as a subspace of X_∞. The following are the main properties of the topological space X_∞.

(1) X_∞ *is compact*. To prove this, let $\{G_i\}$ be an open cover of X_∞. We must produce a finite subcover. If X_∞ occurs among the G_i's, then $\{G_i\}$ clearly has a finite subcover, namely, $\{X_\infty\}$. We may therefore assume that each G_i is a set of type (i) or type (ii). At least one G_i, say G_{i_0}, must contain the point ∞, and this set is necessarily of type (ii). Its complement G_{i_0}' is thus a compact subspace of X which is contained in the union of some class of open subsets of X of the form $G_i \cap X$, so it is contained in the union of some finite subclass of these sets, say $\{G_1 \cap X, G_2 \cap X, \ldots, G_n \cap X\}$. It is now easy to see that the class $\{G_{i_0}, G_1, G_2, \ldots, G_n\}$ is a finite subcover of the original open cover of X_∞, so X_∞ is compact.

(2) X_∞ *is Hausdorff*. X is Hausdorff, so any pair of distinct points in X_∞ both of which lie in X can be separated by open subsets of X, and thus can be separated by open subsets of X_∞ of type (i). It therefore suffices to show that any point x in X and the point ∞ can be separated by open subsets of X_∞. X is locally compact, so x has a neighborhood G whose closure $\bar{G}$ in X is compact. It is now clear that G and $\bar{G}'$ are disjoint open subsets of X_∞ such that $x \in G$ and $\infty \in \bar{G}'$, so X_∞ is Hausdorff.

The compact Hausdorff space X_∞ associated with the locally compact Hausdorff space X in the manner described above is called the *one-point compactification* of X, and the point ∞ is called the *point at infinity*. We know that compact spaces are locally compact, so these ideas apply without change when X is a compact Hausdorff space. It is easy to see that the locally compact Hausdorff space X is compact $\Leftrightarrow \infty$ is an isolated point of X_∞.   It may seem useless to consider the one-point compactification of a compact Hausdorff space, but we shall see in the next section that it enables us to weaken the hypotheses of the Stone-Weierstrass theorems.

The one-point compactification is useful mainly in simplifying the proofs of theorems about locally compact Hausdorff spaces.   As an example, any space X of this kind is easily seen to be completely regular; for X is a subspace of X_∞, which is compact Hausdorff and therefore completely regular, and every subspace of a completely regular space is completely regular.   Accordingly, if x is a point of X, and G a neighborhood of x which does not equal the full space, then there exists a continuous real function f defined on X, all of whose values lie in the closed unit interval $[0,1]$, such that $f(x) = 1$ and $f(G') = 0$.   This fact can easily be generalized, again by using the one-point compactification, to the case in which the point x is replaced by an arbitrary compact subspace of X.

Theorem A.   *Let X be a locally compact Hausdorff space, let C be a compact subspace of X, and let G be a neighborhood of C which does not equal the full space.   Then there exists a continuous real function f defined on X, all of whose values lie in the closed unit interval $[0,1]$, such that $f(C) = 1$ and $f(G') = 0$.*

PROOF.   Let X_∞ be the one-point compactification of X.   Then C and G' are disjoint closed subspaces of X_∞, and by Urysohn's lemma there exists a continuous real function g defined on X_∞, all of whose values lie in $[0,1]$, such that $g(C) = 1$ and $g(G') = 0$.   If f is the restriction of g to X, then f has the required properties.

This result is an important tool in the theory of measure and integration on locally compact Hausdorff spaces.

Problems

1.   Let X be a locally compact Hausdorff space, and C_1 and C_2 disjoint compact subspaces of X.   Show that C_1 and C_2 have disjoint neighborhoods whose closures are compact.

2.   Show that a Hausdorff space is locally compact $\Leftrightarrow$ each of its points is an interior point of some compact subspace.

3.   Let f be a mapping of a locally compact space X onto a Hausdorff space Y.   If f is both continuous and open, show that Y is also locally compact.

4.   Show that if the product of a non-empty class of Hausdorff spaces is locally compact, then each coordinate space is also locally compact.

38. THE EXTENDED STONE-WEIERSTRASS THEOREMS

Let X be a locally compact Hausdorff space. Our present purpose is to generalize the theorems of Sec. 36 to this context.

A real or complex function f defined on X is said to *vanish at infinity* if for each $\epsilon > 0$ there exists a compact subspace C of X such that $|f(x)| < \epsilon$ for every x outside of C. On the real line, for instance, the functions f and g defined by $f(x) = e^{-x^2}$ and $g(x) = (x^2 + 1)^{-1}$ have this property, but the non-zero constant functions do not. It is easy to see that if X is compact, then *every* real or complex function defined on X vanishes at infinity, so in this case the requirement that a function vanish at infinity is no restriction at all.

We denote by $\mathfrak{C}_0(X,R)$ the set of all continuous real functions defined on X which vanish at infinity. $\mathfrak{C}_0(X,C)$ is defined similarly. If f is a function in one of these sets, then since $|f(x)| < \epsilon$ outside of some compact subspace C of X, and f is bounded on C, f is necessarily bounded on all of X. It follows from this that $\mathfrak{C}_0(X,R) \subseteq \mathfrak{C}(X,R)$ and $\mathfrak{C}_0(X,C) \subseteq \mathfrak{C}(X,C)$. Further, the remark in the preceding paragraph shows that when X is compact we have equality in each case.

Lemma. $\mathfrak{C}_0(X,R)$ *and* $\mathfrak{C}_0(X,C)$ *are closed subalgebras of* $\mathfrak{C}(X,R)$ *and* $\mathfrak{C}(X,C)$.

PROOF. We first show that $\mathfrak{C}_0(X,R)$ is a closed subset of $\mathfrak{C}(X,R)$. It suffices to show that if f is a function in $\mathfrak{C}(X,R)$ which is in the closure of $\mathfrak{C}_0(X,R)$, then f vanishes at infinity. Let $\epsilon > 0$ be given. Since f is in the closure of $\mathfrak{C}_0(X,R)$, there exists a function g in $\mathfrak{C}_0(X,R)$ such that $\|f - g\| < \epsilon/2$, and this implies that $|f(x) - g(x)| < \epsilon/2$ for all x. The function g vanishes at infinity, so there exists a compact subspace C of X such that $|g(x)| < \epsilon/2$ for all x outside of C. It now follows at once that

$$|f(x)| = |[f(x) - g(x)] + g(x)| \leq |f(x) - g(x)| + |g(x)| < \epsilon/2 + \epsilon/2 = \epsilon$$

for all x outside of C, so f vanishes at infinity. The same argument shows that $\mathfrak{C}_0(X,C)$ is a closed subset of $\mathfrak{C}(X,C)$.

We next show that if f and g are in $\mathfrak{C}_0(X,R)$, then $f + g$ is also in $\mathfrak{C}_0(X,R)$, that is, that $f + g$ vanishes at infinity. Let $\epsilon > 0$ be given. Since f vanishes at infinity, there exists a compact subspace C_1 of X outside of which $|f(x)| < \epsilon/2$. Similarly, there exists a compact subspace C_2 of X outside of which $|g(x)| < \epsilon/2$. $C = C_1 \cup C_2$ is then a compact subspace of X outside of which

$$|(f + g)(x)| = |f(x) + g(x)| \leq |f(x)| + |g(x)| < \epsilon/2 + \epsilon/2 = \epsilon,$$

so $f + g$ vanishes at infinity. We can prove in much the same way that

$\mathcal{C}_0(X,R)$ is also closed with respect to scalar multiplication and multiplication; and since $\mathcal{C}_0(X,R)$ is non-empty (it contains the function which is identically zero), it is clearly a subalgebra of $\mathcal{C}(X,R)$. Similarly, $\mathcal{C}_0(X,C)$ is a subalgebra of $\mathcal{C}(X,C)$.

This lemma permits us to regard $\mathcal{C}_0(X,R)$ and $\mathcal{C}_0(X,C)$ as algebras of functions in their own right. We next establish a natural and useful connection between continuous functions defined on X which vanish at infinity and continuous functions defined on X_∞ which vanish at the point ∞, where of course X_∞ is the one-point compactification of X. It is important here to be quite clear about the distinction between these concepts. For a function on X to vanish at infinity means precisely what the above definition says. Such a function need not have 0 as a value. On the other hand, to say that a function on X_∞ vanishes at the point ∞ is to say that this function assumes the value 0 at the point ∞.

Lemma. *$\mathcal{C}_0(X,R)$ equals the set of all restrictions to X of those functions in $\mathcal{C}(X_\infty,R)$ which vanish at the point ∞. Similarly, $\mathcal{C}_0(X,C)$ equals the set of all restrictions to X of the functions in $\mathcal{C}(X_\infty,C)$ which vanish at the point ∞.*

PROOF. Let g be a function in $\mathcal{C}(X_\infty,R)$ which vanishes at the point ∞. Since g is continuous at ∞, for each $\epsilon > 0$ there exists a neighborhood G of ∞ such that $|g(x)| < \epsilon$ for all x in G. By the definition of a neighborhood of ∞ given in Sec. 37, G is the full space X_∞ or the complement in X_∞ of a compact subspace of X. In either case, there clearly exists a compact subspace C of X such that $|g(x)| < \epsilon$ for every point x in X and outside of C. In other words, the restriction f of g to X vanishes at infinity, so is a function in $\mathcal{C}_0(X,R)$. We must also show, conversely, that every function f in $\mathcal{C}_0(X,R)$ arises in this way from some function g in $\mathcal{C}(X_\infty,R)$ which vanishes at the point ∞. All that is necessary is to define g by $g(x) = f(x)$ for every x in X and $g(\infty) = 0$, and to observe that the condition that f vanishes at infinity is precisely what is needed to guarantee that g is continuous at ∞. The proof of the second statement of the lemma is exactly the same.

The machinery given above is intended to make the proofs of the following two theorems relatively simple. They are called the *extended Stone-Weierstrass theorems*.

Theorem A. *Let X be a locally compact Hausdorff space, and let A be a closed subalgebra of $\mathcal{C}_0(X,R)$ which separates points and for each point in X contains a function which does not vanish there. Then A equals $\mathcal{C}_0(X,R)$.*

PROOF. Let X_∞ be the one-point compactification of X. By our second lemma, we can extend every function in A to a function in $\mathcal{C}(X_\infty,R)$ which vanishes at ∞. We denote the set of all these extensions by A_0. Our hypotheses imply that A_0 is a closed subalgebra of $\mathcal{C}(X_\infty,R)$ which separates points and has the property that all its functions vanish at ∞. Let A_1 be the set of functions obtained by adding all constant functions to each function in A_0. It is easy to see that A_1 is a closed subalgebra of $\mathcal{C}(X_\infty,R)$ which separates points and contains a non-zero constant function, so by Theorem 36-A, A_1 equals $\mathcal{C}(X_\infty, R)$. It follows from this that A_0 consists of all functions in $\mathcal{C}(X_\infty,R)$ which vanish at ∞, and another application of our second lemma shows that A equals $\mathcal{C}_0(X,R)$.

Theorem B. *Let X be a locally compact Hausdorff space, and let A be a closed subalgebra of $\mathcal{C}_0(X,C)$ which separates points, for each point in X contains a function which does not vanish there, and contains the conjugate of each of its functions. Then A equals $\mathcal{C}_0(X,C)$.*

PROOF. The proof of Theorem A will serve here almost word for word.

We observe that when X is assumed to be compact in the above two theorems, so that $\mathcal{C}_0(X,R) = \mathcal{C}(X,R)$ and $\mathcal{C}_0(X,C) = \mathcal{C}(X,C)$, then they constitute slightly stronger forms of the Stone-Weierstrass theorems, for they yield the same conclusions under slightly weaker assumptions.

Problems

1. If X is a locally compact Hausdorff space, prove that $\mathcal{C}_0(X,R)$ is a sublattice of $\mathcal{C}(X,R)$.

2. Let X be a locally compact Hausdorff space, and show that the weak topology generated by $\mathcal{C}_0(X,R)$ equals the given topology.

3. Let X be a locally compact Hausdorff space and S a subset of $\mathcal{C}_0(X,R)$ which separates points and for each point in X contains a function which does not vanish there. Show that the weak topology generated by S equals the given topology.

PART TWO

CHAPTER EIGHT

Algebraic Systems

We have seen in the preceding chapters that one of the basic aims of topology and modern analysis is the study of the bounded real and complex functions which are defined and continuous on a topological space X. These functions can of course be studied individually, but this doesn't carry us very far. It is desirable to consider the sets $\mathcal{C}(X,R)$ and $\mathcal{C}(X,C)$ of *all* such functions as mathematical systems with a high level of internal organization, and this program compels us to give serious attention to their structural features. It is at this point that algebra enters the picture; for modern algebra is essentially the result of crystallizing into abstract form, and studying for their own sake, a few simple patterns of structure which underlie many diverse parts of mathematics.

In Secs. 14 and 20 we defined what is meant by a linear space and an algebra, but we did not develop the theory of these systems to any appreciable degree. We used them only descriptively, as a convenient means of calling attention to the fact that the points in the spaces R^n and C^n can be added and multiplied by numbers, and those in $\mathcal{C}(X,R)$ and $\mathcal{C}(X,C)$ can be multiplied together as well. Our work in the rest of this book requires a deeper understanding of these systems and several others, and the purpose of this chapter is to provide a concise but reasonably complete exposition of this necessary background material.

The algebraic systems we discuss below—*groups, rings, linear spaces,* and *algebras*—have been the subject of many books and innumerable research articles. In the few pages we devote to each, we clearly can do little more than explain what each system is, mention several outstanding examples, and develop the theory to the limited extent required by our later work. If the reader finds it desirable to amplify our abbreviated treatment by consulting additional sources, we suggest McCoy [31] and Halmos [17].

39. GROUPS

We begin by considering two familiar algebraic systems, each of which is a group, with a view to pointing out those features common to both which are set forth abstractly in the general concept of a group.

We first observe that the set R of all real numbers, together with the operation of ordinary addition, has the following properties: the sum of any two numbers in R is a number in R (R is closed under addition); if x, y, z are any three numbers in R, then $x + (y + z) = (x + y) + z$ (addition is associative); there is present in R a special number, namely 0, with the property that $x + 0 = 0 + x = x$ for every x in R (R contains an additive identity element); and to each number x in R there corresponds another number in R, its negative $-x$, with the property that $x + (-x) = (-x) + x = 0$ (R contains additive inverses).

It is equally clear that the set P of all positive real numbers, together with the operation of ordinary multiplication, has the following corresponding properties: the product of any two numbers in P is a number in P (P is closed under multiplication); if x, y, z are any three numbers in P, then $x(yz) = (xy)z$ (multiplication is associative); there is present in P a special number, namely 1, with the property that $x1 = 1x = x$ for every x in P (P contains a multiplicative identity element); and to each number x in P there corresponds another number in P, its reciprocal $1/x = x^{-1}$, with the property that $xx^{-1} = x^{-1}x = 1$ (P contains multiplicative inverses).

Each of these systems plainly possesses many properties other than those we have mentioned. We ignore all such properties and concentrate our attention solely on the ones we have listed. Let us now consciously disregard the concrete nature of the elements composing the above sets and the familiar character of the algebraic operations involved. What remains in each case is a non-empty set which is closed under an operation possessing certain formal properties, and apart from notation and terminology, these properties are identical in the two systems. The concept of a group is a distillation of the common structural form of these and many other similar systems.

The definition is as follows. A *group* is a non-empty set G together with an operation (called *multiplication*) which associates with each ordered pair x, y of elements in G a third element in G (called their *product* and written xy) in such a manner that

(1)   multiplication is *associative*, that is, if x, y, z are any three elements in G, then $x(yz) = (xy)z$;

(2)   there exists an element e in G, called the *identity element* (or simply the *identity*), with the property that $xe = ex = x$ for every x in G; and

(3)   to each element x in G there corresponds another element in G, called the *inverse* of x and written x^{-1}, with the property that $xx^{-1} = x^{-1}x = e$.

It should be carefully noted that we do not assume that $xy = yx$ for all elements x and y. A group which satisfies this additional condition is called a *commutative group* or an *Abelian group* (after the Norwegian mathematician Abel). If G consists of a finite number of elements, then it is called a *finite group* and this number is called its *order;* otherwise, it is called an *infinite group*.

In axiom (2) we speak of *the* identity, as if there were only one identity element in G. This is indeed the case, for if e' is also an element in G such that $xe' = e'x = x$ for every x, then $e' = e'e = e$ shows that e' necessarily equals e. Similarly, in axiom (3) we speak of *the* inverse of x, as if each element had only one inverse. This also is true, for if x' is another element in G such that $xx' = x'x = e$, then

$$x' = x'e = x'(xx^{-1}) = (x'x)x^{-1} = ex^{-1} = x^{-1}$$

shows that x' equals x^{-1}.

If we have succeeded in disengaging ourselves from our intuitive ideas, we must admit that we know nothing whatever about the actual nature of either the set G or the operation. Both are completely abstract, and it is essential to understand that our knowledge of G and its operation is strictly confined to the information contained in the above axioms.[1] As our examples below will show, the elements of a group need not be numbers at all, and its operation can perfectly well be some bizarre rule of combination which bears no relation to the usual operations of elementary algebra. In its essence, the study of groups is the study of a single algebraic operation in its purest form, and the theory of groups is the body of theorems—together with their applications—which can be deduced from the given axioms. This theory is richer in content than can easily be imagined by anyone who has not delved into it for himself, and the applications reach throughout mathematics and even beyond, into such strikingly diverse fields as geometry, the theory of the solva-

---

[1] Some writers emphasize this by referring to G as an *abstract group*. This concept is then contrasted with that of a *concrete group*, such as the group of all real numbers with addition. The abstract character of an abstract group is sometimes further emphasized by using a noncommittal symbol like $x * y$ in place of xy and by speaking of the *star operation*, or the *group operation*, instead of multiplication.

bility of algebraic equations, crystallography, quantum theory, and the theory of relativity.

Example 1.   We have seen that the real numbers form a group with respect to addition.   In the present example we place this group in a context of several groups of the same type, that is, groups whose elements are numbers, whose operation is ordinary addition, and in which the identity is 0 and the inverse of a number is its negative.

(a)   The single-element set consisting only of the number 0.

(b)   The set I of all integers.   Observe that the set of all non-negative integers meets every requirement for a group except axiom (3), so it is not a group.

(c)   The set of all even integers.   The odd integers do not constitute a group, for the sum of two odd integers is even.

(d)   The set of all rational numbers.

(e)   The set R of all real numbers.

(f)   The set C of all complex numbers.

(g)   The set of all complex numbers whose real and imaginary parts are both integers.

Example 2.   We also saw at the beginning of this section that the positive real numbers form a group with respect to multiplication.   Again, this group is only one among many of a similar kind, some of which are listed below.   In all these the elements are numbers, the operation is ordinary multiplication, the identity is 1, and the inverse of a number is its reciprocal.

(a)   The single-element set consisting only of the number 1.

(b)   The two-element set $\{1, -1\}$.

(c)   The set of all positive rational numbers.

(d)   The set of all positive real numbers.   Observe that the negative real numbers do not form a group, for this set is not closed under multiplication.

(e)   The set of all non-zero real numbers.   The set R of all real numbers contains a number, namely 0, which has no reciprocal, so it is not a group with respect to the operation considered here.

(f)   The set of all non-zero complex numbers.

(g)   The set $\{1, i, -1, -i\}$ of the four fourth roots of unity.

(h)   The set $\{z : z^n = 1\}$ of the n nth roots of unity for a fixed but arbitrary positive integer n.   Groups (a), (b), and (g) are the special cases which correspond to choosing n equal to 1, 2, and 4.   We see by this that there exists a finite group of order n for each positive integer n.

(i)   The unit circle $\{z : |z| = 1\}$ in the complex plane.   This is called **the** *circle group*.

The groups in the next two examples are closely related to our work in the previous chapters. They differ from the groups described above in that their elements are not numbers.

Example 3. (a)   The set R^n of all n-tuples of real numbers, the operation being coordinatewise addition. The identity here is

$$0 = (0, 0, \ldots, 0),$$

and the inverse of the element $x = (x_1, x_2, \ldots, x_n)$ is

$$-x = (-x_1, -x_2, \ldots, -x_n).$$

(b)   The set C^n of all n-tuples of complex numbers with respect to coordinatewise addition.

Example 4. (a)   The set $\mathcal{C}(X,R)$ of all bounded continuous real functions defined on a topological space X. The operation here is pointwise addition, the identity is the function which is identically zero, and the inverse of a function f is the function $-f$ defined by $(-f)(x) = -f(x)$.

(b)   The set $\mathcal{C}(X,C)$ of all bounded continuous complex functions defined on a topological space X, the operation again being pointwise addition.

The following examples are somewhat miscellaneous in character. They should be illuminating to the reader who is not already familiar with these ideas, for several have nothing whatever to do with numbers.

Example 5. (a)   The class of all subsets of a set U, the operation being the formation of symmetric differences. The reader will recall that the symmetric difference of two sets A and B is defined by

$$A \triangle B = (A - B) \cup (B - A);$$

and in Problem 2-3 it was shown that this operation is associative, that the identity is the empty set $\emptyset$, and that the inverse of a set is the set itself. It is interesting to note that if U is non-empty, then this class of sets does not constitute a group with respect to the formation of either unions or intersections.

(b)   Any ring of subsets of a set U (see Problem 2-4), the operation again being the formation of symmetric differences.

Example 6. Let m be a positive integer and define I_m to be the set of all non-negative integers less than $m : I_m = \{0, 1, \ldots, m - 1\}$. If a and b are two numbers in I_m, we define their "sum" $a + b$ to be the remainder obtained when their ordinary sum is divided by m. If m is 7, for instance, then $I_7 = \{0, 1, 2, 3, 4, 5, 6\}$ and we have $2 + 3 = 5$, $5 + 2 = 0$, and

$4 + 5 = 2$.   Figure 28 is a complete addition table for I_7: to find the sum of any two numbers in the set, look for the first number down the left-hand side, look for the second across the top, and observe their sum in the corresponding place within the table.

	0	1	2	3	4	5	6
0	0	1	2	3	4	5	6
1	1	2	3	4	5	6	0
2	2	3	4	5	6	0	1
3	3	4	5	6	0	1	2
4	4	5	6	0	1	2	3
5	5	6	0	1	2	3	4
6	6	0	1	2	3	4	5

Fig. 28.   The addition table for I_7.

Example 7.   The set of all one-to-one mappings of a non-empty set X onto itself. The operation here is the multiplication of mappings defined at the end of Sec. 3: if f and g are two such mappings and x is an arbitrary element in X, then $(fg)(x) = f(g(x))$.   The fact that this system forms a group was shown in Problems 3-1, 3-2, and 3-5.

Example 8.   Consider the special case of the previous example in which the set X is taken to be a finite set with n elements, e.g., the set $\{1, 2, \ldots, n\}$. A one-to-one mapping of this set onto itself is usually called a *permutation,* for it can be regarded as a rearrangement of the elements of the set. If n is 4, for instance, the permutation which sends 1 to 3, 2 to 1, 3 to 4, and 4 to 2 can be written in the convenient form

$$p = \begin{pmatrix} 1234 \\ 3142 \end{pmatrix},$$

where below each of the integers 1, 2, 3, 4 is placed its image under the mapping p.   If

$$q = \begin{pmatrix} 1234 \\ 4132 \end{pmatrix}$$

is another such permutation, then their product pq (first q, then p) takes 1 to 4 then 4 to 2, 2 to 1 then 1 to 3, 3 to 3 then 3 to 4, and 4 to 2 then 2 to 1.   This result can be written

$$pq = \begin{pmatrix} 1234 \\ 2341 \end{pmatrix}.$$

The group of all permutations of n elements is denoted by S_n and called the *symmetric group of degree n.*   The detailed structure of symmetric groups is of fundamental importance in the theory of the solvability of algebraic equations.

Example 9.   Our final example is the *group of symmetries of a square.* Imagine that Fig. 29 represents a cardboard square placed on a plane with fixed axes in such a way that its center is at the origin and its sides are parallel to the axes.   This square is carried onto itself by the following rigid motions: the identity motion I, which leaves fixed each point of the square; the counterclockwise rotations R, R', and R'' about the center through angles of 90, 180, and 270 degrees; the reflections H and V about the horizontal and vertical axes; and the reflections D and D' about the indicated diagonals. Each of these rigid motions is related to a certain aspect of the symmetry of the square, and they are therefore called *symmetries*. We multiply two symmetries by performing them in succession, beginning with the one on the right. Accordingly, RV is the result of first reflecting the square about the vertical axis, then rotating it counterclockwise through 90 degrees. If we trace the effect of these motions by following the manner in which the numbered vertices are shifted about, we see that RV has the same result as D, so $RV = D$. These eight symmetries, together with the operation we have described, are easily seen to form a group. Associativity is a special case of Problem 3-1; I is evidently the identity; and it is clear that H, V, D, and D' are their own inverses and that $R^{-1} = R''$, $R'^{-1} = R'$, and $R''^{-1} = R$. In much

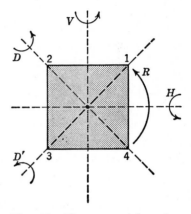

Fig. 29.   The symmetries of a square.

the same way, we can define the group of symmetries of an isosceles triangle, an equilateral triangle, a rectangle, a regular pentagon, etc., and in each case the group describes in a precise fashion the "symmetry characteristics" of the figure. We can go even further and consider the group of symmetries of a regular solid in ordinary three-dimensional space. Groups of this kind have interesting and important applications in crystallography.

We now return briefly to the consideration of a general group G. One of the more elementary facts about G is that certain simple equations are always solvable.

(4)   If a and b are any two elements of G, then the equations $ax = b$ and $ya = b$ have solutions x and y in G.

To prove (4), we have only to observe that $x = a^{-1}b$ and $y = ba^{-1}$ are in fact solutions, since $a(a^{-1}b) = (aa^{-1})b = eb = b$ and

$$(ba^{-1})a = b(a^{-1}a) = be = b.$$

Not only are the equations in (4) solvable in G, but their solutions are unique. This is a direct consequence of the following cancellation law.

(5)   If a is any element in G, then $ax = ax' \Rightarrow x = x'$ and
$ya = y'a \Rightarrow y = y'$.

We prove the first of these statements by multiplying $ax = ax'$ by a^{-1} on the left.   This gives $a^{-1}(ax) = a^{-1}(ax')$, from which we get $(a^{-1}a)x = (a^{-1}a)x'$, $ex = ex'$, and $x = x'$.   The second is proved similarly.

It is sometimes useful to know that (4) is capable of replacing axioms (2) and (3) in the definition of a group.   This amounts to the assertion that if G is a non-empty set which is closed under an associative multiplication with property (4), then G is a group.   To prove this, we must show that G has an identity element and that each element in G has an inverse.   We reason as follows.   Let c be an element in G, and e a solution of $yc = c$.   If a is any element in G and x is a solution of $cx = a$, then $ea = e(cx) = (ec)x = cx = a$, so e acts as an identity on the left. We still must show that $ae = a$.   For any element b in G, denote a solution of $yb = e$ by b^{-1} and call it a left inverse of b.   In particular, $a^{-1}a = e$. It is clear that $(a^{-1}a)a^{-1} = ea^{-1} = a^{-1}$, so $a^{-1}(aa^{-1}) = a^{-1}$; and if we multiply both sides of this on the left by a left inverse of a^{-1}, we get $aa^{-1} = e$.   It is now easy to see that $ae = a$, for

$$ae = a(a^{-1}a) = (aa^{-1})a = ea = a.$$

As the reader will observe, we have not only shown that e is an identity element, but we have also shown that each element has an inverse in the required sense.

A *subgroup* of a group G is a non-empty subset H of G which is itself a group with respect to the operation in G.   It is easy to see that the identity e' in H equals the identity e in G; for $e'e' = e' = e'e$, and by the cancellation law in G we have $e' = e$.   Also, if x is an element in H and x' is its inverse in H, so that $xx' = x'x = e$, then x' equals the inverse x^{-1} of x in G; for $xx' = e$ and $xx^{-1} = e$ yield $xx' = xx^{-1}$, and another application of the cancellation law in G gives $x' = x^{-1}$.   By these remarks, we see that a non-empty subset H of G is a subgroup of $G \Leftrightarrow$ it is closed under multiplication, it contains the identity e of G, and it contains the inverse x^{-1} of each of its elements x.   Since $xx^{-1} = e$, it is equally clear that a non-empty subset H of G is a subgroup of $G \Leftrightarrow$ it is closed under multiplication and the formation of inverses.

Many of the groups described above are subgroups of other groups. For instance, in Example 1 it is easy to see that (a) is a subgroup of (c), (c) of (b), (b) of (d), (d) of (e), and (e) of (f).   The subgroups of Example 7 are of particular importance, and are called *transformation groups*. If the underlying set is finite, as in Example 8, a transformation group is often called a *permutation group*.   In the case of Example 9, for instance, each symmetry can be regarded as a permutation of the numbers which label the vertices of the square; e.g., the reflection H about the horizontal axis interchanges 1 and 4, and also 2 and 3, so we may put

$$H = \begin{pmatrix} 1234 \\ 4321 \end{pmatrix}.$$

The group of symmetries of a square, being a subgroup of the symmetric group S_4 of degree 4, is thus a permutation group. It is obvious that any group G has $\{e\}$ and G itself as trivial subgroups.

Every group in our list of examples is Abelian, with the exception of the last three. We ask the reader to show in Problem 9 that Example 7 is non-Abelian whenever the set X contains more than two elements. It will follow from this that the symmetric group S_n is non-Abelian whenever $n \geq 3$. This can easily be seen for S_4 by computing the product qp, where p and q are the permutations given in Example 8:

$$qp = \begin{pmatrix} 1234 \\ 3421 \end{pmatrix}.$$

Since $pq \neq qp$, S_4 is non-Abelian. We saw in Example 9 that $RV = D$. If we now compute VR, we get $VR = D'$, so $RV \neq VR$ and the group of symmetries of a square is also non-Abelian.

When the theory of groups is studied for its own sake, the emphasis is usually placed on non-Abelian groups. The present section, however, is intended mainly to provide a proper foundation for our work in the rest of this chapter, and Abelian groups are the ones of greatest importance for us. In the Abelian case, the multiplicative notation used above is often replaced by additive notation, in which the product xy is written $x + y$ and called the *sum* of x and y. Correspondingly, the identity is denoted by 0 instead of e and is called the *zero element* (or simply *zero*); and the inverse of x is denoted by $-x$ instead of x^{-1} and is called the *negative* of x. Also, the operation of *subtraction* is defined by

$$x - y = x + (-y),$$

and the element $x - y$ is called the *difference* between x and y. An Abelian group in which this additive notation is used is called an *additive Abelian group*. It is clear that a subgroup of an additive Abelian group is a non-empty subset which is closed under addition and the formation of negatives.

Problems

1. Let G be a group, and show that $(xy)^{-1} = y^{-1}x^{-1}$ for any two elements x and y in G. Show also that $(x^{-1})^{-1} = x$ for any element x in G.

2. Let G be a finite non-empty set which is closed under an associative multiplication with property (5). Show that G is a group. (*Hint:* prove property (4) by considering the mappings of G into itself defined by $f(x) = ax$ and $g(y) = ya$.)

3. Prove that a group with the property that $x^2 = e$ for every element x is necessarily Abelian (needless to say, x^2 is the conventional symbol for the product of x with itself: $x^2 = xx$).

4. Prove that a group of order n with $n \leq 4$ is necessarily Abelian.

5. Let H be a non-empty subset of a group G, and show that H is a subgroup of $G \Leftrightarrow xy^{-1}$ is in H whenever x and y are. We see from this that a non-empty subset of an additive Abelian group is a subgroup $\Leftrightarrow$ it is closed under subtraction.

6. Let G be a group, and let C be the subset of G defined by

$$C = \{a : ax = xa \text{ for every } x \, \varepsilon \, G\}.$$

    Prove that C is a subgroup of G. C is called the *center* of G.

7. Let m be a positive integer, consider the set

$$I_m = \{0, 1, \ldots, m - 1\},$$

    and define the "product" of any two numbers in it to be the remainder left when their ordinary product is divided by m. Construct a multiplication table similar to Fig. 28 for the non-zero elements of I_6. Does this set with this operation form a group? Compute a similar table for the non-zero elements of I_7. Does this system form a group?

8. Introduce a symbol for each element of the symmetric group S_3 of degree 3, and construct a multiplication table for this group. Show that $n!$ is the order of S_n.

9. Show that the group of all one-to-one mappings of a non-empty set X onto itself is non-Abelian if X has more than two elements.

10. Construct a multiplication table for the group of symmetries of a square.

11. Let G and G' be groups. A mapping f of G into G' is called a *homomorphism* if $f(xy) = f(x)f(y)$ for all elements x and y in G. Assume that f is a homomorphism of G into G', and prove the following facts:
    (a) $f(e) = e'$, where e and e' are the identity elements in G and G';
    (b) $f(x^{-1}) = f(x)^{-1}$;
    (c) $f(G)$ is a subgroup of G';
    (d) $f^{-1}(\{e'\})$ is a subgroup of G.
    If a homomorphism is one-to-one, it is called an *isomorphism*. If there exists an isomorphism of G onto G', then G is said to be *isomorphic to G'*. To say that one group is isomorphic to another is to say that they have the same number of elements and the same group

structure, and differ only with respect to such inessentials as notation and terminology.   The reader will observe that the function f defined on the real line by $f(x) = a^x$, where a is a fixed real number greater than 1, is an isomorphism of the group of all real numbers with addition onto the group of positive real numbers with multiplication, so that these two systems *as groups* are abstractly identical.   Now let G be an arbitrary group, and let f be the mapping defined on G by $f(a) = M_a$, where M_a is the mapping of G into itself given by $M_a(x) = ax$.   Show that f is an isomorphism of G into the group of one-to-one mappings of G onto itself.   This fact is called *Cayley's theorem*, and it shows that from the abstract point of view the theory of groups is coextensive with the theory of transformation groups.

40. RINGS

We have seen that the set I of all integers is an additive Abelian group with respect to the operation of ordinary addition.   It is just as important to observe that I is also closed under ordinary multiplication and that multiplication is linked to addition in a way which enriches the structure of the system as a whole.   The theory of rings is the theory of such systems.

A *ring* is an additive Abelian group R which is closed under a second operation called *multiplication*—the *product* of two elements x and y in R is written xy—in such a manner that

(1)   multiplication is *associative*, that is, if x, y, z are any three elements in R, then $x(yz) = (xy)z$; and

(2)   multiplication is *distributive*, that is, if x, y, z are any three elements in R, then $x(y + z) = xy + xz$ and $(x + y)z = xz + yz$.

In other words, a ring is an additive Abelian group whose elements can be multiplied as well as added, and in which multiplication behaves reasonably with respect to itself and addition.   We note particularly that multiplication is not assumed to be commutative.

Many of the additive Abelian groups listed in the previous section are also rings with respect to natural multiplications.

Example 1.   Each of the following rings consists of numbers, and addition and multiplication are understood to have their ordinary meanings.

(a)   The single-element set containing only the number 0.

(b)   The set I of all integers.

(c)   The set of all even integers.

(d)   The set of all rational numbers.

(e)   The set R of all real numbers.

(*f*)   The set C of all complex numbers.

(*g*)   The set of all complex numbers whose real and imaginary parts are both integers.

Example 2.   (*a*) $\mathcal{C}(X,R)$, with pointwise addition and multiplication.
(*b*) $\mathcal{C}(X,C)$, with pointwise addition and multiplication.

Example 3.   Any ring of subsets of a set U, with addition and multiplication defined by $A + B = A \bigtriangleup B$ and $AB = A \cap B$ (see Problems 2-3 and 2-4). The fact that a ring of sets is a ring in our present sense is the reason for the name *ring of sets*.

Example 4.   Let m be a positive integer, and I_m the set of all non-negative integers less than m: $I_m = \{0, 1, \ldots, m - 1\}$. If a and b are two numbers in I_m, we define their "sum" $a + b$ and "product" ab to be the remainders obtained when their ordinary sum and product are divided by m. If m is 6, for instance, then $I_6 = \{0, 1, 2, 3, 4, 5\}$, and we have $3 + 4 = 1$ and $2 \cdot 3 = 0$. I_m with these operations is called the *ring of integers mod m*.

We now consider a general ring R. Many familiar facts from elementary algebra are valid in R. Nevertheless, each must be proved on its own merits from the axioms or previous theorems, for one never knows when something which appears to be "obvious" will turn out to be false.

We have already defined subtraction in any additive Abelian group by $x - y = x + (-y)$, and it is easy to show that such statements as $-(x - y) = y - x$ and $x = y \Leftrightarrow x - y = 0$ are true. Problem 39-1 assures us that $-(-x) = x$. And so on. Properties of this kind relate to the additive structure of R and are comparatively trivial. It is only when we consider multiplication, and its relation to addition, that we begin to encounter some interesting situations.

We illustrate this by proving that $x0 = 0$ for any element x in R. First, we have

$$x0 + x0 = x(0 + 0) = x0.$$

Our next step is to add $-x0$ (the negative of $x0$) to both sides of this on the right, which gives

$$(x0 + x0) + (-x0) = x0 + (-x0);$$

and by the associativity of addition we can write this in the form

$$x0 + (x0 + (-x0)) = x0 + (-x0).$$

Since the sum of any element and its negative is 0, this collapses to

$$x0 + 0 = 0,$$

which yields $$x0 = 0.$$

Similarly, $0x = 0$ for any x. We see in this way that the product of two elements in a ring is zero whenever either factor is zero. We have given the details of the proof of this seemingly obvious fact because, surprisingly enough, its converse is false. It can perfectly well happen (and it often does happen) that the product of two non-zero elements in a ring is zero. The simplest examples of this phenomenon are found in the rings of integers mod m where m is greater than 1 and is not a prime number. We have already seen, for instance, that in I_6 the product of the two non-zero elements 2 and 3 is 0. An element z in a ring such that either $zx = 0$ for some non-zero x or $yz = 0$ for some non-zero y is called a *divisor of zero*. In any ring with non-zero elements, the element 0 itself is a divisor of zero.

By using distributivity and the fact that the product of two elements in the ring R is zero when either factor is zero, it is easy to verify the following familiar rules of calculation: $x(-y) = (-x)y = -xy$, $(-x)(-y) = xy$, $x(y - z) = xy - xz$, $(x - y)z = xz - yz$. As a simple consequence of the last two of these rules, we have the following cancellation law: if a is not a divisor of zero, then either of the relations $ax = ay$ or $xa = ya$ implies that $x = y$.

R is called a *commutative ring* if $xy = yx$ for all elements x and y. Every ring in the above list of examples is commutative. We shall encounter some non-commutative rings of very great importance in Secs. 44 and 45.

If the ring R contains a non-zero element 1 with the property that $x1 = 1x = x$ for every x, then 1 is called an *identity element* (or an *identity*), and R is called a *ring with identity*. If a ring has an identity, then it has only one. In Example 1, only (a) and (c) have no identity. In both rings described in Example 2, the identity is the function which is identically 1. A ring of subsets of a set U has an identity $\Leftrightarrow$ there exists a non-empty set in the ring which contains every set in the ring; in particular, if U is non-empty and the ring is a Boolean algebra of subsets of U, then the set U itself is the identity. The ring I_m has an identity $\Leftrightarrow m > 1$.

Let R be a ring with identity. If x is an element in R, then it may happen that there is present in R an element y such that $xy = yx = 1$. In this case there is only one such element, and it is written x^{-1} and called the *inverse* of x. If an element x in R has an inverse, then x is said to be *regular*. Elements which are not regular are called *singular*. Regular elements are often called *invertible* elements, or *non-singular* elements. The element 0 is always singular in a ring with identity, and the element 1 is always regular. In Example 1b, 1 and -1 are the only regular elements; in 1d to 1f, all non-zero elements are regular; and in 1g, the regular elements are 1, i, -1, and $-i$.

A ring with identity is called a *division ring* if all its non-zero elements are regular. A *field* is a commutative division ring. The rational numbers constitute a field, as do the real numbers and the complex numbers. Roughly speaking, fields are the "number systems" of mathematics.

Problems

1. In the ring of even integers, why is 2 neither regular nor singular?
2. Consider the ring of all subsets of a non-empty set U, with the operations defined in Example 3. What are the regular elements in this ring? What are the singular elements? What are the divisors of zero? Under what conditions is this ring a field?
3. In each of the following rings of functions defined on the closed unit interval [0,1], describe the regular elements, the singular elements, and the divisors of zero:
   (a) all real functions;
   (b) all continuous real functions;
   (c) all bounded continuous real functions.
   What changes are necessary in these descriptions if [0,1] is replaced by (0,1)?
4. Let R be a ring with identity, and show that any divisor of zero in R is singular.
5. Let R be a ring with identity, and show that R is a division ring $\Leftrightarrow$ the non-zero elements of R form a group with respect to multiplication.
6. Show that the ring I_m is a field $\Leftrightarrow$ m is a prime number. (*Hint:* in showing that I_m is a field if m is prime, show first that in this case I_m has no non-zero divisors of zero, so that the non-zero elements of I_m are closed under multiplication and the cancellation law $ax = ay \Rightarrow x = y$ holds for these elements; now apply Problem 39-2 and Problem 5 above.)

41. THE STRUCTURE OF RINGS

Let R be a ring. A non-empty subset S of R is called a *subring* of R if the elements of S form a ring with respect to the operations defined in R. This is equivalent to the requirement that S be closed under the formation of sums, negatives, and products.

We concentrate our attention on a special type of subring. An *ideal* in R is a subring I of R which has the following further property:

$$i \,\varepsilon\, I \Rightarrow xi \text{ and } ix \,\varepsilon\, I \text{ for every element } x \,\varepsilon\, R.$$

It is in this sense that an ideal in R can be described as a subring of R which is closed with respect to multiplication on both sides by every element of R. If the ideal I is a proper subset of R, then it is called a *proper ideal*. The *trivial ideals* in R are the *zero ideal* $\{0\}$ consisting of the zero element alone, and the full ring R itself. We see from this that every ring with non-zero elements has at least two distinct ideals.

In order to clarify the concept of an ideal, we mention a few specific examples. We begin by considering the ring of all integers. The even integers (i.e., all integral multiples of 2) obviously form an ideal in this ring. So also do all integral multiples of 3, of 4, and so on. In general, if m is any positive integer, then the set

$$\bar{m} = \{\ldots, -2m, -m, 0, m, 2m, \ldots\}$$

of all integral multiples of m is a non-zero ideal. We next consider the ring $\mathcal{C}[0,1]$ of all bounded continuous real functions defined on the closed unit interval. If X is a subset of $[0,1]$, then the set

$$I(X) = \{f : f(x) = 0 \text{ for every } x \in X\}$$

is an ideal in this ring. It is easy to see that $I(X)$ equals the full ring when X is the empty set and equals the zero ideal when $X = [0,1]$. As a final example, we consider the ring of all subsets of an infinite set U, and we observe that the class of all finite subsets of U is a proper ideal in this ring.

Some rings have a multitude of non-trivial ideals, while others have none at all. In general, the structure of a ring is very closely connected with the ideals in it. The following theorem illustrates this point.

Theorem A. *If R is a commutative ring with identity, then R is a field $\Leftrightarrow$ it has no non-trivial ideals.*

PROOF. We first assume that R is a field, and we show that it has no non-trivial ideals. It suffices to show that if I is a non-zero ideal in R, then $I = R$. Since I is non-zero, it must contain some element $a \neq 0$. R is a field, so a has an inverse a^{-1}, and I (being an ideal) contains $1 = a^{-1}a$. Since I contains 1, it also contains $x = x1$ for every x in R, and therefore $I = R$.

We now assume that R has no non-trivial ideals, and we prove that R is a field by showing that if x is a non-zero element in R, then x has an inverse. The set $I = \{yx : y \in R\}$ of all multiples of x by elements of R is easily seen to be an ideal. Since I contains $x = 1x$, it is a non-zero ideal, and it consequently equals R. We conclude from this that I contains 1, and therefore that there is an element y in R such that $yx = 1$. This shows that x has an inverse, so R is a field.

The real significance of the ideals in a ring is that they enable us to construct other rings which are associated with the first in a natural way. We explain how this is done.

Let I be an ideal in a ring R. We use I to define an equivalence relation in R as follows: two elements x and y in R are said to be *congruent modulo I*, written $x \equiv y \pmod{I}$, if $x - y$ is in I. Since only one ideal is under consideration, we abbreviate this symbolism to $x \equiv y$. It is easy to verify that we actually do have an equivalence relation here, that is, that the following three conditions are satisfied:

(1) $x \equiv x$ for every x;

(2) $x \equiv y \Rightarrow y \equiv x$;

(3) $x \equiv y$ and $y \equiv z \Rightarrow x \equiv z$.

Furthermore, congruences can be added and multiplied, as if they were ordinary equations:

(4) $x_1 \equiv x_2$ and $y_1 \equiv y_2 \Rightarrow x_1 + y_1 \equiv x_2 + y_2$ and $x_1 y_1 \equiv x_2 y_2$.

The hypothesis of (4) is that $x_1 - x_2$ and $y_1 - y_2$ are elements of I, and since I is an ideal, the conclusions follow at once from

$$(x_1 + y_1) - (x_2 + y_2) = (x_1 - x_2) + (y_1 - y_2)$$

and
$$x_1 y_1 - x_2 y_2 = x_1 y_1 - x_1 y_2 + x_1 y_2 - x_2 y_2$$
$$= x_1(y_1 - y_2) + (x_1 - x_2) y_2.$$

It will be necessary to use property (4) at a critical stage in our discussion below, and the reader will see there that this property is the main reason why the ideals in a ring are so much more important than its subrings.

According to the general theory of Sec. 5, this equivalence relation has associated with it a partition of R into equivalence sets—called *cosets* in this context—which are non-empty and disjoint and whose union is the full ring R. What is the structure of these cosets? In order to answer this question, we let x be an element of R. The coset $[x]$ containing x is by definition the set of all elements y such that $y \equiv x$; that is, $[x] = \{y : y \equiv x\}$. But

$$\{y : y \equiv x\} = \{y : y - x \in I\}$$
$$= \{y : y - x = i \text{ for some } i \in I\}$$
$$= \{y : y = x + i \text{ for some } i \in I\}$$
$$= \{x + i : i \in I\}.$$

A natural notation for the set last written is $x + I$, which we understand to signify the set of all sums of x and elements of I. The structure of the coset $[x]$ containing x is fully exhibited by the fact that $[x] = x + I$. Sometimes it is convenient to denote this coset by $[x]$ and sometimes by $x + I$. We recall that the same coset can perfectly well arise from another element, say x_1, and that $[x] = [x_1]$ means that $x \equiv x_1$, that is, that $x - x_1$ is in I. The elements x and x_1 are called *representatives* of the coset which contains them.

Our next step is to construct a new ring, which we denote by R/I and call the *quotient ring* of R with respect to I. The elements of the ring R/I are the distinct cosets of the form $[x]$ (or $x + I$). All that remains is to define the manner in which these cosets are to be added and multiplied and to verify the fact that we do indeed obtain a ring. The definitions are as follows:

$$[x] + [y] = [x + y]$$
and
$$[x] \cdot [y] = [xy].$$

In other words, we add and multiply two cosets $[x]$ and $[y]$ by first adding and multiplying the representatives x and y, and then by forming the cosets which contain $x + y$ and xy. It is necessary to make certain that these are legitimate definitions, that is, that the resulting cosets $[x + y]$ and $[xy]$ do not depend on the particular representatives x and y chosen for the cosets $[x]$ and $[y]$. To this end, we take two other representatives of the same two cosets, i.e., two other elements x_1 and y_1 of R such that $x_1 \equiv x$ and $y_1 \equiv y$. We want to satisfy ourselves that

$$[x_1 + y_1] = [x + y]$$

and $[x_1y_1] = [xy]$, or equivalently, that $x_1 + y_1 \equiv x + y$ and $x_1y_1 \equiv xy$. Since this is precisely the content of property (4) above, we do have valid definitions for our ring operations in R/I. We omit the detailed verification of the fact that R/I with these operations actually is a ring, remarking only that the zero element of this ring is $[0] = 0 + I = I$ and that the negative of a typical element $[x] = x + I$ is $[-x] = (-x) + I$. It is easy to see that R/I is commutative if R is, and that if R has an identity 1 and I is a proper ideal, then R/I has an identity $1 + I$.

We summarize the results of this discussion in the following theorem.

Theorem B. *Let I be an ideal in a ring R, and let the coset of an element x in R be defined by $x + I = \{x + i : i \in I\}$. Then the distinct cosets form a partition of R; and if addition and multiplication are defined by*

$$(x + I) + (y + I) = (x + y) + I$$
and
$$(x + I)(y + I) = xy + I,$$

then these cosets constitute a ring denoted by R/I and called the quotient ring of R with respect to I, in which the zero element is $0 + I = I$ and the negative of $x + I$ is $(-x) + I$. Further, if R is commutative, then R/I is also commutative; and if R has an identity 1 and I is a proper ideal, then R/I has an identity $1 + I$.

We now give a brief account of homomorphisms and of the manner in which ideals, quotient rings, and homomorphisms are all related to one another.

Let R and R' be two rings. A *homomorphism* of R into R' is a mapping f of R into R' with the following two properties:

$$f(x + y) = f(x) + f(y)$$
and $$f(xy) = f(x)f(y).$$

A homomorphism of one ring into another is thus a mapping of the first ring into the second which preserves the ring operations. It is easy to see that f preserves zero in the sense that $f(0) = 0$;[1] for

$$f(0) + f(0) = f(0 + 0) = f(0),$$

and subtracting $f(0)$ from both sides yields our result. Similarly, f preserves negatives, for $f(-x) = -f(x)$ follows from

$$f(x) + f(-x) = f(x + (-x)) = f(0) = 0.$$

The image $f(R)$ of R under f is clearly a subring of R'. This subring $f(R)$—which is R' itself when f is onto—is called a *homomorphic image* of R. If the homomorphism f is one-to-one, then it is called an *isomorphism*, and the subring $f(R)$ is called an *isomorphic image* of R. An isomorphic image of R can be thought of as a ring which is essentially identical with R, for it differs from R only in the matter of notation. The properties of R are reflected with complete precision in an isomorphic image and with somewhat less precision in a homomorphic image.

Let f be a homomorphism of R into R'. The *kernel K* of this homomorphism is the inverse image in R of the zero ideal in R':

$$K = \{x : x \in R \text{ and } f(x) = 0\}.$$

It is easy to see that K is an ideal in R, and also that K is the zero ideal in $R \Leftrightarrow f$ is an isomorphism. We leave these verifications to the reader. In a sense, the size of the kernel K is a measure of the extent to which f fails to be an isomorphism.

What is the real significance of homomorphisms, homomorphic images, and kernels? A full answer to this question would carry us into the utmost reaches of the general theory of rings, where we have no intention of treading. We will, however, attempt a brief and necessarily vague partial answer. Suppose that R is a ring whose features are unfamiliar, whose structure is unknown. The question confronting us is, What is the nature of R? And, as is often the case in mathematics, if we can state adequately what this question means, we will have taken a long step toward answering it. Suppose now that R' is a homomorphic image of R and that R' is a well-known ring which is intuitively familiar

---

[1] We use the symbol 0 here to designate two different elements: the zero in R and the zero in R'. This convention is customary, and it saves more trouble than it causes.

and thoroughly understood. R' then provides a picture of the structure of R. The details of this picture may be blurred and fragmentary, but we can usually glean from them a few hints as to the nature of R itself. If we have available many homomorphic images of R, it is often possible to correlate the hints we get from these many sources in such a way as to build up a fully detailed and completely precise picture of the original ring R. This is the overall strategy in the structure theory (or representation theory) of rings.[1] The relevance of ideals to this strategic pattern depends on the following fact: all possible homomorphic images of R can be constructed by means of the ideals in R. We next describe how this is accomplished.

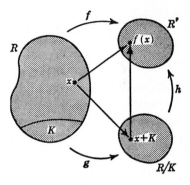

Fig. 30

Let R be a ring, and let f be a homomorphism of R onto a ring R'. Let K be the kernel of f. Since K is an ideal in R, we can form the quotient ring R/K. We now observe that R/K is a homomorphic image of R under the homomorphism g—called the *natural homomorphism* —defined by

$$g(x) = x + K.$$

The fact that g is a homomorphism follows directly from the definition of the ring operations in R/K:

$$g(x + y) = (x + y) + K = (x + K) + (y + K)$$
$$= g(x) + g(y)$$
and $$g(xy) = xy + K = (x + K)(y + K)$$
$$= g(x)g(y).$$

Finally, we show that R/K and R' are essentially identical by producing an isomorphism of R/K onto R'. Let a mapping h be defined on R/K by $h(x + K) = f(x)$. We leave it to the reader to verify that h is a well-defined mapping of R/K onto R' and is also an isomorphism. Figure 30 gives a schematic representation of this situation. Since R/K and R' are isomorphic, we can replace R' in any discussion by its replica R/K. It is

[1] The procedure described here is loosely similar to a familiar technique from three-dimensional analytic geometry, in which the form of a curved surface is studied by means of its cross sections. The information obtainable from any given cross section is meager, but an intelligent consideration of all the successive cross sections can yield a satisfactory mental image of the surface as a whole.

therefore unnecessary to go beyond the ring R to find all its homomorphic images.

If I is an ideal in a ring R, then its properties relative to all of R are reflected in corresponding properties of the quotient ring R/I, and many aspects of the study of R depend on the presence in it of ideals whose corresponding quotient rings are simple and familiar.

In order to illustrate this fundamental principle, we introduce the following concept. An ideal I in a ring R is said to be a *maximal ideal* if it is a proper ideal which is not properly contained in any other proper ideal. Our next theorem is an immediate consequence of this concept and Theorem A.

Theorem C. *If R is a commutative ring with identity, then an ideal I in R is maximal $\Leftrightarrow$ R/I is a field.*

PROOF. We first observe that if I is maximal, then R/I is a commutative ring with identity in which there are no non-trivial ideals, so by Theorem A it follows that R/I is a field. We now assume that I is not maximal, and we show that R/I is not a field. There are two possibilities: (a) that $I = R$, and (b) that there exists an ideal J such that $I \subset J \subset R$. In case (a), R/I has no non-zero elements, so it cannot be a field. In case (b), R/I is a commutative ring with identity which contains the non-trivial ideal J/I, so again it cannot be a field.

The commutative rings we study in later chapters have a great many distinct maximal ideals, and this theorem will serve us well in our program of analyzing the structure of these rings.

Problems

1. Let R be a ring with identity which is not necessarily commutative. In view of Theorem A, it is natural to conjecture that R is a division ring $\Leftrightarrow$ it has no non-trivial ideals. Try to prove this conjecture by the method used in the proof of Theorem A. At what precise point does this attempted proof break down? How much of the conjecture can you prove?

2. Let R be the ring of all real functions defined on the closed unit interval $[0,1]$. If X is a subset of $[0,1]$, show that the ideal $I(X)$ in R defined by $I(X) = \{f : f(x) = 0 \text{ for every } x \in X\}$ is maximal $\Leftrightarrow$ X consists of a single point.

3. Let I be the ring of integers and m a positive integer. It is easy to see that if x is any integer, then x can be represented uniquely in the form $x = qm + r$, where q and r are integers and r is in the set $\{0, 1, \ldots, m - 1\}$. Use this fact to show that a non-zero ideal in I is necessarily of the form $\bar{m} = \{\ldots, -2m, -m, 0, m, 2m, \ldots\}$

for some positive integer m. Show that the mapping f defined on I by $f(x) = r$ is a homomorphism of I onto the ring I_m of integers mod m. Show that the kernel of this homomorphism is the ideal $\bar{m}$, so that the quotient ring $I/\bar{m}$ is isomorphic to I_m, and conclude from this that $\bar{m}$ is maximal $\Leftrightarrow m$ is a prime number.

42. LINEAR SPACES

We introduced linear spaces in Sec. 14, and we also mentioned a few of their simpler properties. Our present purpose is to develop the theory of these systems in somewhat greater detail.

We begin by restating the definition in terms of concepts now available to us. The reader will recall that by the *scalars* we mean either the system of real numbers or the system of complex numbers. A *linear space* (or *vector space*) is an additive Abelian group L (whose elements are called *vectors*) with the property that any scalar α and any vector x can be combined by an operation called *scalar multiplication* to yield a vector αx in such a manner that

(1) $\alpha(x + y) = \alpha x + \alpha y$;

(2) $(\alpha + \beta)x = \alpha x + \beta x$;

(3) $(\alpha\beta)x = \alpha(\beta x)$;

(4) $1 \cdot x = x$.

A linear space is thus an additive Abelian group whose elements can be multiplied by numbers in a reasonable way, but not necessarily by one another (as in the case of rings). The two primary operations in a linear space—addition and scalar multiplication—are called the *linear operations*, and its zero element is usually referred to as the *origin*.

A linear space is called a *real* linear space or a *complex* linear space according as the scalars are the real numbers or the complex numbers. The advantage of calling the numerical coefficients scalars is that we avoid committing ourselves to either the real case or the complex case and are free to develop the theory for both simultaneously.[1] In later chapters we shall be concerned exclusively with complex linear spaces, but for the present we prefer to leave the door open.

Before proceeding to the general theory of linear spaces, we list a few examples.

---

[1] In some approaches to the theory of linear spaces, the system of scalars is allowed to be an arbitrary field. This degree of generality is unnecessary for our purposes; and since unnecessary generality is undesirable, we limit ourselves accordingly. Even further, the system of scalars can be taken to be an arbitrary ring. In this case, one speaks of a *module* instead of a linear space. Modules are of great importance in the structure theory of rings.

Example 1. The set R of all real numbers, with ordinary addition and multiplication taken as the linear operations, is a real linear space.

Example 2. The set R^n of all n-tuples of real numbers is a real linear space under the following coordinatewise linear operations: if

$$x = (x_1, x_2, \ldots, x_n) \quad \text{and} \quad y = (y_1, y_2, \ldots, y_n),$$

then
$$x + y = (x_1 + y_1, x_2 + y_2, \ldots, x_n + y_n)$$
and
$$\alpha x = (\alpha x_1, \alpha x_2, \ldots, \alpha x_n).$$

This reduces to Example 1 when $n = 1$.

Example 3. The set $\mathcal{C}(X,R)$ of all bounded continuous real functions defined on a topological space X is a real linear space under the following pointwise linear operations: if f and g are functions in $\mathcal{C}(X,R)$, then $f + g$ and αf are defined by

$$(f + g)(x) = f(x) + g(x)$$
and
$$(\alpha f)(x) = \alpha f(x).$$

Example 4. The set C of all complex numbers is a complex linear space under ordinary addition and multiplication.

Example 5. The set C^n of all n-tuples of complex numbers is a complex linear space with respect to the coordinatewise linear operations defined in Example 2. This reduces to Example 4 when $n = 1$.

Example 6. The set $\mathcal{C}(X,C)$ of all bounded continuous complex functions defined on a topological space X is a complex linear space with respect to the pointwise linear operations defined in Example 3.

Example 7. Let P be the set of all polynomials, with real coefficients, defined on the closed unit interval $[0,1]$. We specifically include all non-zero constant polynomials (which have degree 0) and the polynomial which is identically zero (this has no degree at all). If the linear operations are taken to be the usual addition of two polynomials and the multiplication of a polynomial by a real number, then P is a real linear space.

Example 8. For a given positive integer n, let P_n be the subset of P consisting of the polynomial which is identically zero and all polynomials of degree less than n. P_n is a real linear space with respect to the linear operations defined in P.

Example 9. A linear space may consist solely of the vector 0, with scalar multiplication defined by $\alpha \cdot 0 = 0$ for all α. We refer to this as the *zero space*, and we always denote it by $\{0\}$.

These examples are typical of the spaces which will concern us and give ample scope for the illustration of all the important phenomena. There are a number of other linear spaces of great interest, and we mention some of these from time to time in later chapters. For the present, however, the above list will suffice.

We saw in Sec. 14 that in any linear space we have $\alpha \cdot 0 = 0$, $0 \cdot x = 0$, and $(-1)x = -x$. It is also easy to show that $\alpha x = 0 \Rightarrow \alpha = 0$ or $x = 0$; for if $\alpha \neq 0$, then multiplying both sides of $\alpha x = 0$ by α^{-1} yields $\alpha^{-1}(\alpha x) = \alpha^{-1} \cdot 0$, $(\alpha^{-1}\alpha)x = 0$, $1 \cdot x = 0$, and finally, $x = 0$.

We now turn to the general theory of an arbitrary linear space L.

A non-empty subset M of L is called a *subspace* (or a *linear subspace*) of L if M is a linear space in its own right with respect to the linear operations defined in L. This is clearly equivalent to the condition that M contain all sums, negatives, and scalar multiples of its elements; and since $-x = (-1)x$, this in turn is equivalent to the condition that M be closed under addition and scalar multiplication. If the subspace M is a proper subset of L, then it is called a *proper subspace* of L. The zero space $\{0\}$ and the full space L itself are always subspaces of L. Among our examples, P_n is a subspace of P for each positive integer n, and P is a subspace of $\mathcal{C}[0,1]$. Also, the following are easily seen to be subspaces of R^3:

$$M_1 = \{(x_1, 0, 0)\}, \quad M_2 = \{(0, x_2, 0)\}, \quad M_3 = \{(0, 0, x_3)\},$$
and $\quad M_4 = \{(0, x_2, x_3)\}, \; M_5 = \{(x_1, 0, x_3)\}, \; M_6 = \{(x_1, x_2, 0)\}.$

The subspaces M_1, M_2, M_3 are usually called the *coordinate axes* in solid analytic geometry, and M_4, M_5, M_6 are called the *coordinate planes*. The most general non-zero proper subspace of R^3 is a line or a plane through the origin.

If M is a subspace of L, then—just as in the case of an ideal in a ring—we can use M to define an equivalence relation in L as follows: $x \equiv y \pmod{M}$ means that $x - y$ is in M. The discussion leading up to Theorem 41-B can be repeated without essential change (but with considerable simplification) to yield the concept of the *quotient space* L/M of L with respect to M. We give a formal statement of the basic facts in the following theorem.

Theorem A. *Let M be a subspace of a linear space L, and let the coset of an element x in L be defined by $x + M = \{x + m : m \in M\}$. Then the distinct cosets form a partition of L; and if addition and scalar multiplication are defined by*

$$(x + M) + (y + M) = (x + y) + M$$
and $\qquad\qquad \alpha(x + M) = \alpha x + M,$

then these cosets constitute a linear space denoted by L/M and called the quotient space of L with respect to M. The origin in L/M is the coset $0 + M = M$, and the negative of $x + M$ is $(-x) + M$.

The proof of this theorem is routine, and we leave the details to the reader.

It is worth remarking that the concept of a quotient space has a simple geometric interpretation. To bring this out most clearly, we let L be the linear space R^2 and M the subspace indicated in Fig. 31. If we

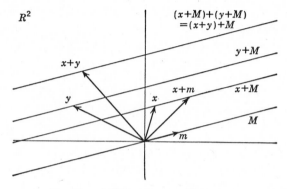

Fig. 31.   Addition in a quotient space.

think of the vectors in L as the heads of arrows whose tails are at the origin, then the non-zero proper subspace M is a straight line through the origin, a typical coset $x + M$ is a line parallel to M, and L/M consists of all lines parallel to M. We add two cosets $x + M$ and $y + M$ by adding x and y and by forming the line $(x + y) + M$ through the head of $x + y$ and parallel to M. Scalar multiplication is carried out similarly.

The subspaces of our linear space L can be characterized conveniently as follows. If $\{x_1, x_2, \ldots, x_n\}$ is a finite non-empty set of vectors in L, then the vector

$$x = \alpha_1 x_1 + \alpha_2 x_2 + \cdots + \alpha_n x_n$$

is called a *linear combination* of $x_1, x_2, \ldots, x_n$. It is evident that a subspace of L is simply a non-empty subset of L which is closed under the formation of linear combinations. If S is an arbitrary non-empty subset of L, then the set of all linear combinations of vectors in S is clearly a subspace of L; we denote this subspace by $[S]$, and we call it the subspace *spanned by S*. Since $[S]$ is a subspace which contains S and is contained in every subspace which contains S, we may think of $[S]$ as the smallest subspace which contains S. If M is a subspace of L, then a non-empty subset S of M is said to *span M* if $[S] = M$.

Suppose now that M and N are subspaces of L, and consider the set

$M + N$ of all sums of the form $x + y$, where $x \in M$ and $y \in N$. Since M and N are subspaces, it is easy to see that $M + N$ is the subspace spanned by all vectors in M and N together, i.e., that $M + N = [M \cup N]$. If it happens that $M + N = L$, then we say that L is the *sum* of the subspaces M and N. This means that each vector in L is expressible as the sum of a vector in M and a vector in N. The case in which even more is true—namely, that each vector z in L is expressible *uniquely* in the form $z = x + y$, with $x \in M$ and $y \in N$—will be of particular importance for us. In this case we say that L is the *direct sum* of the subspaces M and N, and we symbolize this statement by writing $L = M \oplus N$.

Theorem B. *Let a linear space L be the sum of two subspaces M and N, so that $L = M + N$. Then $L = M \oplus N \Leftrightarrow M \cap N = \{0\}$.*

PROOF. We begin by assuming that $L = M \oplus N$, and we deduce a contradiction from the further assumption that there is a non-zero vector z in $M \cap N$. It suffices to observe that z is expressible in two different ways as the sum of a vector x in M and a vector y in N, for $z = z + 0$ (here $x = z$ and $y = 0$) and $z = 0 + z$ (here $x = 0$ and $y = z$). This contradicts the uniqueness required by the assumption that $L = M \oplus N$.

We now assume that $M \cap N = \{0\}$, and we show that it follows from this that $L = M + N$ can be strengthened to $L = M \oplus N$. Since $L = M + N$, each z in L can be written in the form $z = x + y$ with $x \in M$ and $y \in N$. We wish to show that this decomposition is unique. If we have two such decompositions of z, so that $z = x_1 + y_1 = x_2 + y_2$, then $x_1 - x_2 = y_2 - y_1$. The left side of this is in M, the right side is in N, and they are equal; it therefore follows from $M \cap N = \{0\}$ that both sides are 0, that $x_1 = x_2$ and $y_1 = y_2$, and that the decomposition of z is unique.

The condition in this theorem—that the subspaces M and N have only the origin in common—is often expressed by saying that M and N are *disjoint*. There is fortunately little danger of confusing this with the set-theoretical notion of disjointness, for a subspace of a linear space always contains the vector 0, so the intersection of any two must also contain this vector and they can never be disjoint in the set-theoretical sense.

The concept of a direct sum can easily be broadened to allow for three or more subspaces. If $M_1, M_2, \ldots, M_n$ ($n > 2$) are subspaces of L, then the statement that L is the *direct sum* of the M_i's—written

$$L = M_1 \oplus M_2 \oplus \cdots \oplus M_n$$

—means that each vector z in L can be represented uniquely in the form $z = x_1 + x_2 + \cdots + x_n$, where $x_i \in M_i$ for every i. The reader will observe that R^3 can be represented in various ways as direct sums of the coordinate axes and coordinate planes mentioned above:

$$R^3 = M_1 \oplus M_2 \oplus M_3 = M_1 \oplus M_4 = M_2 \oplus M_5 = M_3 \oplus M_6.$$

We shall often have occasion in the following chapters to study problems which are intimately concerned with the representation of a linear space as the direct sum of certain of its subspaces.

Problems

1. Each of the following conditions determines a subset of the real linear space R^3 of all triples $x = (x_1, x_2, x_3)$ of real numbers: (a) x_1 is an integer; (b) $x_1 = 0$ or $x_2 = 0$; (c) $x_1 + 2x_2 = 0$; (d) $x_1 + 2x_2 = 1$. Which of these subsets are subspaces of R^3?
2. Each of the following conditions determines a subset of the real linear space $\mathcal{C}[-1,1]$ of all bounded continuous real functions $y = f(x)$ defined on $[-1,1]$: (a) f is differentiable; (b) f is a polynomial of degree 3; (c) f is an even function, in the sense that $f(-x) = f(x)$ for all x; (d) f is an odd function, in the sense that $f(-x) = -f(x)$ for all x; (e) $f(0) = 0$; (f) $f(0) = 1$; (g) $f(x) \geq 0$ for all x. Which of these subsets are subspaces of $\mathcal{C}[-1,1]$?
3. In the preceding problem, show that $\mathcal{C}[-1,1]$ is the direct sum of the subspaces defined by conditions (c) and (d). (*Hint:* observe that $f(x) = [f(x) + f(-x)]/2 + [f(x) - f(-x)]/2$.)
4. Let a linear space L be the sum of certain subspaces $M_1, M_2, \ldots, M_n$ $(n > 2)$, and show that L is the direct sum of these subspaces $\Leftrightarrow$ each M_i is disjoint from the subspace spanned by all the others. The latter condition clearly implies that each M_i is disjoint from each of the others. Show that the converse of this statement is false by exhibiting three subspaces M_1, M_2, M_3 of R^3 such that

$$M_1 \cap M_2 = M_1 \cap M_3 = M_2 \cap M_3 = \{0\}$$

and $M_1 \cap (M_2 + M_3) \neq \{0\}$.

43. THE DIMENSION OF A LINEAR SPACE

Let L be a linear space, and let $S = \{x_1, x_2, \ldots, x_n\}$ be a finite non-empty set of vectors in L. S is said to be *linearly dependent* if there exist scalars $\alpha_1, \alpha_2, \ldots, \alpha_n$, not all of which are 0, such that

$$\alpha_1 x_1 + \alpha_2 x_2 + \cdots + \alpha_n x_n = 0. \tag{1}$$

If S is not linearly dependent, then it is called *linearly independent;* and this clearly means that if Eq. (1) holds for certain scalar coefficients α_1, α_2, . . . , α_n, then all these scalars are necessarily 0. In other words, S is linearly independent if the trivial linear combination of its vectors (with all scalar coefficients equal to 0) is the only one which equals 0, and it is linearly dependent if some non-trivial linear combination of its vectors equals 0. In either case, as we know, the vectors in the subspace $[S]$ spanned by S are precisely the linear combinations

$$x = \alpha_1 x_1 + \alpha_2 x_2 + \cdots + \alpha_n x_n \tag{2}$$

of the x_i's. The significance of the linear independence of S rests on the fact that if S is linearly independent, then each vector x in $[S]$ is *uniquely* expressible in this form; for if we also have

$$x = \beta_1 x_1 + \beta_2 x_2 + \cdots + \beta_n x_n, \tag{3}$$

then subtracting (3) from (2) yields

$$(\alpha_1 - \beta_1)x_1 + (\alpha_2 - \beta_2)x_2 + \cdots + (\alpha_n - \beta_n)x_n = 0,$$

from which—by the linear independence of S—we obtain $\alpha_i - \beta_i = 0$ or $\alpha_i = \beta_i$ for every i. Further, the linear independence of S not only implies this uniqueness, but is also implied by it, for the statement that the vector 0 in $[S]$ is uniquely expressible in the form

$$0 = 0 \cdot x_1 + 0 \cdot x_2 + \cdots + 0 \cdot x_n$$

is exactly what is meant by the linear independence of S.

It is necessary to extend these concepts to cover the case of an arbitrary non-empty set of vectors in L. We shall say that such a set is *linearly independent* if every finite non-empty subset is linearly independent in the sense of the above paragraph; otherwise, it is said to be *linearly dependent*. Just as in the finite case, an arbitrary non-empty subset S of L is linearly independent $\Leftrightarrow$ each vector in the subspace $[S]$ spanned by S is uniquely expressible as a linear combination of the vectors in S. We are particularly interested in linearly independent sets which span the whole space L. Such a set is called a *basis* for L. It is important to observe that if S is a linearly independent subset of L, then S is a basis for $L \Leftrightarrow$ it is *maximal* with respect to being linearly independent, in the sense that every subset of L which properly contains S is linearly dependent.

Our first theorem assures us that if a linearly independent set is not already a basis, then it can always be enlarged to form a basis.

Theorem A. *If S is a linearly independent set of vectors in a linear space L, then there exists a basis B for L such that $S \subseteq B$.*

PROOF. Consider the class **P** of all linearly independent subsets of L which contain S. **P** is clearly a partially ordered set with respect to set inclusion. It suffices to show that **P** contains a maximal set B, for such a maximal set will automatically be a basis for L such that $S \subseteq B$. By Zorn's lemma, it suffices to show that every chain in **P** has an upper bound in **P**. But this is evident from the fact that the union of all the sets in any chain of linearly independent sets which contain S is itself a linearly independent set which contains S.

A linearly independent set is non-empty by definition, and it clearly cannot contain the vector 0. We see from this that if our linear space L is the zero space $\{0\}$, then no subset of L is linearly independent and L has no basis. On the other hand, if $L \neq \{0\}$ and x is a non-zero vector in L, then the single-element set $\{x\}$ is linearly independent and Theorem A guarantees that L has a basis which contains $\{x\}$. This proves

Theorem B. *Every non-zero linear space has a basis.*

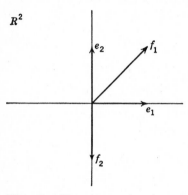

R^2

Fig. 32. Two bases $\{e_1, e_2\}$ and $\{f_1, f_2\}$ for R^2.

Since any single-element set consisting of a non-zero vector can be enlarged to form a basis, it is evident that any given non-zero linear space has a great many different bases. In R^2, for instance, the vectors $e_1 = (1,0)$ and $e_2 = (0,1)$ form a basis, as do $f_1 = (1,1)$ and $f_2 = (0,-1)$ (see Fig. 32). If we think of a vector as an arrow whose tail is the origin, it is fairly clear on geometrical grounds that in this space any two non-zero vectors form a basis if they are not collinear. We bring order out of this apparent chaos by proving in several stages that any two bases in a non-zero linear space have the same number of elements. Our next theorem is the first step in this process.

Theorem C. *Let $S = \{x_1, x_2, \ldots, x_n\}$ be a finite non-empty set of vectors in a linear space L. If $n = 1$, then S is linearly dependent $\Leftrightarrow x_1 = 0$. If $n > 1$ and $x_1 \neq 0$, then S is linearly dependent $\Leftrightarrow$ some one of the vectors $x_2, \ldots, x_n$ is a linear combination of the vectors in S which precede it.*

PROOF. The first statement is obvious, so we assume that $n > 1$ and that $x_1 \neq 0$. It is easy to see that if one of the vectors $x_2, \ldots, x_n$ is a linear combination of the preceding ones, then the equation expressing this fact can be rewritten in the form of Eq. (1) in such a way that the coefficient of the vector in question is 1, so S is linearly dependent. We now assume that S is linearly dependent, so that Eq. (1) holds with at least one non-zero coefficient. If α_i is the last non-zero coefficient, then

$i > 1$ (since $x_1 \neq 0$) and Eq. (1) can be rewritten in such a way that x_i is exhibited as a linear combination of $x_1, \ldots, x_{i-1}$ (with coefficients $-\alpha_1/\alpha_i, \ldots, -\alpha_{i-1}/\alpha_i$).

We next prove the following restricted form of our main theorem.

Theorem D. *Let L be a non-zero linear space. If L has a finite basis $B_1 = \{e_i\} = \{e_1, e_2, \ldots, e_n\}$ with n elements, then any other basis $B_2 = \{f_j\}$ is also finite and also has n elements.*

PROOF. To show that B_2 is finite, we assume that it is not, and we deduce a contradiction from this assumption. We first observe that each e_i is a linear combination of certain f_j's and that all the f_j's which occur in this way constitute a finite subset S of B_2. Since B_2 is assumed to be infinite, there exists a vector f_{j_0} in B_2 which is not in S. But f_{j_0} is a linear combination of the e_i's, and therefore of the vectors in S. This shows that $S \cup \{f_{j_0}\}$ is a linearly dependent subset of B_2, which contradicts the fact that B_2 is a basis.

Since the basis B_2 is finite, it can be written in the form

$$B_2 = \{f_j\} = \{f_1, f_2, \ldots, f_m\}$$

for some positive integer m. We must now show that m and n are equal, and this we do as follows. Since the e_i's span L, f_1 is a linear combination of the e_i's, and the set $S_1 = \{f_1, e_1, e_2, \ldots, e_n\}$ is linearly dependent. We know by Theorem C that one of the e_i's, say e_{i_0}, is a linear combination of the vectors in S_1 which precede it. If we delete e_{i_0} from S_1, then the remaining set $S_2 = \{f_1, e_1, \ldots, e_{i_0-1}, e_{i_0+1}, \ldots, e_n\}$ still spans L. Just as before, f_2 is a linear combination of the vectors in S_2, so the set $S_3 = \{f_1, f_2, e_1, \ldots, e_{i_0-1}, e_{i_0+1}, \ldots, e_n\}$ is linearly dependent. Another application of Theorem C shows that some vector in S_3 is a linear combination of the preceding ones; and since the f_j's are linearly independent, this vector must be one of the e_i's. If we delete this vector, then the remaining set again spans L. If we continue in this way, it is clear that we cannot run out of e_i's before the f_j's are exhausted; for if we do, then the remaining f_j's are linear combinations of those already used, which contradicts the linear independence of the f_j's. This shows that n is not less than m, or equivalently, that $m \leq n$. If we reverse the roles of the e_i's and f_j's, then precisely the same reasoning yields $n \leq m$, from which we conclude that $m = n$.

We are now in a position to prove our main theorem in its full generality.

Theorem E. *Let L be a non-zero linear space. If $B_1 = \{e_i\}$ and $B_2 = \{f_j\}$ are any two bases for L, then B_1 and B_2 have the same number of elements (that is, the same cardinal number).*

PROOF.   If either B_1 or B_2 is finite, then by the preceding theorem the other is also finite, and they have the same number of elements.   We may therefore confine our attention to the case in which both are infinite.

Since B_2 is a basis, each e_i can be expressed uniquely as a linear combination (with non-zero coefficients) of certain f_j's:

$$e_i = \alpha_1 f_{j_1} + \alpha_2 f_{j_2} + \cdots + \alpha_n f_{j_n}.$$

Further, every f_j occurs in at least one such expression, for if a certain one, say f_{j_0}, does not, then since B_1 is a basis, f_{j_0} is a linear combination of certain e_i's, and therefore of certain f_j's $\neq f_{j_0}$—which contradicts the fact that the f_j's are linearly independent.   This process associates with each e_i a finite non-empty set F_{e_i} of f_j's, and in such a way that $B_2 = \bigcup_{e_i \varepsilon B_1} F_{e_i}$.   Let n_1 and n_2 be the cardinal numbers of B_1 and B_2, and let n be the cardinal number of the indicated union.   It follows from the above set equality that $n_2 = n$, and Problem 8-10 shows that $n \leq n_1$, so $n_2 \leq n_1$.   If we reverse the roles of the e_i's and f_j's, then in the same manner we obtain $n_1 \leq n_2$, from which we conclude that $n_1 = n_2$.

These theorems enable us to define the dimension of an arbitrary linear space L.   If $L = \{0\}$, then it is said to be 0-*dimensional*, or to have *dimension* 0; and if $L \neq \{0\}$, then its *dimension* is the number of elements in any basis.   A linear space is called *finite-dimensional* if its dimension is 0 or a positive integer, and *infinite-dimensional* otherwise. We can now justify the usual practice of calling R^n and C^n n-dimensional spaces by exhibiting the following n vectors as a basis for both spaces:

$$
\begin{aligned}
e_1 &= (1, 0, 0, \ldots, 0), \\
e_2 &= (0, 1, 0, \ldots, 0), \\
e_3 &= (0, 0, 1, \ldots, 0), \\
&\cdots \cdots \cdots \cdots \cdots \\
e_n &= (0, 0, 0, \ldots, 1).
\end{aligned}
$$

It is easy to see that P_n is also n-dimensional, for the polynomials 1, x, $x^2, \ldots, x^{n-1}$ constitute a basis for this space.   Similarly, the set $\{1, x, x^2, \ldots, x^n, \ldots\}$ is a basis for P, so this space is infinite-dimensional.   More precisely, the dimension of P is $\aleph_0$.

The existence of a basis for an arbitrary non-zero linear space, and the fact that the number of elements in a basis is a constant determined only by the space, can also be used to give a simple but complete structure theory for these spaces.   We proceed as follows.

Let L and L' be linear spaces with the same system of scalars.   An *isomorphism* of L onto L' is a one-to-one mapping f of L onto L' such that $f(x + y) = f(x) + f(y)$ and $f(\alpha x) = \alpha f(x)$; and if there exists such an isomorphism, then L is said to be *isomorphic to L'*.   To say that one linear space is isomorphic to another is to say, in effect, that they are abstractly identical with respect to their structure as linear spaces.

Now let L be a non-zero finite-dimensional linear space of dimension n, and let $B = \{e_1, e_2, \ldots, e_n\}$ be a basis for L whose elements are written in a definite order as indicated by the subscripts. Each vector x in L is uniquely expressible in the form

$$x = \alpha_1 e_1 + \alpha_2 e_2 + \cdots + \alpha_n e_n,$$

so the n-tuple $(\alpha_1, \alpha_2, \ldots, \alpha_n)$ of scalars is uniquely determined by x. If we define a mapping f by $f(x) = (\alpha_1, \alpha_2, \ldots, \alpha_n)$, then it is easy to see that f is an isomorphism of L onto R^n or C^n according as L was real or complex to begin with. It should be recognized that the isomorphism f is by no means unique, for if some other basis is chosen for L, or if the order of the elements in the basis B is altered, then the resulting isomorphism of L onto R^n or C^n will clearly be different from f. We summarize these remarks in

Theorem F.   *Let L be a non-zero finite-dimensional linear space of dimension n.   If L is real, then it is isomorphic to R^n; and if it is complex, then it is isomorphic to C^n.*

This theorem can easily be extended to the case of an arbitrary non-zero linear space. We begin by describing the concrete linear spaces which will replace R^n and C^n in our generalized form of Theorem F. Let X be an arbitrary non-empty set, and denote by $L(X)$ the set of all scalar-valued functions defined on X which vanish outside finite sets. Addition and scalar multiplication for such functions are understood to be defined pointwise, and $L(X)$ is obviously a non-zero linear space which is real or complex according as the functions considered are real or complex. Our purpose is to show that these spaces are universal models for non-zero linear spaces, in the sense that an arbitrary non-zero linear space L is isomorphic to some $L(X)$. We start by choosing a basis $B = \{e_i\}$ for L, and we let B be the set X. We next establish an isomorphism of L onto $L(B)$ by making correspond to each vector x in L a scalar-valued function f_x defined on B. If $x = 0$, then f_x is defined by $f_x(e_i) = 0$ for every e_i in B. If $x \neq 0$, it is uniquely expressible in the form

$$x = \alpha_1 e_{i_1} + \alpha_2 e_{i_2} + \cdots + \alpha_n e_{i_n}$$

with non-zero coefficients; and f_x is defined by $f_x(e_i) = 0$ outside the set $\{e_{i_1}, e_{i_2}, \ldots, e_{i_n}\}$ and by $f_x(e_{i_j}) = \alpha_j$ inside this set. It is trivial to verify that the mapping we have described is an isomorphism of L onto $L(B)$. This discussion yields

Theorem G. *Let L be a non-zero linear space. If B is a basis for L, then L is isomorphic to the linear space $L(B)$ of all scalar-valued functions defined on B which vanish outside finite sets.*

Theorems F and G are of considerable interest in that they reveal what simple things linear spaces really are. The reader may well feel, in the light of these results, that the concept of an abstract linear space has served its purpose and should now be abandoned, and that all further study of linear spaces should be directed specifically at the $L(X)$'s, or in the finite-dimensional case, at R^n and C^n. There are at least two reasons why this is not a useful point of view. One of these lies in the fact that the above isomorphisms were established by arbitrarily choosing one particular basis B for L in preference to all the others, whereas most of the important ideas in the theory of linear spaces are independent of any specially chosen basis and are best treated, when this is possible, without reference to any basis whatever. A second reason is that almost all the linear spaces of greatest interest carry additional algebraic or topological structure, which need not be related in any significant manner to the above isomorphisms.

Problems

1. Let L be a non-zero finite-dimensional linear space of dimension n. Show that every set of $n + 1$ vectors in L is linearly dependent. Show that a set of n vectors in L is a basis $\Leftrightarrow$ it is linearly independent $\Leftrightarrow$ it spans L.

2. Show that the vectors $(1, 0, 0)$, $(1, 1, 0)$, $(1, 1, 1)$ form a basis for R^3. Show that if $\{e_1, e_2, e_3\}$ is a basis for R^3, then $\{e_1 + e_2, e_1 + e_3, e_2 + e_3\}$ is also a basis.

3. Let M be a subspace of a linear space L, and show that there exists a subspace N such that $L = M \oplus N$. Give an example for the case in which $L = R^2$ to show that N need not be uniquely determined by M.

4. If M and N are subspaces of a linear space L, and if $L = M \oplus N$, show that the mapping $y \to y + M$ which sends each y in N to $y + M$ in L/M is an isomorphism of N onto L/M.

5. Denote the dimension of a linear space L by $d(L)$. If L is finite-dimensional, and if M and N are subspaces of L, prove the following:
    (a) $d(M) \leq d(L)$, and $d(M) = d(L) \Leftrightarrow M = L$;
    (b) $d(M) + d(N) = d(M + N) + d(M \cap N)$;
    (c) if $L = M + N$, then $L = M \oplus N \Leftrightarrow d(L) = d(M) + d(N)$;
    (d) $d(L/M) = d(L) - d(M)$.

6. If L and L' are linear spaces, show that L is isomorphic to $L' \Leftrightarrow$ they have the same scalars and the same dimension.

44. LINEAR TRANSFORMATIONS

Let L and L' be linear spaces with the same system of scalars. A mapping T of L into L' is called a *linear transformation* if

$$T(x + y) = T(x) + T(y) \quad \text{and} \quad T(\alpha x) = \alpha T(x),$$

or equivalently, if

$$T(\alpha x + \beta y) = \alpha T(x) + \beta T(y).$$

A linear transformation of one linear space into another is thus a homomorphism of the first space into the second, for it is a mapping which preserves the linear operations. T also preserves the origin and negatives, for $T(0) = T(0 \cdot 0) = 0 \cdot T(0) = 0$ and

$$T(-x) = T((-1)x) = (-1)T(x) = -T(x).$$

The importance of linear spaces lies mainly in the linear transformations they carry, for vast tracts of algebra and analysis, when placed in their proper context, reduce to the study of linear transformations of one linear space into another. The theory of matrices, for instance, is one small corner of this subject, as are the theory of certain types of differential and integral equations and the theory of integration in its most elegant modern form.

In the following examples we leave it to the reader to show that each mapping described actually is a linear transformation.

Example 1. We consider the linear space R^2, and each linear transformation mentioned is a mapping of R^2 into itself.

(a) $T_1((x_1,x_2)) = (\alpha x_1, \alpha x_2)$, where α is a real number. The effect of T_1 is to multiply each vector in R^2 by the scalar α.

(b) $T_2((x_1,x_2)) = (x_2,x_1)$. T_2 reflects R^2 about the diagonal line $x_1 = x_2$.

(c) $T_3((x_1,x_2)) = (x_1,0)$. T_3 projects R^2 onto the x_1 axis.

(d) $T_4((x_1,x_2)) = (0,x_2)$. T_4 projects R^2 onto the x_2 axis.

Example 2. Consider the linear space P of all polynomials $p(x)$, with real coefficients, defined on $[0,1]$. The mapping D defined by

$$D(p) = \frac{dp}{dx}$$

is clearly a linear transformation of P into itself.

Example 3. The mapping I defined by

$$I(f) = \int_0^1 f(x)\, dx$$

is easily seen to be a linear transformation of $\mathcal{C}[0,1]$ into the real linear space R of all real numbers.

We return to our consideration of the linear spaces L and L' and of the linear transformations of L into L'. If T and U are two such transformations, then they can be added in a natural way to yield $T + U$, which is defined by

$$(T + U)(x) = T(x) + U(x). \tag{1}$$

Similarly, any such transformation T can be multiplied by any scalar α, in accordance with

$$(\alpha T)(x) = \alpha T(x). \tag{2}$$

Simple computations show readily that $T + U$ and αT are themselves linear transformations of L into L', and it is easily proved that these definitions convert the set of all such linear transformations into a linear space. The *zero transformation* 0 (i.e., the zero element of this linear space) and the *negative* $-T$ of a transformation T are defined by $0(x) = 0$ and $(-T)(x) = -T(x)$. In summary, we have

Theorem A. *Let L and L' be two linear spaces with the same system of scalars. Then the set of all linear transformations of L into L' is itself a linear space with respect to the linear operations defined by Eqs. (1) and (2).*

The most interesting and significant applications of these ideas occur in the special cases in which (1) L' equals L, and (2) L' equals the linear space of all scalars of L. We now develop a few of the simpler concepts which arise in case (1). Case (2) will be treated in some detail in the next chapter.

We assume, then, that we have a single linear space L, and we consider the linear space of all linear transformations of L into itself. We usually speak of these as linear transformations *on L*. The most important feature of this situation is that if T and U are any two linear transformations on L, then we can define their *product TU* by means of

$$(TU)(x) = T(U(x)). \tag{3}$$

This is precisely the multiplication of mappings discussed at the end of Sec. 3, and Problem 3-1 assures us that this operation is associative:

$$T(UV) = (TU)V. \tag{4}$$

Furthermore, multiplication is related to addition by the distributive laws

$$T(U + V) = TU + TV \tag{5}$$

and
$$(T + U)V = TV + UV, \tag{6}$$

and to scalar multiplication by

$$\alpha(TU) = (\alpha T)U = T(\alpha U). \tag{7}$$

The proofs of these facts are easy. As an illustration, we prove (6) by the following computation:

$$[(T + U)V](x) = (T + U)(V(x))$$
$$= T(V(x)) + U(V(x))$$
$$= (TV)(x) + (UV)(x)$$
$$= (TV + UV)(x).$$

Examples can readily be found to show that multiplication is in general non-commutative. For instance, if we define a linear transformation M on the space P of polynomials $p(x)$ by $M(p) = xp$, then

$$(MD)(p) = M(D(p)) = xD(p) = x\frac{dp}{dx}$$

and $$(DM)(p) = D(M(p)) = D(xp) = x\frac{dp}{dx} + p,$$

so $MD \neq DM$. Also, it is quite possible for the product of two non-zero linear transformations to be 0. Examples 1c and 1d demonstrate this, for the transformations T_3 and T_4 are both different from 0, and yet $T_3T_4 = 0$.

We have so far seen only one specific linear transformation on the arbitrary linear space L, namely, the zero transformation 0. Another is the *identity transformation I*, defined by $I(x) = x$. We observe that $I \neq 0 \Leftrightarrow L \neq \{0\}$, and that

$$TI = IT = T \tag{8}$$

for every linear transformation T on L. If α is any scalar, then the linear transformation αI is called a *scalar multiplication*, for

$$(\alpha I)(x) = \alpha I(x) = \alpha x$$

shows that the effect of αI is to multiply each vector in L by α.

A linear transformation T on L is called *non-singular* if it is one-to-one and onto, and *singular* otherwise. If T is non-singular, then by Sec. 3 its *inverse* T^{-1} exists as a mapping and satisfies the following equation:

$$TT^{-1} = T^{-1}T = I. \tag{9}$$

It is not difficult to show that when T is non-singular, then the mapping T^{-1} is also a linear transformation on L.

A particularly important type of linear transformation on L arises as follows. Let L be the direct sum of the subspaces M and N, so that $L = M \oplus N$. This means, of course, that each vector z in L can be written uniquely in the form $z = x + y$ with x in M and y in N. Since x is uniquely determined by z, we can define a mapping E of L into itself by $E(z) = x$.　E is easily seen to be a linear transformation on L, and it is called the *projection on M along N*.　Figure 33 indicates the geometric reason for this terminology.　The most significant property of E is that it is *idempotent*, in the sense that $E^2 = E$; for since $x = x + 0$ is the

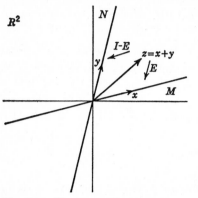

representation of x as the sum of a vector in M and a vector in N, we have

$$E^2(z) = (EE)(z) = E(E(z))$$
$$= E(x) = x = E(z).$$

This property of idempotence is characteristic of projections, as our next theorem shows.

Theorem B. *If E is a linear transformation on a linear space L, then E is idempotent $\Leftrightarrow$ there exist subspaces M and N of L such that $L = M \oplus N$ and E is the projection on M along N.*

Fig. 33.　The projections E on M along N and $I - E$ on N along M.

PROOF.　In view of the above remarks, it suffices to show that if E is idempotent, then it is the projection on M along N for suitable M and N. We define M and N by $M = \{E(z): z \in L\}$ and $N = \{z: E(z) = 0\}$. Both are clearly subspaces, and we must show that $L = M \oplus N$.　By Theorem 42-B, it suffices to show that M and N span L and are disjoint. That M and N span L follows from the fact that each z in L can be written in the form

$$z = E(z) + (I - E)(z); \tag{10}$$

for $E(z)$ is obviously in M, and

$$E((I - E)(z)) = (E(I - E))(z) = (E - E^2)(z) = (E - E)(z)$$
$$= 0(z) = 0$$

shows that $(I - E)(z)$ is in N.　To see that M and N are disjoint, we have only to notice that if a vector $E(z)$ in M is also in N, so that $E(E(z)) = 0$, then $E(E(z)) = E^2(z) = E(z)$ shows that $E(z) = 0$.　This proves that $L = M \oplus N$, and it follows from Eq. (10) that E is precisely the projection on M along N.

The unsymmetric way in which M and N are treated in this discussion can easily be balanced by considering the mapping which makes correspond to each $z = x + y$ the vector y (instead of x).　This linear transformation (it is clearly $I - E$) is the projection on N along M.　In the light of our theorem, we define a *projection* on L to be an idempotent linear transformation on L. If E is a linear transformation on L, then the equation

$$(I - E)^2 = (I - E)(I - E) = I - E - E + E^2$$

shows that E is a projection $\Leftrightarrow I - E$ is a projection; and we know that if E is the projection on M along N, then $I - E$ is the projection on N along M, and conversely. We make one further comment on these matters: if M is a given subspace of L, then by Problem 43-3 there certainly exists a projection on M and along some N; but since there may be many different subspaces N such that $L = M \oplus N$, there may also be many different projections on M (and along various N's).

Problems

1. Show that the mappings defined by Eqs. (1) to (3) are linear transformations.
2. Show that the linear transformations T_2 and T_3 defined in Example 1 do not commute; that is, show that $T_2T_3 \neq T_3T_2$.
3. If D and M are the linear transformations on the space P defined in the text, show that $DM = MD + I$ and $(MD)^2 = M^2D^2 + MD$.
4. Let T be a linear transformation on a linear space L, and show that T is non-singular $\Leftrightarrow$ there exists a linear transformation T' on L such that $TT' = T'T = I$.
5. Let T be a linear transformation on a linear space L, and prove that T is non-singular $\Leftrightarrow T(B)$ is a basis for L whenever B is.
6. Prove that a linear transformation on a finite-dimensional linear space is non-singular $\Leftrightarrow$ it is one-to-one $\Leftrightarrow$ it is onto.
7. Show that the set of all non-singular linear transformations on a linear space L is a group with respect to multiplication. If L is finite-dimensional with dimension $n > 0$, this group is called the *full linear group* of degree n.
8. If L and L' are non-zero linear spaces (both real or both complex), prove that there exists a non-zero linear transformation of L into L'.
9. Let L be a linear space, and let x and y be vectors in L such that $x \neq 0$. Prove that there exists a linear transformation T on L such that $T(x) = y$. If y is not a scalar multiple of x, prove that there exists a linear transformation T' on L such that $T'(x) = 0$ and $T'(y) \neq 0$.
10. Let L and L' be linear spaces with the same scalars, and let T be a linear transformation of L into L'. The *null space* of T, namely, $\{x : T(x) = 0\}$, and its *range*, $\{T(x) : x \in L\}$, are clearly subspaces of L and L'. The *nullity* of T, denoted by $n(T)$, is the dimension

of its null space, and its *rank* $r(T)$ is the dimension of its range.   If L is finite-dimensional, prove that $n(T) + r(T) = d(L)$.

11.   If E is a projection on a linear space L, show that its range equals the set of all vectors which are fixed under E; i.e., show that $\{E(z): z \in L\} = \{z: E(z) = z\}$.

45. ALGEBRAS

A linear space A is called an *algebra* (see Sec. 20) if its vectors can be multiplied in such a way that A is also a ring in which scalar multiplication is related to multiplication by the following property:

$$\alpha(xy) = (\alpha x)y = x(\alpha y).$$

The concept of an algebra is therefore a natural combination of the concepts of a linear space and a ring.   Figure 34 illustrates the manner in

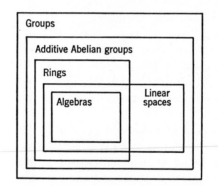

Fig. 34.   The major algebraic systems.

which the major algebraic systems defined in this chapter are related to one another.

Since an algebra is a linear space, all the ideas developed in Secs. 42 and 43 are immediately applicable.   Some algebras are real and some are complex, and every algebra has a well-defined dimension.   Furthermore, since an algebra is also a ring, it may be commutative or non-commutative, and may or may not have an identity; and if it does have an identity, then we can speak of its regular and singular elements.   A *division algebra* is an algebra with identity which, as a ring, is a division ring.   A *subalgebra* of an algebra A is a non-empty subset A_0 of A which is an algebra in its own right with respect to the operations in A.   This condition evidently means that A_0 is closed under addition, scalar multiplication, and multiplication.

Example 1. (*a*) The real linear space R of all real numbers (see Example 42-1) is a commutative real algebra with identity if multiplication is defined in the ordinary way. The reader will observe that scalar multiplication is indistinguishable from ring multiplication in this system.

(*b*) The complex linear space C of all complex numbers defined in Example 42-4 is a commutative complex algebra with identity if multiplication is defined as usual. Again we see that scalar multiplication and ring multiplication are the same.

Example 2. (*a*) The real linear space $\mathcal{C}(X,R)$ (see Example 42-3) is a commutative real algebra with identity if multiplication is defined pointwise.

(*b*) The complex linear space $\mathcal{C}(X,C)$ defined in Example 42-6 is a commutative complex algebra with identity with respect to pointwise multiplication.

Example 3. Let L be a linear space. We know by Theorem 44-A that the set of all linear transformations on L is a linear space with respect to the linear operations defined by Eqs. 44-(1) and 44-(2). The discussion following this theorem can be summed up as follows. If multiplication is defined by Eq. 44-(3), then this linear space is an algebra which is real or complex according as L is real or complex. This algebra has an identity (the identity transformation) $\Leftrightarrow L \neq \{0\}$, and in general it is non-commutative and has non-zero divisors of zero.

An *ideal* I in an algebra A is a non-empty subset of A which is both a subspace when A is considered as a linear space and an ideal when A is considered as a ring. By Theorems 42-A and 41-B, A/I is both a linear space and a ring. It is easy to see that A/I is actually an algebra, called the *quotient algebra* of A with respect to I. An ideal in A in our present sense is sometimes called an *algebra ideal*, as opposed to what we might call a *ring ideal*, that is, an ideal in A when A is considered as a ring. By our definition, an algebra ideal is a ring ideal which is also a subspace. In the cases of interest to us, the distinction between these two types of ideals disappears. For if A has an identity 1, and if I is a ring ideal in A, then the fact that $i \varepsilon I \Rightarrow \alpha i = \alpha(1i) = (\alpha 1)i \varepsilon I$ for every scalar α shows that I is closed under scalar multiplication, and is therefore an algebra ideal.

We shall return to the subject of algebras and their ideals in later chapters.

Problems

1. Let T be a non-singular linear transformation on a linear space L, and let A be the algebra of all linear transformations on L. Show

that T is non-singular (i.e., regular) as an element of the algebra $A \Leftrightarrow L \neq \{0\}$.

2. If A is an algebra, show that the subset of A defined by $C = \{x:xy = yx \text{ for every } y \in A\}$ is a subalgebra of A. C is called the *center* of A (see Problem 39-6).

3. Let A be an algebra of linear transformations on a linear space L. If A contains the identity transformation, prove that the center of A contains all scalar multiplications. If A is the algebra of all linear transformations on L, prove that the center of A consists precisely of the scalar multiplications (*Hint:* see Problem 44-9).

4. Let A and A' be algebras which are both real or both complex. As usual, we define a *homomorphism* of A into A' to be a mapping f of A into A' which preserves all the operations, in the sense that $f(x + y) = f(x) + f(y), f(\alpha x) = \alpha f(x)$, and $f(xy) = f(x)f(y)$. An *isomorphism* is a one-to-one homomorphism, and A is said to be *isomorphic to A'* if there exists an isomorphism of A onto A'. Now let A be an arbitrary algebra with identity, and prove that the mapping f defined on A by $f(x) = M_x$, where $M_x(y) = xy$, is an isomorphism of A into the algebra of all linear transformations on A. This fact is analogous to Cayley's theorem (see Problem 39-11). The isomorphism f is called the *regular representation* of A (by linear transformations on itself).

CHAPTER NINE

Banach Spaces

We have already seen, in Sec. 14, that a Banach space is a linear space which is also, in a special way, a complete metric space. This combination of algebraic and metric structures opens up the possibility of studying linear transformations of one Banach space into another which have the additional property of being continuous.

Most of our work in this chapter centers around three fundamental theorems relating to continuous linear transformations. The *Hahn-Banach theorem* guarantees that a Banach space is richly supplied with continuous linear functionals, and makes possible an adequate theory of conjugate spaces. The *open mapping theorem* enables us to give a satisfactory description of the projections on a Banach space, and has the important closed graph theorem as one of its consequences. We use the *uniform boundedness theorem* in our discussion of the conjugate of an operator on a Banach space, and this in turn provides the setting for our treatment in the next chapter of the adjoint of an operator on a Hilbert space.

Virtually all this theory had its origins in analysis. Our present interest, however, lies in the study of form and structure, not in exploring the many applications of these ideas to specific problems. This chapter is therefore strongly oriented toward the algebraic and topological aspects of the matters at hand.

46. THE DEFINITION AND SOME EXAMPLES

We begin by restating the definition of a Banach space.

A *normed linear space* is a linear space N in which to each vector x there corresponds a real number, denoted by $\|x\|$ and called the *norm* of x, in such a manner that

(1) $\|x\| \geq 0$, and $\|x\| = 0 \Leftrightarrow x = 0$;

(2) $\|x + y\| \leq \|x\| + \|y\|$;

(3) $\|\alpha x\| = |\alpha| \, \|x\|$.

The non-negative real number $\|x\|$ is to be thought of as the length of the vector x. If we regard $\|x\|$ as a real function defined on N, this function is called the *norm* on N. It is easy to verify that the normed linear space N is a metric space with respect to the metric d defined by $d(x,y) = \|x - y\|$. A *Banach space* is a complete normed linear space. Our main interest in this chapter is in Banach spaces, but there are several points in the body of the theory at which it is convenient to have the basic definitions and some of the simpler facts formulated in terms of normed linear spaces. For this reason, and also to emphasize the role of completeness in theorems which require this assumption, we work in the more general context whenever possible. The reader will find that the deeper theorems, in which completeness hypotheses are necessary, often make essential use of Baire's theorem.

Several simple but important facts about a normed linear space are based on the following inequality:

$$| \, \|x\| - \|y\| \, | \leq \|x - y\|. \tag{1}$$

To prove this, it suffices to prove that

$$\|x\| - \|y\| \leq \|x - y\|; \tag{2}$$

for it follows from (2) that we also have

$$-(\|x\| - \|y\|) = \|y\| - \|x\| \leq \|y - x\| = \|-(x - y)\| = \|x - y\|,$$

which together with (2) yields (1). We now prove (2) by observing that $\|x\| = \|(x - y) + y\| \leq \|x - y\| + \|y\|$. The main conclusion we draw from (1) is that the norm is a continuous function:

$$x_n \to x \Rightarrow \|x_n\| \to \|x\|.$$

This is clear from the fact that $| \, \|x_n\| - \|x\| \, | \leq \|x_n - x\|$, since $x_n \to x$ means that $\|x_n - x\| \to 0$. In the same vein, we can prove that addition and scalar multiplication are jointly continuous (see Problem 22-5), for

$$x_n \to x \text{ and } y_n \to y \Rightarrow x_n + y_n \to x + y$$

and
$$\alpha_n \to \alpha \text{ and } x_n \to x \Rightarrow \alpha_n x_n \to \alpha x.$$

These assertions follow from

$$\|(x_n + y_n) - (x + y)\| = \|(x_n - x) + (y_n - y)\|$$
$$\leq \|x_n - x\| + \|y_n - y\|$$

and

$$\|\alpha_n x_n - \alpha x\| = \|\alpha_n(x_n - x) + (\alpha_n - \alpha)x\|$$
$$\leq |\alpha_n| \, \|x_n - x\| + |\alpha_n - \alpha| \, \|x\|.$$

Our first theorem exhibits one of the most useful ways of forming new normed linear spaces out of old ones.

Theorem A. *Let M be a closed linear subspace of a normed linear space N. If the norm of a coset $x + M$ in the quotient space N/M is defined by*

$$\|x + M\| = \inf \{\|x + m\| : m \in M\}, \tag{3}$$

then N/M is a normed linear space. Further, if N is a Banach space, then so is N/M.

PROOF. We first verify that (3) defines a norm in the required sense. It is obvious that $\|x + M\| \geq 0$; and since M is closed, it is easy to see that $\|x + M\| = 0 \Leftrightarrow$ there exists a sequence $\{m_k\}$ in M such that $\|x + m_k\| \to 0 \Leftrightarrow x$ is in $M \Leftrightarrow x + M = M =$ the zero element of N/M. Next, we have $\|(x + M) + (y + M)\| = \|(x + y) + M\| = \inf \{\|x + y + m\| : m \in M\} = \inf \{\|x + y + m + m'\| : m$ and $m' \in M\} = \inf \{\|(x + m) + (y + m')\| : m$ and $m' \in M\} \leq \inf \{\|x + m\| + \|y + m'\| : m$ and $m' \in M\} = \inf \{\|x + m\| : m \in M\} + \inf \{\|y + m'\| : m' \in M\} = \|x + M\| + \|y + M\|$. The proof of $\|\alpha(x + M)\| = |\alpha| \, \|x + M\|$ is similar.

Finally, we assume that N is complete, and we show that N/M is also complete. If we start with a Cauchy sequence in N/M, then by Problem 12-2 it suffices to show that this sequence has a convergent subsequence. It is clearly possible to find a subsequence $\{x_n + M\}$ of the original Cauchy sequence such that $\|(x_1 + M) - (x_2 + M)\| < \frac{1}{2}$, $\|(x_2 + M) - (x_3 + M)\| < \frac{1}{4}$, and, in general, $\|(x_n + M) - (x_{n+1} + M)\| < 1/2^n$. We prove that this sequence is convergent in N/M. We begin by choosing any vector y_1 in $x_1 + M$, and we select y_2 in $x_2 + M$ such that $\|y_1 - y_2\| < \frac{1}{2}$. We next select a vector y_3 in $x_3 + M$ such that $\|y_2 - y_3\| < \frac{1}{4}$. Continuing in this way, we obtain a sequence $\{y_n\}$ in N such that $\|y_n - y_{n+1}\| < 1/2^n$. If $m < n$, then

$$\|y_m - y_n\| = \|(y_m - y_{m+1}) + (y_{m+1} - y_{m+2}) + \cdots$$
$$+ (y_{n-1} - y_n)\| \leq \|y_m - y_{m+1}\| + \|y_{m+1} - y_{m+2}\| + \cdots$$
$$+ \|y_{n-1} - y_n\| < 1/2^m + 1/2^{m+1} + \cdots + 1/2^{n-1} < 1/2^{m-1},$$

so $\{y_n\}$ is a Cauchy sequence in N. Since N is complete, there exists a vector y in N such that $y_n \to y$. It now follows from $\|(x_n + M) - (y + M)\| \leq \|y_n - y\|$ that $x_n + M \to y + M$, so N/M is complete.

In the following sections and chapters, we shall often have occasion to consider the quotient space of a normed linear space with respect to a closed linear subspace. In accordance with our theorem, a quotient space of this kind can always be regarded as a normed linear space in its own right.

We now describe some of the main examples of Banach spaces. In each of these, the linear operations are understood to be defined either coordinatewise or pointwise, whichever is appropriate in the circumstances.

Example 1.   The spaces R and C—the real numbers and the complex numbers—are the simplest of all normed linear spaces. The norm of a number x is of course defined by $\|x\| = |x|$, and each space is a Banach space.

Example 2.   The linear spaces R^n and C^n of all n-tuples $x = (x_1, x_2, \ldots, x_n)$ of real and complex numbers can be made into normed linear spaces in an infinite variety of ways, as we shall see below. If the norm is defined by

$$\|x\| = \Big(\sum_{i=1}^{n} |x_i|^2 \Big)^{1/2}, \tag{4}$$

then we get the n-dimensional Euclidean and unitary spaces familiar to us from our earlier work. We denoted these spaces by R^n and C^n in Part 1 of this book, and we know by the theorems of Sec. 15 that both are Banach spaces.

Each of the following examples consists of n-tuples of scalars, sequences of scalars, or scalar-valued functions defined on some non-empty set, where the scalars are the real numbers or the complex numbers. We do not normally specify which system of scalars is to be used, and it should be emphasized that both possibilities are allowed unless the contrary is clearly stated. Also, we make no distinction in notation between the real case and the complex case. When it turns out to be necessary to distinguish these two cases, we do so verbally, by referring, for instance, to "the complex space —." These conventions are in accord with the standard usage preferred by most mathematicians, and they enable us to avoid a good deal of cumbersome notation and many unnecessary case distinctions.

Example 3.   Let p be a real number such that $1 \le p < \infty$. We denote by l_p^n the space of all n-tuples $x = (x_1, x_2, \ldots, x_n)$ of scalars, with the norm defined by

$$\|x\|_p = \Big(\sum_{i=1}^{n} |x_i|^p \Big)^{1/p}. \tag{5}$$

Formula (4) is obviously the special case of (5) which corresponds to $p = 2$, so the real and complex spaces l_2^n are the n-dimensional Euclidean and unitary spaces R^n and C^n. It is easy to see that (5) satisfies conditions (1) and (3) required by the definition of a norm. In Problem 4 we outline a proof of the fact that (5) also satisfies condition (2), that is, that $\|x + y\|_p \leq \|x\|_p + \|y\|_p$. The completeness of l_p^n follows from substantially the same reasoning as that used in the proof of Theorem 15-A, so l_p^n is a Banach space.

Example 4. We again consider a real number p with the property that $1 \leq p < \infty$, and we denote by l_p the space of all sequences

$$x = \{x_1, x_2, \ldots, x_n, \ldots\}$$

of scalars such that $\Sigma_{n=1}^{\infty} |x_n|^p < \infty$, with the norm defined by

$$\|x\|_p = \Big(\sum_{n=1}^{\infty} |x_n|^p \Big)^{1/p}. \tag{6}$$

The reader will observe that the real and complex spaces l_2 are precisely the infinite-dimensional Euclidean and unitary spaces R^∞ and C^∞ defined in Problem 15-4. The proof of the fact that l_p actually is a Banach space requires arguments similar to those used in Problems 15-3 and 15-4.

The Banach spaces discussed in these examples are all special cases of the important L_p spaces studied in the theory of measure and integration. A detailed treatment of these spaces is outside the scope of this book, but we can describe them loosely as follows. An L_p space essentially consists of all measurable functions f defined on a measure space X with measure m which are such that $|f(x)|^p$ is integrable, with

$$\|f\|_p = \left(\int |f(x)|^p \, dm(x) \right)^{1/p} \tag{7}$$

taken as the norm. In order to include the spaces l_p^n and l_p within the theory of L_p spaces, we have only to consider the sets $\{1, 2, \ldots, n\}$ and $\{1, 2, \ldots, n, \ldots\}$ as measure spaces in which each point has measure 1, and to regard n-tuples and sequences of scalars as functions defined on these sets. Since integration is a generalized type of summation, formulas (5) and (6) are special cases of formula (7).[1]

Example 5. Just as in Example 3, we start with the linear space of all n-tuples $x = (x_1, x_2, \ldots, x_n)$ of scalars, but this time we define the

---

[1] Several remarks and examples relating to L_p spaces are scattered about in this and the next chapter. This fragmentary material is not essential for an understanding of these chapters, and may be disregarded by any reader without the necessary background. Brief sketches of the relevant ideas can be found in Taylor [41, chap. 7] and Loomis [27, chap. 3]. For more extended treatments, see Halmos [18], Zaanen [45], or Kolmogorov and Fomin [26, vol. 2].

norm by

$$\|x\| = \max \{|x_1|, |x_2|, \ldots, |x_n|\}. \tag{8}$$

This Banach space is commonly denoted by l_∞^n, and the symbol $\|x\|_\infty$ is occasionally used for the norm given by (8). The reason for this practice lies in the interesting fact that

$$\|x\|_\infty = \lim \|x\|_p \quad \text{as } p \to \infty,$$

that is, that

$$\max \{|x_i|\} = \lim \Big(\sum_{i=1}^{n} |x_i|^p \Big)^{1/p} \quad \text{as } p \to \infty. \tag{9}$$

We briefly inspect the case $n = 2$ to see why this is true. Let $x = (x_1, x_2)$ be an ordered pair of real numbers with x_1 and $x_2 \geq 0$. It is clear that $\|x\|_\infty = \max \{x_1, x_2\} \leq (x_1{}^p + x_2{}^p)^{1/p} = \|x\|_p$. If $x_1 = x_2$, then $\lim \|x\|_p = \lim (2x_2{}^p)^{1/p} = \lim 2^{1/p} x_2 = x_2 = \|x\|_\infty$. And if $x_1 < x_2$, then

$$\lim \|x\|_p = \lim (x_1{}^p + x_2{}^p)^{1/p} = \lim ([(x_1/x_2)^p + 1]x_2{}^p)^{1/p}$$
$$= \lim [(x_1/x_2)^p + 1]^{1/p} x_2 = x_2 = \|x\|_\infty.$$

Example 6. Consider the linear space of all bounded sequences $x = \{x_1, x_2, \ldots, x_n, \ldots\}$ of scalars. By analogy with Example 5, we define the norm by

$$\|x\| = \sup |x_n|, \tag{10}$$

and we denote the resulting Banach space by l_∞. The set c of all convergent sequences is easily seen to be a closed linear subspace of l_∞ and is therefore itself a Banach space. Another Banach space in this family is the subset c_0 of c which consists of all convergent sequences with limit 0.

Example 7. The Banach space of primary interest to us is the space $\mathcal{C}(X)$ of all bounded continuous scalar-valued functions defined on a topological space X, with the norm given by

$$\|f\| = \sup |f(x)|.^1 \tag{11}$$

This norm is sometimes called the *uniform norm*, because the statement that f_n converges to f with respect to this norm means that f_n converges to f uniformly on X. The fact that this space is complete amounts to the fact that if f is the uniform limit of a sequence of bounded continuous functions, then f itself is bounded and continuous. If, as above, we consider n-tuples and sequences as functions defined on $\{1, 2, \ldots, n\}$ and $\{1, 2, \ldots, n, \ldots\}$, then the spaces l_∞^n and l_∞ are the special cases of $\mathcal{C}(X)$ which correspond to choosing X to be the sets just mentioned, each with the discrete topology.

---

[1] The real space $\mathcal{C}(X)$ and the complex space $\mathcal{C}(X)$ are, of course, the spaces previously denoted by $\mathcal{C}(X,R)$ and $\mathcal{C}(X,C)$.

Many important properties of a Banach space are closely linked to the shape of its *closed unit sphere,* that is, the set $S = \{x : \|x\| \leq 1\}$. One basic property of S is that it is always *convex,* in the sense (see Problem 32-5) that if x and y are any two vectors in S, then the vector $z = \alpha x + \beta y$ is also in S, where α and β are non-negative real numbers such that $\alpha + \beta = 1$; for $\|z\| = \|\alpha x + \beta y\| \leq \alpha\|x\| + \beta\|y\| \leq \alpha + \beta = 1$. In this connection, it is illuminating to consider the shape of S for certain simple examples. Let our underlying linear space be the real linear

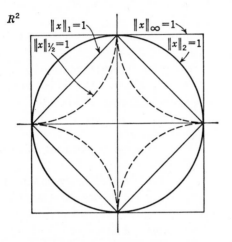

Fig. 35.   Some closed unit spheres.

space R^2 of all ordered pairs $x = (x_1, x_2)$ of real numbers. As we have seen, there are many different norms which can be defined on R^2, among which are the following: $\|x\|_1 = |x_1| + |x_2|$; $\|x\|_2 = (|x_1|^2 + |x_2|^2)^{1/2}$; and $\|x\|_\infty = \max\{|x_1|, |x_2|\}$. Figure 35 illustrates the closed unit sphere which corresponds to each of these norms. In the first case, S is the square with vertices $(1,0)$, $(0,1)$, $(-1,0)$, $(0,-1)$; in the second, it is the circular disc of radius 1; and in the third, it is the square with vertices $(1,1)$, $(-1,1)$, $(-1,-1)$, $(1,-1)$. If we consider the norm defined by

$$\|x\|_p = (|x_1|^p + |x_2|^p)^{1/p}, \tag{12}$$

where $1 \leq p < \infty$, and if we allow p to increase from 1 to ∞, then the corresponding S's swell continuously from the first square mentioned to the second. We note that S is truly "spherical" $\Leftrightarrow p = 2$. These considerations also show quite clearly why we always assume that $p \geq 1$; for if we were to define $\|x\|_p$ by formula (12) with $p < 1$, then $S = \{x : \|x\|_p \leq 1\}$ would not be convex (see the star-shaped inner portion of Fig. 35). For $p < 1$, therefore, formula (12) does not yield a norm.

In the above examples, we have exhibited several different types of Banach spaces, and there are yet others which we have not mentioned. Amid this diversity of possibilities, it is well to realize that any Banach space can be regarded—from the point of view of its linear and norm structures alone—as a closed linear subspace of $\mathcal{C}(X)$ for a suitable compact Hausdorff space X. We prove this below, in our discussion of the natural imbedding of a Banach space in its second conjugate space.

Problems

1. Let N be a non-zero normed linear space, and prove that N is a Banach space $\Leftrightarrow \{x : \|x\| = 1\}$ is complete.
2. Let a Banach space B be the direct sum of the linear subspaces M and N, so that $B = M \oplus N$. If $z = x + y$ is the unique expression of a vector z in B as the sum of vectors x and y in M and N, then a new norm can be defined on the linear space B by $\|z\|' = \|x\| + \|y\|$. Prove that this actually is a norm. If B' symbolizes the linear space B equipped with this new norm, prove that B' is a Banach space if M and N are closed in B.
3. Prove Eq. (9) for the case of an arbitrary positive integer n.
4. In this problem we sketch the proofs—and we ask the reader to fill in the details—of some important inequalities relating to n-tuples $x = (x_1, x_2, \ldots, x_n)$ and $y = (y_1, y_2, \ldots, y_n)$ of scalars. Whenever p occurs alone, and nothing is said to the contrary, we assume that $1 \le p < \infty$; and whenever p and q occur together, we assume that both are greater than 1 and that $1/p + 1/q = 1$.
   (a) Show that a and $b \ge 0 \Rightarrow a^{1/p} b^{1/q} \le a/p + b/q$. (If $a = 0$ or $b = 0$, the conclusion is clear, so assume that both are positive. If $k \in (0,1)$, define $f(t)$ for $t \ge 1$ by $f(t) = k(t - 1) - t^k + 1$. Note that $f(1) = 0$ and $f'(t) \ge 0$, and conclude that $t^k \le kt + (1 - k)$. If $a \ge b$, put $t = a/b$ and $k = 1/p$; if $a < b$, put $t = b/a$ and $k = 1/q$; and in each case, draw the required conclusion.)
   (b) Prove *Hölder's inequality*: $\sum_{i=1}^n |x_i y_i| \le \|x\|_p \|y\|_q$. (If $x = 0$ or $y = 0$, the inequality is obvious, so assume that both are $\ne 0$. Put $a_i = (|x_i|/\|x\|_p)^p$ and $b_i = (|y_i|/\|y\|_q)^q$, and use part (a) to obtain $|x_i y_i|/\|x\|_p \|y\|_q \le a_i/p + b_i/q$. Add these inequalities for $i = 1, 2, \ldots, n$, and conclude that

   $$\left(\sum_{i=1}^n |x_i y_i| \right) / \|x\|_p \|y\|_q \le 1/p + 1/q = 1.)$$

   (c) Prove *Minkowski's inequality*: $\|x + y\|_p \le \|x\|_p + \|y\|_p$. (The inequality is evident when $p = 1$, so assume that $p > 1$.

Use Hölder's inequality to obtain

$$\|x + y\|_p{}^p = \sum_{i=1}^{n} |x_i + y_i|^p = \sum_{i=1}^{n} |x_i + y_i| \, |x_i + y_i|^{p-1}$$

$$\leq \sum_{i=1}^{n} |x_i| \, |x_i + y_i|^{p-1} + \sum_{i=1}^{n} |y_i| \, |x_i + y_i|^{p-1}$$

$$\leq (\|x\|_p + \|y\|_p) \, \|x + y\|_p{}^{p/q}.)$$

When $p = q = 2$, Hölder's inequality becomes Cauchy's inequality as stated and proved in Sec. 15. The Hölder and Minkowski inequalities can easily be extended from finite sums to series. For readers with some knowledge of the theory of measure and integration, we remark that these inequalities can also be stated in the following much more general forms: if f is in L_p and g is in L_q, then their pointwise product fg is in L_1 and

$$\|fg\|_1 \leq \|f\|_p\|g\|_q;$$

and if f and g are both in L_p, then $f + g$ is also in L_p and

$$\|f + g\|_p \leq \|f\|_p + \|g\|_p.$$

It is to be understood that f and g are measurable functions defined on an arbitrary measure space and that the norms occurring in these inequalities are those defined by formula (7).

47. CONTINUOUS LINEAR TRANSFORMATIONS

Let N and N' be normed linear spaces with the same scalars, and let T be a linear transformation of N into N'.[1] When we say that T is continuous, we mean that it is continuous as a mapping of the metric space N into the metric space N'. By Theorem 13-B, this amounts to the condition that $x_n \to x$ in $N \Rightarrow T(x_n) \to T(x)$ in N'. Our main purpose in this section is to convert the requirement of continuity into several more useful equivalent forms and to show that the set of all continuous linear transformations of N into N' can itself be made into a normed linear space in a natural way.

Theorem A. *Let N and N' be normed linear spaces and T a linear transformation of N into N'. Then the following conditions on T are all equivalent to one another:*

(1) *T is continuous;*
(2) *T is continuous at the origin, in the sense that $x_n \to 0 \Rightarrow T(x_n) \to 0$;*

---

[1] In the future, whenever we mention two normed linear spaces with a view to considering linear transformations of one into the other, we shall always assume—without necessarily saying so explicitly—that they have the same scalars.

(3)   *there exists a real number $K \geq 0$ with the property that $\|T(x)\|$ $\leq K\|x\|$ for every $x \in N$;*

(4)   *if $S = \{x : \|x\| \leq 1\}$ is the closed unit sphere in N, then its image $T(S)$ is a bounded set in N'.*

PROOF.   (1) $\Leftrightarrow$ (2). If T is continuous, then since $T(0) = 0$ it is certainly continuous at the origin.   On the other hand, if T is continuous at the origin, then $x_n \to x \Leftrightarrow x_n - x \to 0 \Rightarrow T(x_n - x) \to 0 \Leftrightarrow T(x_n) - T(x) \to 0 \Leftrightarrow T(x_n) \to T(x)$, so T is continuous.

(2) $\Leftrightarrow$ (3).   It is obvious that (3) $\Rightarrow$ (2), for if such a K exists, then $x_n \to 0$ clearly implies that $T(x_n) \to 0$.   To show that (2) $\Rightarrow$ (3), we assume that there is no such K.   It follows from this that for each positive integer n we can find a vector x_n such that $\|T(x_n)\| > n\|x_n\|$, or equivalently, such that $\|T(x_n/n\|x_n\|)\| > 1$.   If we now put

$$y_n = x_n/n\|x_n\|,$$

then it is easy to see that $y_n \to 0$ but $T(y_n) \nrightarrow 0$, so T is not continuous at the origin.

(3) $\Leftrightarrow$ (4).   Since a non-empty subset of a normed linear space is bounded $\Leftrightarrow$ it is contained in a closed sphere centered on the origin, it is evident that (3) $\Rightarrow$ (4); for if $\|x\| \leq 1$, then $\|T(x)\| \leq K$.   To show that (4) $\Rightarrow$ (3), we assume that $T(S)$ is contained in a closed sphere of radius K centered on the origin.   If $x = 0$, then $T(x) = 0$, and clearly $\|T(x)\| \leq K\|x\|$; and if $x \neq 0$, then $x/\|x\| \in S$, and therefore $\|T(x/\|x\|)\| \leq K$, so again we have $\|T(x)\| \leq K\|x\|$.

If the linear transformation T in this theorem satisfies condition (3), so that there exists a real number $K \geq 0$ with the property that

$$\|T(x)\| \leq K\|x\|$$

for every x, then K is called a *bound* for T, and such a T is often referred to as a *bounded linear transformation*.   According to our theorem, T is bounded $\Leftrightarrow$ it is continuous, so these two adjectives can be used interchangeably.   We now assume that T is continuous, so that it satisfies condition (4), and we define its *norm* by

$$\|T\| = \sup \{\|T(x)\| : \|x\| \leq 1\}. \qquad (1)$$

When $N \neq \{0\}$, this formula can clearly be written in the equivalent form

$$\|T\| = \sup \{\|T(x)\| : \|x\| = 1\}. \qquad (2)$$

It is apparent from the proof of Theorem A that the set of all bounds for T equals the set of all radii of closed spheres centered on the origin which contain $T(S)$.   This yields yet another expression for the norm of T, namely,

$$\|T\| = \inf \{K : K \geq 0 \text{ and } \|T(x)\| \leq K\|x\| \text{ for all } x\} ; \qquad (3)$$

and from this we see at once that

$$\|T(x)\| \leq \|T\| \, \|x\| \qquad (4)$$

for all x.

We now denote the set of all continuous (or bounded) linear transformations of N into N' by $\mathcal{B}(N,N')$, where the letter "$\mathcal{B}$" is intended to suggest the adjective "bounded." It is a routine matter to verify that this set is a linear space with respect to the pointwise linear operations defined by Eqs. 44-(1) and 44-(2) and to show that formula (1) actually does define a norm on this linear space. We summarize and extend these remarks in

Theorem B. *If N and N' are normed linear spaces, then the set $\mathcal{B}(N,N')$ of all continuous linear transformations of N into N' is itself a normed linear space with respect to the pointwise linear operations and the norm defined by (1). Further, if N' is a Banach space, then $\mathcal{B}(N,N')$ is also a Banach space.*

PROOF. We leave to the reader the simple task of showing that $\mathcal{B}(N,N')$ is a normed linear space, and we prove that this space is complete when N' is.

Let $\{T_n\}$ be a Cauchy sequence in $\mathcal{B}(N,N')$. If x is an arbitrary vector in N, then $\|T_m(x) - T_n(x)\| = \|(T_m - T_n)(x)\| \leq \|T_m - T_n\| \, \|x\|$ shows that $\{T_n(x)\}$ is a Cauchy sequence in N'; and since N' is complete, there exists a vector in N'—we denote it by $T(x)$—such that $T_n(x) \to T(x)$. This defines a mapping T of N into N', and by the joint continuity of addition and scalar multiplication, T is easily seen to be a linear transformation. To conclude the proof, we have only to show that T is continuous and that $T_n \to T$ with respect to the norm on $\mathcal{B}(N,N')$. By the inequality 46-(1), the norms of the terms of a Cauchy sequence in a normed linear space form a bounded set of numbers, so

$$\|T(x)\| = \|\lim T_n(x)\| = \lim \|T_n(x)\| \leq \sup (\|T_n\| \, \|x\|) = (\sup \|T_n\|)\|x\|$$

shows that T has a bound and is therefore continuous. It remains to be proved that $\|T_n - T\| \to 0$. Let $\epsilon > 0$ be given, and let n_0 be a positive integer such that $m, n \geq n_0 \Rightarrow \|T_m - T_n\| < \epsilon$. If $\|x\| \leq 1$ and $m, n \geq n_0$, then

$$\|T_m(x) - T_n(x)\| = \|(T_m - T_n)(x)\| \leq \|T_m - T_n\| \, \|x\|$$
$$\leq \|T_m - T_n\| < \epsilon.$$

We now hold m fixed and allow n to approach ∞, and we see that $\|T_m(x) - T_n(x)\| \to \|T_m(x) - T(x)\|$, from which we conclude that

$\|T_m(x) - T(x)\| \leq \epsilon$ for all $m \geq n_0$ and all x such that $\|x\| \leq 1$. This shows that $\|T_m - T\| \leq \epsilon$ for all $m \geq n_0$, and the proof is complete.

Let N be a normed linear space. We call a continuous linear transformation of N into itself an *operator* on N, and we denote the normed linear space of all operators on N by $\mathfrak{B}(N)$ instead of $\mathfrak{B}(N,N)$. Theorem B shows that $\mathfrak{B}(N)$ is a Banach space when N is. Furthermore, if operators are multiplied in accordance with formula 44-(3), then $\mathfrak{B}(N)$ is an algebra in which multiplication is related to the norm by

$$\|TT'\| \leq \|T\| \, \|T'\|. \tag{5}$$

This relation is proved by the following computation:

$$\|TT'\| = \sup \{\|(TT')(x)\| : \|x\| \leq 1\} = \sup \{\|T(T'(x))\| : \|x\| \leq 1\}$$
$$\leq \sup \{\|T\| \, \|T'(x)\| : \|x\| \leq 1\} = \|T\| \sup \{\|T'(x)\| : \|x\| \leq 1\}$$
$$= \|T\| \, \|T'\|.$$

We know from the previous section that addition and scalar multiplication in $\mathfrak{B}(N)$ are jointly continuous, as they are in any normed linear space. Property (5) permits us to conclude that multiplication is also jointly continuous:

$$T_n \to T \text{ and } T'_n \to T' \Rightarrow T_n T'_n \to TT'.$$

This follows at once from

$$\|T_n T'_n - TT'\| = \|T_n(T'_n - T') + (T_n - T)T'\| \leq \|T_n\| \, \|T'_n - T'\|$$
$$+ \|T_n - T\| \, \|T'\|.$$

We also remark that when $N \neq \{0\}$, then the identity transformation I is an identity for the algebra $\mathfrak{B}(N)$. In this case, we clearly have

$$\|I\| = 1; \tag{6}$$

for $\|I\| = \sup \{\|I(x)\| : \|x\| \leq 1\} = \sup \{\|x\| : \|x\| \leq 1\} = 1$.

We complete this section with some definitions which will often be useful in our later work. Let N and N' be normed linear spaces. An *isometric isomorphism* of N into N' is a one-to-one linear transformation T of N into N' such that $\|T(x)\| = \|x\|$ for every x in N; and N is said to be *isometrically isomorphic* to N' if there exists an isometric isomorphism of N onto N'. This terminology enables us to give precise meaning to the statement that one normed linear space is essentially the same as another.

Problems

1. If M is a closed linear subspace of a normed linear space N, and if T is the natural mapping of N onto N/M defined by $T(x) = x + M$, show that T is a continuous linear transformation for which $\|T\| \leq 1$.

2. If T is a continuous linear transformation of a normed linear space N into a normed linear space N', and if M is its null space, show that T induces a natural linear transformation T' of N/M into N' and that $\|T'\| = \|T\|$.

3. Let N and N' be normed linear spaces with the same scalars. If N is infinite-dimensional and $N' \neq \{0\}$, show that there exists a linear transformation of N into N' which is not continuous. (We shall see in Problem 7 that if N is finite-dimensional, then every linear transformation of N into N' is automatically continuous.)

4. Let a linear space L be made into a normed linear space in two ways, and let the two norms of a vector x be denoted by $\|x\|$ and $\|x\|'$. These norms are said to be *equivalent* if they generate the same topology on L. Show that this is the case $\Leftrightarrow$ there exist two positive real numbers K_1 and K_2 such that $K_1\|x\| \leq \|x\|' \leq K_2\|x\|$ for all x. (If L is finite-dimensional, then any two norms defined on it are equivalent. See Problem 7.)

5. If n is a fixed positive integer, the spaces l_p^n $(1 \leq p \leq \infty)$ consist of a single underlying linear space with different norms defined on it. Show that these norms are all equivalent to one another. (*Hint:* show that convergence with respect to each norm amounts to coordinatewise convergence.)

6. If N is an arbitrary normed linear space, show that any linear transformation T of l_p^n $(1 \leq p \leq \infty)$ into N is continuous. (*Hint:* if $\{e_1, e_2, \ldots, e_n\}$ is the natural basis for l_p^n, where e_i is the n-tuple with 1 in the ith place and 0's elsewhere, then an arbitrary vector x in l_p^n can be written uniquely in the form

$$x = \alpha_1 e_1 + \alpha_2 e_2 + \cdots + \alpha_n e_n,$$

and from this we get $T(x) = \alpha_1 T(e_1) + \alpha_2 T(e_2) + \cdots + \alpha_n T(e_n)$; now apply the hint given for Problem 5.)

7. Let N be a finite-dimensional normed linear space with dimension $n > 0$, and let $\{e_1, e_2, \ldots, e_n\}$ be a basis for N. Each vector x in N can be written uniquely in the form

$$x = \alpha_1 e_1 + \alpha_2 e_2 + \cdots + \alpha_n e_n.$$

If T is the one-to-one linear transformation of N onto l_1^n defined by $T(x) = (\alpha_1, \alpha_2, \ldots, \alpha_n)$, then T^{-1} is continuous by Problem 6.

(a) Prove that T is continuous. (*Hint:* if T is not continuous, then for some $\epsilon > 0$ there exists a sequence $\{y_n\}$ in N such that $y_n \to 0$ and $\|T(y_n)\| \geq \epsilon$; if $z_n = y_n/\|T(y_n)\|$, then $z_n \to 0$ and $\|T(z_n)\| = 1$; the subset of l_1^n consisting of all vectors of norm 1 is compact, so $\{T(z_n)\}$ has a subsequence which converges to a vector with norm 1; now use the continuity of T^{-1}.)

(b)   Show that every linear transformation of N into an arbitrary normed linear space N' is continuous.

(c)   Show that any other norm defined on N is equivalent to the given norm.

(d)   Show that N is complete, and infer from this that every finite-dimensional linear subspace of an arbitrary normed linear space is closed.

8.   It is a simple consequence of Problem 7 that every finite-dimensional normed linear space is locally compact.   Prove the converse, that is, that a locally compact normed linear space N is finite-dimensional. *Hint:* the closed unit sphere S of N is compact, so there is a finite subset of S, say $\{x_1, x_2, \ldots, x_n\}$, with the property that each point of S is distant by less than $\frac{1}{2}$ from one of the x_i's; let M be the linear subspace of N spanned by the x_i's; and show that $M = N$ (to do this, assume that there exists a vector y not in M, use the fact that M is closed to infer that $d = d(y,M) > 0$, find m_0 in M such that $d \leq \|y - m_0\| \leq 3d/2$, and deduce the contradiction that the vector y_0 in S defined by $y_0 = (y - m_0)/\|y - m_0\|$ is distant from M by at least $\frac{2}{3}$).

48. THE HAHN-BANACH THEOREM

One of the basic principles of strategy in the study of an abstract mathematical system can be stated as follows: consider the set of all structure-preserving mappings of that system into the simplest system of the same type.   This principle is richly fruitful in the structure theory (or representation theory) of groups, rings, and algebras, and we shall see in the next section how it works for normed linear spaces.

We have remarked that the spaces R and C are the simplest of all normed linear spaces.   If N is an arbitrary normed linear space, the above principle leads us to form the set of all continuous linear transformations of N into R or C, according as N is real or complex.   This set—it is $\mathscr{B}(N,R)$ or $\mathscr{B}(N,C)$—is denoted by N^* and is called the *conjugate space* of N.   The elements of N^* are called *continuous linear functionals*, or more briefly, *functionals*.[1]   It follows from our work in the previous section that if these functionals are added and multiplied by scalars

---

[1] The noun "functional" seems to have originated in the theory of integral equations.   It was used to distinguish between a function in the elementary sense defined on a set of numbers and a function (or functional) defined on a set of functions. In this book, we always use the word to mean a scalar-valued continuous linear function defined on a normed linear space.

pointwise, and if the norm of a functional f is defined by

$$\|f\| = \sup \{|f(x)| : \|x\| \leq 1\}$$
$$= \inf \{K : K \geq 0 \text{ and } |f(x)| \leq K\|x\| \text{ for all } x\},$$

then N^* is a Banach space.

When we consider various specific Banach spaces, the problem arises of determining the concrete nature of the functionals associated with these spaces. It is not our aim in this section to explore the ample body of theory which centers around this problem, and in any case, the machinery necessary for such an enterprise (mostly the theory of measure and integration) is not available to us. Nevertheless, for the reader who may have the required background, we mention some of the main facts without proof.

Let X be a measure space with measure m, and let p be a given real number such that $1 < p < \infty$. Consider the Banach space L_p of all measurable functions f defined on X for which $|f(x)|^p$ is integrable. If g is a function in L_q, where $1/p + 1/q = 1$, we define a function F_g on L_p by

$$F_g(f) = \int f(x)g(x) \, dm(x).$$

The Hölder inequality for integrals mentioned at the end of Problem 46-4 shows that

$$|F_g(f)| = \left| \int f(x)g(x) \, dm(x) \right|$$
$$\leq \int |f(x)g(x)| \, dm(x)$$
$$\leq \|f\|_p \|g\|_q.$$

We conclude from this that F_g is a well-defined scalar-valued linear function on L_p with the property that $\|F_g\| \leq \|g\|_q$, and is therefore a functional on L_p. It can be shown that equality holds here, so that

$$\|F_g\| = \|g\|_q.$$

It can also be shown that every functional on L_p arises in this way, so the mapping $g \to F_g$ (which is clearly linear) is an isometric isomorphism of L_q onto $L_p{}^*$. This statement is usually expressed by writing

$$L_p{}^* = L_q, \tag{1}$$

where the equality sign is to be interpreted in the sense just explained.

If we specialize these considerations to n-tuples of scalars, we see that (1) becomes

$$(l_p^n)^* = l_q^n. \tag{2}$$

Further, it can be shown that

$$(l_1^n)^* = l_\infty^n \tag{3}$$

and that

$$(l_\infty^n)^* = l_1^n. \tag{4}$$

We sketch proofs of (2), (3), and (4) in the problems. When we consider sequences of scalars, then for $1 < p < \infty$ we have the following special case of (1):

$$l_p^* = l_q. \tag{5}$$

If $p = 1$, we obtain a natural extension of (3):

$$l_1^* = l_\infty. \tag{6}$$

The corresponding extension of (4) is another matter, for it is false that $l_\infty^* = l_1$. Instead, we have

$$c_0^* = l_1. \tag{7}$$

What is l_∞^*? We saw in Sec. 46 that l_∞ is a special case of $\mathcal{C}(X)$, so this question leads naturally to the problem of determining the nature of the conjugate space $\mathcal{C}^*(X)$. The classic solution of this problem for a space X which is compact Hausdorff (or even normal) is known as the *Riesz representation theorem*, and it depends on some of the deeper parts of the theory of measure and integration (see Dunford and Schwartz [8, pp. 261–265]). The situation is somewhat simpler for the case in which X is an interval $[a,b]$ on the real line, but even here an adequate treatment requires a knowledge of Stieltjes integrals (see Riesz and Sz.-Nagy [35, secs. 49–51]).

Most of the theory of conjugate spaces rests on the Hahn-Banach theorem, which asserts that any functional defined on a linear subspace of a normed linear space can be extended linearly and continuously to the whole space without increasing its norm. The proof is rather complicated, so we begin with a lemma which serves to isolate its most difficult parts.

Lemma. *Let M be a linear subspace of a normed linear space N, and let f be a functional defined on M. If x_0 is a vector not in M, and if*

$$M_0 = M + [x_0]$$

is the linear subspace spanned by M and x_0, then f can be extended to a functional f_0 defined on M_0 such that $\|f_0\| = \|f\|$.

PROOF. We first prove the lemma under the assumption that N is a real normed linear space. We may assume, without loss of generality, that $\|f\| = 1$. Since x_0 is not in M, each vector y in M_0 is uniquely expressible in the form $y = x + \alpha x_0$ with x in M. It is clear that the definition $f_0(x + \alpha x_0) = f_0(x) + \alpha f_0(x_0) = f(x) + \alpha r_0$ extends f linearly to M_0 for every choice of the real number $r_0 = f_0(x_0)$. Since we are trying to arrange matters so that $\|f_0\| = 1$, our problem is to show that r_0 can be chosen in such a way that $|f_0(x + \alpha x_0)| \leq \|x + \alpha x_0\|$ for every x in M and every $\alpha \neq 0$. Since $f_0(x + \alpha x_0) = f(x) + \alpha r_0$, this inequality can be written as

$$- \|x + \alpha x_0\| \leq f(x) + \alpha r_0 \leq \|x + \alpha x_0\|$$

or $\qquad -f(x) - \|x + \alpha x_0\| \leq \alpha r_0 \leq -f(x) + \|x + \alpha x_0\|,$

which in turn is equivalent to

$$-f\left(\frac{x}{\alpha}\right) - \left\|\frac{x}{\alpha} + x_0\right\| \leq r_0 \leq -f\left(\frac{x}{\alpha}\right) + \left\|\frac{x}{\alpha} + x_0\right\|. \tag{8}$$

We now observe that for any two vectors x_1 and x_2 in M we have

$$f(x_2) - f(x_1) = f(x_2 - x_1) \leq |f(x_2 - x_1)| \leq \|f\| \, \|x_2 - x_1\|$$
$$= \|x_2 - x_1\| = \|(x_2 + x_0) - (x_1 + x_0)\| \leq \|x_2 + x_0\| + \|x_1 + x_0\|,$$

so $\qquad -f(x_1) - \|x_1 + x_0\| \leq -f(x_2) + \|x_2 + x_0\|. \tag{9}$

If we define two real numbers a and b by

$$a = \sup \{-f(x) - \|x + x_0\| : x \in M\}$$

and $\qquad b = \inf \{-f(x) + \|x + x_0\| : x \in M\},$

then (9) shows that $a \leq b$. If we now choose r_0 to be any real number such that $a \leq r_0 \leq b$, then the required inequality (8) is satisfied and this part of the proof is complete.

We next use the result of the above paragraph to prove the lemma for the case in which N is complex. Here f is a complex-valued functional defined on M for which $\|f\| = 1$. We begin by remarking that a complex linear space can be regarded as a real linear space by simply restricting the scalars to be real numbers. If g and h are the real and imaginary parts of f, so that $f(x) = g(x) + ih(x)$ for every x in M, then both g and h are easily seen to be real-valued functionals on the real space M; and since $\|f\| = 1$, we have $\|g\| \leq 1$. The equation

$$f(ix) = if(x),$$

together with $f(ix) = g(ix) + ih(ix)$ and

$$if(x) = i(g(x) + ih(x)) = ig(x) - h(x),$$

shows that $h(x) = -g(ix)$, so we can write $f(x) = g(x) - ig(ix)$. By the above paragraph, we can extend g to a real-valued functional g_0 on the real space M_0 in such a way that $\|g_0\| = \|g\|$, and we define f_0 for

x in M_0 by $f_0(x) = g_0(x) - ig_0(ix)$. It is easy to see that f_0 is an extension of f from M to M_0, that $f_0(x + y) = f_0(x) + f_0(y)$, and that $f_0(\alpha x) = \alpha f_0(x)$ for all real α's. The fact that the property last stated is also valid for all complex α's is a direct consequence of

$$f_0(ix) = g_0(ix) - ig_0(i^2 x) = g_0(ix) + ig_0(x) = i(g_0(x) - ig_0(ix)) = if_0(x),$$

so f_0 is linear as a complex-valued function defined on the complex space M_0. All that remains to be proved is that $\|f_0\| = 1$, and we dispose of this by showing that if x is a vector in M_0 for which $\|x\| = 1$, then $|f_0(x)| \leq 1$. If $f_0(x)$ is real, this follows from $f_0(x) = g_0(x)$ and $\|g_0\| \leq 1$. If $f_0(x)$ is complex, then we can write $f_0(x) = re^{i\theta}$ with $r > 0$, so

$$|f_0(x)| = r = e^{-i\theta} f_0(x) = f_0(e^{-i\theta} x);$$

and our conclusion now follows from $\|e^{-i\theta} x\| = \|x\| = 1$ and the fact that $f_0(e^{-i\theta} x)$ is real.

Theorem A (the Hahn-Banach Theorem). *Let M be a linear subspace of a normed linear space N, and let f be a functional defined on M. Then f can be extended to a functional f_0 defined on the whole space N such that $\|f_0\| = \|f\|$.*

PROOF. The set of all extensions of f to functionals g with the same norm defined on subspaces which contain M is clearly a partially ordered set with respect to the following relation: $g_1 \leq g_2$ means that the domain of g_1 is contained in the domain of g_2, and $g_2(x) = g_1(x)$ for all x in the domain of g_1. It is easy to see that the union of any chain of extensions is also an extension and is therefore an upper bound for the chain. Zorn's lemma now implies that there exists a maximal extension f_0. We complete the proof by observing that the domain of f_0 must be the entire space N, for otherwise it could be extended further by our lemma and would not be maximal.

As we stated in the introduction to this chapter, the main force of the Hahn-Banach theorem lies in the guarantee it provides that any Banach space (or normed linear space) has a rich supply of functionals. This property is to be understood in the sense of the following two theorems, on which most of its applications depend.

Theorem B. *If N is a normed linear space and x_0 is a non-zero vector in N, then there exists a functional f_0 in N^* such that $f_0(x_0) = \|x_0\|$ and $\|f_0\| = 1$.*

PROOF. Let $M = \{\alpha x_0\}$ be the linear subspace of N spanned by x_0, and define f on M by $f(\alpha x_0) = \alpha \|x_0\|$. It is clear that f is a functional on M such that $f(x_0) = \|x_0\|$ and $\|f\| = 1$. By the Hahn-Banach theorem, f can be extended to a functional f_0 in N^* with the required properties.

Among other things, this result shows that N^* separates the vectors in N, for if x and y are any two distinct vectors, so that $x - y \neq 0$, then there exists a functional f in N^* such that $f(x - y) \neq 0$, or equivalently, $f(x) \neq f(y)$.

Theorem C. *If M is a closed linear subspace of a normed linear space N and x_0 is a vector not in M, then there exists a functional f_0 in N^* such that $f_0(M) = 0$ and $f_0(x_0) \neq 0$.*

PROOF. The natural mapping T of N onto N/M (see Problem 47-1) is a continuous linear transformation such that $T(M) = 0$ and

$$T(x_0) = x_0 + M \neq 0.$$

By Theorem B, there exists a functional f in $(N/M)^*$ such that

$$f(x_0 + M) \neq 0.$$

If we now define f_0 by $f_0(x) = f(T(x))$, then f_0 is easily seen to have the desired properties.

These theorems play a critical role in the ideas developed in the following sections, and their significance will emerge quite clearly in the proper context.

Problems

1. Let M be a closed linear subspace of a normed linear space N, and let x_0 be a vector not in M. If d is the distance from x_0 to M, show that there exists a functional f_0 in N^* such that $f_0(M) = 0, f_0(x_0) = 1$, and $\|f_0\| = 1/d$.

2. Prove that a normed linear space N is separable if its conjugate space N^* is. (*Hint:* let $\{f_n\}$ be a countable dense set in N^* and $\{x_n\}$ a corresponding set in N such that $\|x_n\| \leq 1$ and $|f_n(x_n)| \geq \|f_n\|/2$; let M be the set of all linear combinations of the x_n's whose coefficients are rational or—if N is complex—have rational real and imaginary parts; and use Theorem C to show that $\bar{M} = N$.) We remark that N^* need not be separable when N is, for l_1 is easily proved to be separable, $l_1^* = l_\infty$, and l_∞ is not separable (see Problem 18-4).

3. In this problem we ask the reader to convince himself of the validity of Eqs. (2) to (4). Let L be the linear space of all n-tuples

$$x = (x_1, x_2, \ldots, x_n)$$

of scalars. If $\{e_1, e_2, \ldots, e_n\}$ is the natural basis described in Problem 47-6, then $x = x_1e_1 + x_2e_2 + \cdots + x_ne_n$; and if f is any scalar-valued linear function defined on L, then the equation $f(x) = x_1 f(e_1) + x_2 f(e_2) + \cdots + x_n f(e_n)$ shows that f determines, and is determined by, the n scalars $y_i = f(e_i)$. The mapping

$$y = (y_1, y_2, \ldots, y_n) \to f,$$

where $f(x) = \Sigma_{i=1}^n x_i y_i$, is clearly an isomorphism of L onto the linear space L' of all f's. When the space L of all x's is made into l_p^n ($1 \le p \le \infty$) by suitably defining its norm, then by Problem 47-6 the space L' of all f's equals its conjugate space $(l_p^n)^*$, where the norm of f is understood to be given by

$$\|f\| = \inf \{K : K \ge 0 \text{ and } |f(x)| \le K\|x\| \text{ for all } x\}.$$

All that remains is to see what norm for the y's makes the mapping $y \to f$ an isometric isomorphism.

(a) If $1 < p < \infty$, then $(l_p^n)^* = l_q^n$. The norm in this case is defined by $\|x\| = (\Sigma_{i=1}^n |x_i|^p)^{1/p}$, and it follows from

$$|f(x)| = \left| \sum_{i=1}^n x_i y_i \right| \le \sum_{i=1}^n |x_i y_i| \le \left(\sum_{i=1}^n |x_i|^p \right)^{1/p} \left(\sum_{i=1}^n |y_i|^q \right)^{1/q}$$

that $\|f\| \le (\Sigma_{i=1}^n |y_i|^q)^{1/q}$. Show that $\|f\| = (\Sigma_{i=1}^n |y_i|^q)^{1/q}$ by considering the vector x defined by $x_i = 0$ if $y_i = 0$ and

$$x_i = |y_i|^q / y_i$$

otherwise.

(b) $(l_1^n)^* = l_\infty^n$. Here we have $\|x\| = \Sigma_{i=1}^n |x_i|$, and it follows from $|f(x)| = |\Sigma_{i=1}^n x_i y_i| \le \Sigma_{i=1}^n |x_i| \, |y_i| \le \max \{|y_i|\} (\Sigma_{i=1}^n |x_i|)$ that $\|f\| \le \max \{|y_i|\}$. Show that $\|f\| = \max \{|y_i|\}$ by considering the vector x defined by $x_i = |y_i|/y_i$ if $|y_i| = \max \{|y_i|\}$ and $x_i = 0$ otherwise.

(c) $(l_\infty^n)^* = l_1^n$. In this case, the norm is defined by

$$\|x\| = \max \{|x_i|\},$$

and it follows from

$$|f(x)| = \left| \sum_{i=1}^n x_i y_i \right| \le \sum_{i=1}^n |x_i| \, |y_i| \le \max \{|x_i|\} \left(\sum_{i=1}^n |y_i| \right)$$

that $\|f\| \le \Sigma_{i=1}^n |y_i|$. Show that $\|f\| = \Sigma_{i=1}^n |y_i|$ by considering the vector x defined by $x_i = 0$ if $y_i = 0$ and $x_i = |y_i|/y_i$ otherwise.

4. The following generalized form of part of the Hahn-Banach theorem is useful in certain problems of measure theory. Prove it by suitably modifying the arguments given in the text. If p is a real function

defined on a real linear space L such that $p(\alpha x) = \alpha p(x)$ for $\alpha \geq 0$ and $p(x + y) \leq p(x) + p(y)$, and if f is a real linear function defined on a linear subspace M such that $f(x) \leq p(x)$ for all x in M, then f can be extended to a real linear function f_0 defined on L such that $f_0(x) \leq p(x)$ for all x in L.

49. THE NATURAL IMBEDDING OF N IN N^{**}

Since the conjugate space N^* of a normed linear space N is itself a normed linear space, it is possible to form the conjugate space $(N^*)^*$ of N^*. We denote this space by N^{**}, and we call it the *second conjugate space* of N.

The importance of N^{**} rests on the fact that each vector x in N gives rise to a functional F_x in N^{**}. If we denote a typical element of N^* by f, then F_x is defined by

$$F_x(f) = f(x).$$

In other words, we invert the usual practice by regarding the symbol $f(x)$ as specifying a function of f for each fixed x, and we emphasize this point of view by writing $f(x)$ in the form $F_x(f)$. A simple manipulation of the definition shows that F_x is linear:

$$\begin{aligned}
F_x(\alpha f + \beta g) &= (\alpha f + \beta g)(x) \\
&= \alpha f(x) + \beta g(x) \\
&= \alpha F_x(f) + \beta F_x(g).
\end{aligned}$$

If we now compute the norm of F_x, we see that

$$\begin{aligned}
\|F_x\| &= \sup \{|F_x(f)| : \|f\| \leq 1\} \\
&= \sup \{|f(x)| : \|f\| \leq 1\} \\
&\leq \sup \{\|f\| \, \|x\| : \|f\| \leq 1\} \\
&\leq \|x\|.
\end{aligned}$$

Theorem 48-B is exactly what is needed to guarantee that equality holds here, so for each x in N we have

$$\|F_x\| = \|x\|.$$

It follows from these observations that $x \to F_x$ is a norm-preserving mapping of N into N^{**}. F_x is called the functional on N^* *induced by* the vector x, and we refer to functionals of this kind as *induced functionals*. We next point out that the mapping $x \to F_x$ is linear and is therefore an isometric isomorphism of N into N^{**}. To verify this, we must show that $F_{x+y}(f) = (F_x + F_y)(f)$ and $F_{\alpha x}(f) = (\alpha F_x)(f)$ for every f in N^*. The first of these relations follows from

$$F_{x+y}(f) = f(x + y)$$
$$= f(x) + f(y)$$
$$= F_x(f) + F_y(f)$$
$$= (F_x + F_y)(f),$$

and the second is proved similarly. The isometric isomorphism $x \to F_x$ is called the *natural imbedding* of N in N^{**}, for it allows us to regard N as part of N^{**} without altering any of its structure as a normed linear space. We write

$$N \subseteq N^{**},$$

where this set inclusion is to be understood in the sense just explained.

A normed linear space N is said to be *reflexive* if $N = N^{**}$. The spaces l_p (and L_p) for $1 < p < \infty$ are reflexive, for $l_p^* = l_q$ and

$$l_p^{**} = l_q^* = l_p.$$

It follows from Problem 48-3 that the spaces l_p^n for $1 \leq p \leq \infty$ are also reflexive. Since N^{**} is complete, N is necessarily complete if it is reflexive. If N is complete, however, it is not necessarily reflexive, as we see from $c_0^* = l_1$ and $c_0^{**} = l_1^* = l_\infty$. If X is a compact Hausdorff space, it can be shown that $\mathcal{C}(X)$ is reflexive $\Leftrightarrow X$ is a finite set.

There is an interesting criterion for reflexivity, which depends on the concept of the *weak topology* on a normed linear space N. This is defined to be the weak topology on N generated by the functions in N^* in the sense of Sec. 19; that is, it is the weakest topology on N with respect to which all the functions in N^* remain continuous. The criterion referred to is the following: if B is a Banach space, and if $S = \{x : \|x\| \leq 1\}$ is its closed unit sphere, then B is reflexive $\Leftrightarrow S$ is compact in the weak topology. This fact is something one should know about Banach spaces, but we shall have no need for it ourselves, so we state it without proof.[1]

Far more important for our purposes is the *weak* topology* on N^*, which is defined to be the weak topology on N^* generated by all the induced functionals F_x in N^{**}. This situation is rather complicated, so we shall try to make clear just what is going on.

First of all, N^* (like N) is a normed linear space, and it therefore has a topology derived from its character as a metric space. This is called the *strong topology*. N^{**} is the set of all scalar-valued linear functions defined on N^* which are continuous with respect to its strong topology. The *weak topology* on N^* (like the weak topology on N) is the weakest topology on N^* with respect to which all the functions in N^{**} are continuous, and clearly this is weaker than its strong topology. So far, as

[1] See Hille and Phillips [20, p. 38] or Dunford and Schwartz [8, p. 425].

we have indicated, these concepts apply equally to N and N^*. However, since N^* is the conjugate space of N, the natural imbedding enables us to consider N as part of N^{**}. We now form the weakest topology on N^* with respect to which all the functions in N—regarded as a subset of N^{**}—remain continuous. This is the *weak* topology*, and it is evidently weaker than the weak topology. The weak* topology can be given a more explicit description, in which its defining subbasic open sets are displayed. Consider a vector x in N and its induced functional F_x in N^{**}. The weak* topology on N^* is the weakest topology under which all such F_x's are continuous. If f_0 is an arbitrary element in N^*, and if $\epsilon > 0$ is given, then the set

$$S(x, f_0, \epsilon) = \{f : f \in N^* \text{ and } |F_x(f) - F_x(f_0)| < \epsilon\}$$
$$= \{f : f \in N^* \text{ and } |f(x) - f_0(x)| < \epsilon\}$$

is an open set (in fact, a neighborhood of f_0) in the weak* topology. Furthermore, the class of all sets of this kind, for all x's, f_0's, and ϵ's, is the defining open subbase for the weak* topology. All finite intersections of these sets constitute an open base for this topology, and the open sets themselves are all unions of these finite intersections.

We remark at this point that N^* is a Hausdorff space with respect to its weak* topology. This follows at once from the fact that if f and g are distinct functionals in N^*, then there must exist a vector x in N such that $f(x) \neq g(x)$; for if we put $\epsilon = |f(x) - g(x)|/3$, then $S(x, f, \epsilon)$ and $S(x, g, \epsilon)$ are disjoint neighborhoods of f and g in the weak* topology.

Let us now consider the closed unit sphere S^* in N^*, that is, the set $S^* = \{f : f \in N^* \text{ and } \|f\| \leq 1\}$.[1] It is an easy consequence of Problem 2 that S^* is compact in the strong topology $\Leftrightarrow N$ is finite-dimensional, so the strong compactness of S^* is a very stringent condition. If N is complete, it follows from Problem 3 and our unproved criterion for reflexivity that S^* is compact in the weak topology $\Leftrightarrow N$ is reflexive, so the weak compactness of S^* is still a fairly substantial restriction. We state these facts to emphasize that the situation is quite different with the weak* topology, for here S^* is always compact.

Theorem A. *If N is a normed linear space, then the closed unit sphere S^* in N^* is a compact Hausdorff space in the weak* topology.*

PROOF. We already know that S^* is a Hausdorff space in this topology, so we confine our attention to proving compactness. With each vector x in N we associate a compact space C_x, where C_x is the closed interval $[-\|x\|, \|x\|]$ or the closed disc $\{z : |z| \leq \|x\|\}$, according as N is real or complex. By Tychonoff's theorem, the product C of all the C_x's is

---

[1] When we use the adjective "closed" in referring to S^*, we intend only to emphasize the inequality $\|f\| \leq 1$, as contrasted with $\|f\| < 1$.

also a compact space. For each x, the values $f(x)$ of all f's in S^* lie in C_x. This enables us to imbed S^* in C by regarding each f in S^* as identical with the array of all its values at the vectors x in N. It is clear from the definitions of the topologies concerned that the weak* topology on S^* equals its topology as a subspace of C; and since C is compact, it suffices to show that S^* is closed as a subspace of C. We show that if g is in $\overline{S^*}$, then g is in S^*. If we consider g to be a function defined on the index set N, then since g is in C we have $|g(x)| \leq \|x\|$ for every x in N. It therefore suffices to show that g is linear as a function defined on N. Let $\epsilon > 0$ be given, and let x and y be any two vectors in N. Every basic neighborhood of g intersects S^*, so there exists an f in S^* such that $|g(x) - f(x)| < \epsilon/3$, $|g(y) - f(y)| < \epsilon/3$, and $|g(x + y) - f(x + y)| < \epsilon/3$. Since f is linear, $f(x + y) - f(x) - f(y) = 0$, and we therefore have

$$|g(x + y) - g(x) - g(y)| = |[g(x + y) - f(x + y)] - [g(x) - f(x)]$$
$$- [g(y) - f(y)]| \leq |g(x + y) - f(x + y)| + |g(x) - f(x)|$$
$$+ |g(y) - f(y)| < \epsilon/3 + \epsilon/3 + \epsilon/3 = \epsilon.$$

The fact that this inequality is true for every $\epsilon > 0$ now implies that $g(x + y) = g(x) + g(y)$. We can show in the same way that

$$g(\alpha x) = \alpha g(x)$$

for every scalar α, so g is linear and the theorem is proved.

We are now in a position to keep the promise made in the last paragraph of Sec. 46, for the following result is an obvious consequence of our preceding work.

Theorem B. *Let N be a normed linear space, and let S^* be the compact Hausdorff space obtained by imposing the weak* topology on the closed unit sphere in N^*. Then the mapping $x \to F_x$, where $F_x(f) = f(x)$ for each f in S^*, is an isometric isomorphism of N into $\mathcal{C}(S^*)$. If N is a Banach space, this mapping is an isometric isomorphism of N onto a closed linear subspace of $\mathcal{C}(S^*)$.*

This theorem shows, in effect, that the most general Banach space is essentially a closed linear subspace of $\mathcal{C}(X)$, where X is a compact Hausdorff space. The purpose of representation theorems in abstract mathematics is to reveal the structures of complex systems in terms of simpler ones, and from this point of view, Theorem B is satisfying to a degree. It must be pointed out, however, that we know next to nothing about the closed linear subspaces of $\mathcal{C}(X)$, though we know a good deal about $\mathcal{C}(X)$ itself. Theorem B is therefore somewhat less revealing than appears at first glance. We shall see in Chaps. 13 and 14 that the corresponding representation theorem for Banach algebras is much more significant and useful.

Problems

1. Let X be a compact Hausdorff space, and justify the assertion that $\mathcal{C}(X)$ is reflexive if X is finite.
2. If N is a finite-dimensional normed linear space of dimension n, show that N^* also has dimension n. Use this to prove that N is reflexive.
3. If B is a Banach space, prove that B is reflexive $\Leftrightarrow B^*$ is reflexive.
4. Prove that if B is a reflexive Banach space, then its closed unit sphere S is weakly compact.
5. Show that a linear subspace of a normed linear space is closed $\Leftrightarrow$ it is weakly closed.

50. THE OPEN MAPPING THEOREM

In this section we have our first encounter with basic theorems which require that the spaces concerned be complete. The following rather technical lemma is the key to these theorems.

Lemma. *If B and B' are Banach spaces, and if T is a continuous linear transformation of B onto B', then the image of each open sphere centered on the origin in B contains an open sphere centered on the origin in B'.*

PROOF. We denote by S_r and S'_r the open spheres with radius r centered on the origin in B and B'. It is easy to see that

$$T(S_r) = T(rS_1) = rT(S_1),$$

so it suffices to show that $T(S_1)$ contains some S'_r.

We begin by proving that $\overline{T(S_1)}$ contains some S'_r. Since T is onto, we see that $B' = \bigcup_{n=1}^{\infty} T(S_n)$. B' is complete, so Baire's theorem implies that some $\overline{T(S_{n_0})}$ has an interior point y_0, which may be assumed to lie in $T(S_{n_0})$. The mapping $y \to y - y_0$ is a homeomorphism of B' onto itself, so $\overline{T(S_{n_0})} - y_0$ has the origin as an interior point. Since y_0 is in $T(S_{n_0})$, we have $T(S_{n_0}) - y_0 \subseteq T(S_{2n_0})$; and from this we obtain $\overline{T(S_{n_0})} - y_0 = \overline{T(S_{n_0}) - y_0} \subseteq \overline{T(S_{2n_0})}$, which shows that the origin is an interior point of $\overline{T(S_{2n_0})}$. Multiplication by any non-zero scalar is a homeomorphism of B' onto itself, so $\overline{T(S_{2n_0})} = \overline{2n_0 T(S_1)} = 2n_0\overline{T(S_1)}$; and it follows from this that the origin is also an interior point of $\overline{T(S_1)}$, so $S'_\epsilon \subseteq \overline{T(S_1)}$ for some positive number ϵ.

We conclude the proof by showing that $S'_\epsilon \subseteq T(S_3)$, which is clearly equivalent to $S'_{\epsilon/3} \subseteq T(S_1)$. Let y be a vector in B' such that $\|y\| < \epsilon$.

Since y is in $\overline{T(S_1)}$, there exists a vector x_1 in B such that $\|x_1\| < 1$ and $\|y - y_1\| < \epsilon/2$, where $y_1 = T(x_1)$. We next observe that $S'_{\epsilon/2} \subseteq \overline{T(S_{\frac{1}{2}})}$, so there exists a vector x_2 in B such that $\|x_2\| < \frac{1}{2}$ and $\|(y - y_1) - y_2\| < \epsilon/4$, where $y_2 = T(x_2)$. Continuing in this way, we obtain a sequence $\{x_n\}$ in B such that $\|x_n\| < 1/2^{n-1}$ and $\|y - (y_1 + y_2 + \cdots + y_n)\| < \epsilon/2^n$, where $y_n = T(x_n)$. If we put

$$s_n = x_1 + x_2 + \cdots + x_n,$$

then it follows from $\|x_n\| < 1/2^{n-1}$ that $\{s_n\}$ is a Cauchy sequence in B for which

$$\|s_n\| \leq \|x_1\| + \|x_2\| + \cdots + \|x_n\| < 1 + \frac{1}{2} + \cdots + 1/2^{n-1} < 2.$$

B is complete, so there exists a vector x in B such that $s_n \to x$; and $\|x\| = \|\lim s_n\| = \lim \|s_n\| \leq 2 < 3$ shows that x is in S_3. All that remains is to notice that the continuity of T yields

$$T(x) = T(\lim s_n) = \lim T(s_n) = \lim (y_1 + y_2 + \cdots + y_n) = y,$$

from which we see that y is in $T(S_3)$.

This makes our main theorem easy to prove.

Theorem A (the Open Mapping Theorem). *If B and B' are Banach spaces, and if T is a continuous linear transformation of B onto B', then T is an open mapping.*

PROOF. We must show that if G is an open set in B, then $T(G)$ is also an open set in B'. If y is a point in $T(G)$, it suffices to produce an open sphere centered on y and contained in $T(G)$. Let x be a point in G such that $T(x) = y$. Since G is open, x is the center of an open sphere—which can be written in the form $x + S_r$—contained in G. Our lemma now implies that $T(S_r)$ contains some S'_{r_1}. It is clear that $y + S'_{r_1}$ is an open sphere centered on y, and the fact that it is contained in $T(G)$ follows at once from $y + S'_{r_1} \subseteq y + T(S_r) = T(x) + T(S_r) = T(x + S_r) \subseteq T(G)$.

Most of the applications of the open mapping theorem depend **more** directly on the following special case, which we state separately **for the** sake of emphasis.

Theorem B. *A one-to-one continuous linear transformation of one Banach space onto another is a homeomorphism. In particular, if a one-to-one linear transformation T of a Banach space onto itself is continuous, then its inverse T^{-1} is automatically continuous.*

As our first application of Theorem B, we give a geometric characterization of the projections on a Banach space. The reader will recall from Sec. 44 that a projection E on a linear space L is simply an idempotent ($E^2 = E$) linear transformation of L into itself. He will also recall that projections on L can be described geometrically as follows:

(1)   a projection E determines a pair of linear subspaces M and N such that $L = M \oplus N$, where $M = \{E(x) : x \in L\}$ and $N = \{x : E(x) = 0\}$ are the range and null space of E;

(2)   a pair of linear subspaces M and N such that $L = M \oplus N$ determines a projection E whose range and null space are M and N (if $z = x + y$ is the unique representation of a vector in L as a sum of vectors in M and N, then E is defined by $E(z) = x$).

These facts show that the study of projections on L is equivalent to the study of pairs of linear subspaces which are disjoint and span L.

In the theory of Banach spaces, however, more is required of a projection than mere linearity and idempotence. A *projection* on a Banach space B is an idempotent operator on B; that is, it is a projection on B in the algebraic sense which is also continuous. Our present task is to assess the effect of the additional requirement of continuity on the geometric descriptions given in (1) and (2) above. The analogue of (1) is easy.

Theorem C.   *If P is a projection on a Banach space B, and if M and N are its range and null space, then M and N are closed linear subspaces of B such that $B = M \oplus N$.*

PROOF. P is an algebraic projection, so (1) gives everything except the fact that M and N are closed. The null space of any continuous linear transformation is closed, so N is obviously closed; and the fact that M is also closed is a consequence of

$$M = \{P(x) : x \in B\} = \{x : P(x) = x\} = \{x : (I - P)(x) = 0\},$$

which exhibits M as the null space of the operator $I - P$.

The analogue of (2) is more difficult, for Theorem B is needed in its proof.

Theorem D.   *Let B be a Banach space, and let M and N be closed linear subspaces of B such that $B = M \oplus N$. If $z = x + y$ is the unique representation of a vector in B as a sum of vectors in M and N, then the mapping P defined by $P(z) = x$ is a projection on B whose range and null space are M and N.*

PROOF.   Everything stated is clear from (2) except the fact that P is continuous, and this we prove as follows. By Problem 46-2, if B' denotes the linear space B equipped with the norm defined by

$$\|z\|' = \|x\| + \|y\|,$$

then B' is a Banach space; and since $\|P(z)\| = \|x\| \leq \|x\| + \|y\| = \|z\|'$, P is clearly continuous as a mapping of B' into B. It therefore suffices to prove that B' and B have the same topology. If T denotes the identity mapping of B' onto B, then

$$\|T(z)\| = \|z\| = \|x + y\| \leq \|x\| + \|y\| = \|z\|'$$

shows that T is continuous as a one-to-one linear transformation of B' onto B. Theorem B now implies that T is a homeomorphism, and the proof is complete.

This theorem raises some interesting and significant questions. Let M be a closed linear subspace of a Banach space B. As we remarked at the end of Sec. 44, there is always at least one algebraic projection defined on B whose range is M, and there may be a great many. However, it might well happen that none of these are continuous, and that consequently none are projections in our present sense. In the light of our theorems, this is equivalent to saying that there might not exist any closed linear subspace N such that $B = M \oplus N$. What sorts of Banach spaces have the property that this awkward situation cannot occur? We shall see in the next chapter that a Hilbert space—which is a special type of Banach space—has this property. We shall also see that this property is closely linked to the satisfying geometric structure which sets Hilbert spaces apart from general Banach spaces.

We now turn to the closed graph theorem. Let B and B' be Banach spaces. If we define a metric on the product $B \times B'$ by

$$d((x_1, y_1), (x_2, y_2)) = \max \{\|x_1 - x_2\|, \|y_1 - y_2\|\},$$

then the resulting topology is easily seen to be the same as the product topology, and convergence with respect to this metric is equivalent to coordinatewise convergence. Now let T be a linear transformation of B into B'. We recall that the graph of T is that subset of $B \times B'$ which consists of all ordered pairs of the form $(x, T(x))$. Problem 26-6 shows that if T is continuous, then its graph is closed as a subset of $B \times B'$. In the present context, the converse is also true.

Theorem E (the Closed Graph Theorem). *If B and B' are Banach spaces, and if T is a linear transformation of B into B', then T is continuous $\Leftrightarrow$ its graph is closed.*

PROOF. In view of the above remarks, we may confine our attention to proving that T is continuous if its graph is closed. We denote by B_1 the linear space B renormed by $\|x\|_1 = \|x\| + \|T(x)\|$. Since

$$\|T(x)\| \leq \|x\| + \|T(x)\| = \|x\|_1,$$

T is continuous as a mapping of B_1 into B'. It therefore suffices to show that B and B_1 have the same topology. The identity mapping of B_1 onto B is clearly continuous, for $\|x\| \leq \|x\| + \|T(x)\| = \|x\|_1$. If we can show that B_1 is complete, then Theorem B will guarantee that this mapping is a homeomorphism, and this will conclude the proof. Let $\{x_n\}$ be a Cauchy sequence in B_1. It follows that $\{x_n\}$ and $\{T(x_n)\}$ are also Cauchy sequences in B and B'; and since both of these spaces are complete, there exist vectors x and y in B and B' such that $\|x_n - x\| \to 0$ and $\|T(x_n) - y\| \to 0$. Our assumption that the graph of T is closed in $B \times B'$ implies that (x,y) lies on this graph, so $T(x) = y$. The completeness of B_1 now follows from

$$\|x_n - x\|_1 = \|x_n - x\| + \|T(x_n - x)\| = \|x_n - x\| + \|T(x_n) - T(x)\|$$
$$= \|x_n - x\| + \|T(x_n) - y\| \to 0.$$

The closed graph theorem has a number of interesting applications to problems in analysis, but since our concern here is mainly with matters of algebra and topology, we do not pause to illustrate its uses in this direction.[1]

Problems

1. Let a Banach space B be made into a Banach space B' by means of a new norm, and show that the topologies generated by these norms are the same if either is stronger than the other.
2. In the text, we used Theorem B to prove the closed graph theorem. Show that Theorem B is a consequence of the closed graph theorem.
3. Let T be a linear transformation of a Banach space B into a Banach space B'. If $\{f_i\}$ is a set of functionals in B'^* which separates the vectors in B', and if f_iT is continuous for each f_i, prove that T is continuous.

51. THE CONJUGATE OF AN OPERATOR

We shall see in this section that each operator T on a normed linear space N induces a corresponding operator, denoted by T^* and called the *conjugate* of T, on the conjugate space N^*. Our first task is to define T^*, and our second is to investigate the properties of the mapping $T \to T^*$. We base our discussion on the following theorem.

Theorem A (the Uniform Boundedness Theorem). *Let B be a Banach space and N a normed linear space. If $\{T_i\}$ is a non-empty set of con-*

[1] See Taylor [41, pp. 181–185].

tinuous linear transformations of B into N with the property that $\{T_i(x)\}$ is a bounded subset of N for each vector x in B, then $\{\|T_i\|\}$ is a bounded set of numbers; that is, $\{T_i\}$ is bounded as a subset of $\mathcal{B}(B,N)$.

PROOF.   For each positive integer n, the set

$$F_n = \{x : x \in B \text{ and } \|T_i(x)\| \leq n \text{ for all } i\}$$

is clearly a closed subset of B, and by our assumption we have

$$B = \bigcup_{n=1}^{\infty} F_n.$$

Since B is complete, Baire's theorem shows that one of the F_n's, say F_{n_0}, has non-empty interior, and thus contains a closed sphere S_0 with center x_0 and radius $r_0 > 0$. This says, in effect, that each vector in every set $T_i(S_0)$ has norm less than or equal to n_0; and for the sake of brevity, we express this fact by writing $\|T_i(S_0)\| \leq n_0$. It is clear that $S_0 - x_0$ is the closed sphere with radius r_0 centered on the origin, so $(S_0 - x_0)/r_0$ is the closed unit sphere S. Since x_0 is in S_0, it is evident that $\|T_i(S_0 - x_0)\| \leq 2n_0$. This yields $\|T_i(S)\| \leq 2n_0/r_0$, so $\|T_i\| \leq 2n_0/r_0$ for every i, and the proof is complete.

This theorem is often called the *Banach-Steinhaus theorem*, and it has several significant applications to analysis. See, for example, Zygmund [46, vol. 1, pp. 165–168] or Gál [11]. For the purposes we have in view, our main interest is in the following simple consequence of it.

Theorem B.   *A non-empty subset X of a normed linear space N is bounded $\Leftrightarrow f(X)$ is a bounded set of numbers for each f in N^*.*

PROOF.   Since $|f(x)| \leq \|f\| \|x\|$, it is obvious that if X is bounded, then $f(X)$ is also bounded for each f.

In order to prove the other half of the theorem, it is convenient to exhibit the vectors in X by writing $X = \{x_i\}$. We now use the natural imbedding to pass from X to the corresponding subset $\{F_{x_i}\}$ of N^{**}. Our assumption that $f(X) = \{f(x_i)\}$ is bounded for each f is clearly equivalent to the assumption that $\{F_{x_i}(f)\}$ is bounded for each f, and since N^* is complete, Theorem A shows that $\{F_{x_i}\}$ is a bounded subset of N^{**}. We know that the natural imbedding preserves norms, so X is evidently a bounded subset of N.

We now turn to the problem of defining the conjugate of an operator on a normed linear space N.

Let L be the linear space of all scalar-valued linear functions defined on N. The conjugate space N^* is clearly a linear subspace of L. Let T be a linear transformation of N into itself which is not necessarily continuous. We use T to define a linear transformation T' of L into itself, as follows.   If f is in L, then $T'(f)$ is defined by

$$[T'(f)](x) = f(T(x)).\tag{1}$$

We leave it to the reader to verify that $T'(f)$ actually is linear as a function defined on N, and also that T' is linear as a mapping of L into itself.

The following natural question now presents itself. Under what circumstances does T' map N^* into N^*? This question has a simple and elegant answer: $T'(N^*) \subseteq N^* \Leftrightarrow T$ is continuous. If we keep Theorem B in mind, the proof of this statement is very easy; for if S is the closed unit sphere in N, then T is continuous $\Leftrightarrow T(S)$ is bounded $\Leftrightarrow f(T(S))$ is bounded for each f in $N^* \Leftrightarrow [T'(f)](S)$ is bounded for each f in $N^* \Leftrightarrow T'(f)$ is in N^* for each f in N^*.

We now assume that the linear transformation T is continuous and is therefore an operator on N. The preceding developments allow us to consider the restriction of T' to a mapping of N^* into itself.   We denote this restriction by T^*, and we call it the *conjugate* of T. The action of T^* is given by

$$[T^*(f)](x) = f(T(x)),\tag{2}$$

in which—in contrast to (1)—f is understood to be a functional on N, and not merely a scalar-valued linear function. T^* is clearly linear, and the following computation shows that it is continuous:

$$
\begin{aligned}
\|T^*\| &= \sup \{\|T^*(f)\| : \|f\| \le 1\}\\
&= \sup \{|[T^*(f)](x)| : \|f\| \text{ and } \|x\| \le 1\}\\
&= \sup \{|f(T(x))| : \|f\| \text{ and } \|x\| \le 1\}\\
&\le \sup \{\|f\|\, \|T\|\, \|x\| : \|f\| \text{ and } \|x\| \le 1\}\\
&\le \|T\|.
\end{aligned}
$$

Since $\|T\| = \sup \{\|T(x)\| : \|x\| \le 1\}$, we see at once from Theorem 48-B that equality holds here, that is, that

$$\|T^*\| = \|T\|.\tag{3}$$

The mapping $T \to T^*$ is thus a norm-preserving mapping of $\mathcal{B}(N)$ into $\mathcal{B}(N^*)$.

We continue in this vein by observing that the mapping $T \to T^*$ also has the following pleasant algebraic properties:

$$(\alpha T_1 + \beta T_2)^* = \alpha T_1{}^* + \beta T_2{}^*,\tag{4}$$
$$(T_1 T_2)^* = T_2{}^* T_1{}^*,\tag{5}$$

and
$$I^* = I.\tag{6}$$

The proofs of these facts are easy consequences of the definitions.   We illustrate the principles involved by proving (5). It must be shown that

$(T_1T_2)^*(f) = (T_2{}^*T_1{}^*)(f)$ for each f in N^*, and this means that

$$[(T_1T_2)^*(f)](x) = [(T_2{}^*T_1{}^*)(f)](x)$$

for each f in N^* and each x in N. A simple computation now shows that

$$[(T_1T_2)^*(f)](x) = f((T_1T_2)(x)) = f(T_1(T_2(x))) = [T_1{}^*(f)](T_2(x))$$
$$= [T_2{}^*(T_1{}^*(f))](x) = [(T_2{}^*T_1{}^*)(f)](x).$$

It may be helpful to the reader to have the following summary of the results of this discussion.

Theorem C. *If T is an operator on a normed linear space N, then its conjugate T^* defined by Eq. (2) is an operator on N^*, and the mapping $T \to T^*$ is an isometric isomorphism of $\mathfrak{B}(N)$ into $\mathfrak{B}(N^*)$ which reverses products and preserves the identity transformation.*

The general significance of the ideas developed here can be understood only in the light of the theory of operators on Hilbert spaces. Some preliminary comments on these matters are given in the introduction to the next chapter.

Problems

1. Let B be a Banach space and N a normed linear space. If $\{T_n\}$ is a sequence in $\mathfrak{B}(B,N)$ such that $T(x) = \lim T_n(x)$ exists for each x in B, prove that T is a continuous linear transformation.

2. Let T be an operator on a normed linear space N. If N is considered to be part of N^{**} by means of the natural imbedding, show that T^{**} is an extension of T. Observe that if N is reflexive, then $T^{**} = T$.

3. Let T be an operator on a Banach space B. Show that T has an inverse $T^{-1} \Leftrightarrow T^*$ has an inverse $(T^*)^{-1}$, and that in this case $(T^*)^{-1} = (T^{-1})^*$.

CHAPTER TEN

Hilbert Spaces

One of the principal applications of the theory of Banach algebras developed in Chaps. 12 to 14 is to the study of operators on Hilbert spaces. Our purpose in this chapter is to present enough of the elementary theory of Hilbert spaces and their operators to provide an adequate foundation for the deeper theory discussed in these later chapters.

We shall see from the formal definition given in the next section that a Hilbert space is a special type of Banach space, one which possesses additional structure enabling us to tell when two vectors are orthogonal (or perpendicular). The first part of the chapter is concerned solely with the geometric implications of this additional structure.

As we have said before, the objects of greatest interest in connection with any linear space are the linear transformations on that space. In our treatment of Banach spaces, we took advantage of the metric structure of such a space by focusing our attention on its operators. A Banach space, however, is still a bit too general to yield a really rich theory of operators. One fact that did emerge, which is of great significance for our present work, is that with each operator T on a Banach space B there is associated an operator T^* (its conjugate) on the conjugate space B^*. We shall see below that one of the central properties of a Hilbert space H is that there is a natural correspondence between H and its conjugate space H^*. If T is an operator on H, this correspondence makes it possible to regard the conjugate T^* as acting on H itself (instead of H^*), where it can be compared with T. These ideas lead to the concept of the adjoint of an operator on a Hilbert space, and they make it easy to understand the importance of operators (such as self-adjoint and normal operators) which are related in simple ways to their adjoints.

52. THE DEFINITION AND SOME SIMPLE PROPERTIES

The Banach spaces studied in the previous chapter are little more than linear spaces provided with a reasonable notion of the length of a vector. The main geometric concept missing in an abstract space of this type is that of the angle between two vectors. The theory of Hilbert spaces does not hinge on angles in general, but rather on some means of telling when two vectors are orthogonal.

In order to see how to introduce this concept, we begin by considering the three-dimensional Euclidean space R^3. A vector in R^3 is of course an ordered triple $x = (x_1, x_2, x_3)$ of real numbers, and its norm is defined by

$$\|x\| = (|x_1|^2 + |x_2|^2 + |x_3|^2)^{1/2}.$$

In elementary vector algebra, the inner product of x and another vector $y = (y_1, y_2, y_3)$ is defined by

$$(x,y) = x_1 y_1 + x_2 y_2 + x_3 y_3,[1]$$

and this inner product is related to the norm by

$$(x,x) = \|x\|^2.$$

We assume that the reader is familiar with the equation

$$(x,y) = \|x\| \, \|y\| \cos \theta,$$

where θ is the angle between x and y, and also with the fact that x and y are orthogonal precisely when $(x,y) = 0$.

Most of these ideas can readily be adapted to the three-dimensional unitary space C^3. For any two vectors $x = (x_1, x_2, x_3)$ and $y = (y_1, y_2, y_3)$ in this space, we define their inner product by

$$(x,y) = x_1 \overline{y_1} + x_2 \overline{y_2} + x_3 \overline{y_3}. \tag{1}$$

Complex conjugates are introduced here to guarantee that the relation

$$(x,x) = \|x\|^2$$

remains true. It is clear that the inner product defined by (1) is linear as a function of x for each fixed y, and is also conjugate-symmetric, in the

---

[1] The term *dot product* and the notation $x \cdot y$ are used in most introductory treatments of vectors.

sense that $\overline{(x,y)} = (y,x)$. In this case, it is no longer possible to think of (x,y) as representing the product of the norms of x and y and the cosine of the angle between them, for (x,y) is in general a complex number. Nevertheless, if the condition $(x,y) = 0$ is taken as the definition of orthogonality, then this concept is just as useful here as it is in the real case.

With these ideas as a background, we are now in a position to give our basic definition. A *Hilbert space* is a complex Banach space whose norm arises from an *inner product*, that is, in which there is defined a complex function (x,y) of vectors x and y with the following properties:

(1) $(\alpha x + \beta y, z) = \alpha(x,z) + \beta(y,z)$;
(2) $\overline{(x,y)} = (y,x)$;
(3) $(x,x) = \|x\|^2$.

It is evident that the further relation

$$(x, \alpha y + \beta z) = \bar{\alpha}(x,y) + \bar{\beta}(x,z)$$

is a direct consequence of properties (1) and (2).

The reader may wonder why we restrict our attention to complex spaces. Why not consider real spaces as well? As a matter of fact, we could easily do so, and many writers adopt this approach. There are a few places in this chapter where complex scalars are necessary, but the theorems involved are not crucial, and we could get along with real scalars without too much difficulty. It is only in the complex case, however, that the theory of operators on a Hilbert space assumes a really satisfactory form. This will appear with particular clarity in the next chapter, where we make essential use of the fact that every polynomial equation of the nth degree with complex coefficients has exactly n complex roots (some of which, of course, may be repeated). For this and other reasons, we limit ourselves to the complex case throughout the rest of this book.

The following are the main examples of Hilbert spaces. In accordance with the above remarks, the scalars in each example are understood to be the complex numbers.

Example 1.   The space l_2^n, with the inner product of two vectors

$$x = (x_1, x_2, \ldots, x_n) \text{ and } y = (y_1, y_2, \ldots, y_n)$$

defined by

$$(x,y) = \sum_{i=1}^{n} x_i \overline{y_i}.$$

It is obvious that conditions (1) to (3) are satisfied.

Example 2. The space l_2, with the inner product of the vectors

$$x = \{x_1, x_2, \ldots, x_n, \ldots\} \text{ and } y = \{y_1, y_2, \ldots, y_n, \ldots\}$$

defined by

$$(x,y) = \sum_{n=1}^{\infty} x_n \overline{y_n}.$$

The fact that this series converges—and thus defines a complex number—for each x and y in l_2 is an easy consequence of Cauchy's inequality.

Example 3. The space L_2 associated with a measure space X with measure m, with the inner product of two functions f and g defined by

$$(f,g) = \int f(x)\overline{g(x)}\, dm(x).$$

This Hilbert space is of course not part of the official content of this book, but we mention it anyway in case the reader has some knowledge of these matters.

As our first theorem, we prove a fundamental relation known as the *Schwarz inequality.*

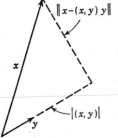

Fig. 36. Schwarz's inequality.

Theorem A. *If x and y are any two vectors in a Hilbert space, then $|(x,y)| \le \|x\|\,\|y\|$.*

PROOF. When $y = 0$, the result is clear, for both sides vanish. When $y \neq 0$, the inequality is equivalent to $|(x, y/\|y\|)| \le \|x\|$. We may therefore confine our attention to proving that if $\|y\| = 1$, then we have $|(x,y)| \le \|x\|$ for all x. This is a direct consequence of the fact that

$$0 \le \|x - (x,y)y\|^2 = (x - (x,y)y,\, x - (x,y)y)$$
$$= (x,x) - (x,y)\overline{(x,y)} - (x,y)\overline{(x,y)} + (x,y)\overline{(x,y)}(y,y)$$
$$= (x,x) - (x,y)\overline{(x,y)} = \|x\|^2 - |(x,y)|^2.$$

An inspection of Fig. 36 will reveal the geometric motivation for this computation.

It follows easily from Schwarz's inequality that the inner product in a Hilbert space is jointly continuous:

$$x_n \to x \text{ and } y_n \to y \Rightarrow (x_n, y_n) \to (x,y).$$

To prove this, it suffices to observe that

$$|(x_n,y_n) - (x,y)| = |(x_n,y_n) - (x_n,y) + (x_n,y) - (x,y)| \le |(x_n,y_n)$$
$$- (x_n,y)| + |(x_n,y) - (x,y)| = |(x_n, y_n - y)|$$
$$+ |(x_n - x, y)| \le \|x_n\|\,\|y_n - y\| + \|x_n - x\|\,\|y\|.$$

A well-known theorem of elementary geometry states that the sum of the squares of the sides of a parallelogram equals the sum of the squares of its diagonals. This fact has an analogue in the present context, for in any Hilbert space the so-called *parallelogram law* holds:

$$\|x + y\|^2 + \|x - y\|^2 = 2\|x\|^2 + 2\|y\|^2.$$

This is readily proved by writing out the expression on the left in terms of inner products:

$$\begin{aligned}
\|x + y\|^2 + \|x - y\|^2 &= (x + y,\, x + y) + (x - y,\, x - y)\\
&= (x,x) + (x,y) + (y,x) + (y,y) + (x,x) - (x,y) - (y,x) + (y,y)\\
&= 2(x,x) + 2(y,y) = 2\|x\|^2 + 2\|y\|^2.
\end{aligned}$$

The parallelogram law has the following important consequence for our work in the next section.

Theorem B. *A closed convex subset C of a Hilbert space H contains a unique vector of smallest norm.*

PROOF. We recall from the definition in Problem 32-5 that since C is convex, it is non-empty and contains $(x + y)/2$ whenever it contains x and y. Let $d = \inf\{\|x\| : x \in C\}$. There clearly exists a sequence $\{x_n\}$ of vectors in C such that $\|x_n\| \to d$. By the convexity of C, $(x_m + x_n)/2$ is in C and $\|(x_m + x_n)/2\| \geq d$, so $\|x_m + x_n\| \geq 2d$. Using the parallelogram law, we obtain

$$\begin{aligned}
\|x_m - x_n\|^2 &= 2\|x_m\|^2 + 2\|x_n\|^2 - \|x_m + x_n\|^2\\
&\leq 2\|x_m\|^2 + 2\|x_n\|^2 - 4d^2;
\end{aligned}$$

and since $2\|x_m\|^2 + 2\|x_n\|^2 - 4d^2 \to 2d^2 + 2d^2 - 4d^2 = 0$, it follows that $\{x_n\}$ is a Cauchy sequence in C. Since H is complete and C is closed, C is complete, and there exists a vector x in C such that $x_n \to x$. It is clear by the fact that $\|x\| = \|\lim x_n\| = \lim \|x_n\| = d$ that x is a vector in C with smallest norm. To see that x is unique, suppose that x' is a vector in C other than x which also has norm d. Then $(x + x')/2$ is also in C, and another application of the parallelogram law yields

$$\begin{aligned}
\left\|\frac{x + x'}{2}\right\|^2 &= \frac{\|x\|^2}{2} + \frac{\|x'\|^2}{2} - \left\|\frac{x - x'}{2}\right\|^2\\
&< \frac{\|x\|^2}{2} + \frac{\|x'\|^2}{2} = d^2,
\end{aligned}$$

which contradicts the definition of d.

The parallelogram law has another interesting application, which depends on the fact that in any Hilbert space the inner product is related to the norm by the following identity:

$$4(x,y) = \|x + y\|^2 - \|x - y\|^2 + i\|x + iy\|^2 - i\|x - iy\|^2. \quad (2)$$

This is easily verified by converting the expression on the right into inner products.

Theorem C. *If B is a complex Banach space whose norm obeys the parallelogram law, and if an inner product is defined on B by (2), then B is a Hilbert space.*

PROOF: All that is necessary is to make sure that the inner product defined by (2) has the three properties required by the definition of a Hilbert space.   This is easy in the case of properties (2) and (3).   Property (1) is best treated by splitting it into two parts:

$$(x + y, z) = (x,z) + (y,z),$$

and $(\alpha x,y) = \alpha(x,y)$.   The first requires the parallelogram law, and the second follows from the first.   We ask the reader (in Problem 6) to work out the details.

        This result has no implications at all for our future work.   However, it does provide a satisfying geometric insight into the place Hilbert spaces occupy among all complex Banach spaces: they are precisely those in which the parallelogram law is true.

Problems

1. Show that the series which defines the inner product in Example 2 is convergent.
2. The *Hilbert cube* is the subset of l_2 consisting of all sequences

$$x = \{x_1, x_2, \ldots, x_n, \ldots\}$$

such that $|x_n| \leq 1/n$ for all n.   Show that this set is compact as a subspace of l_2.
3. For the special Hilbert space l_2^n, use Cauchy's inequality to prove Schwarz's inequality.
4. Show that the parallelogram law is not true in l_1^n ($n > 1$).
5. In a Hilbert space, show that if $\|x\| = \|y\| = 1$, and if $\epsilon > 0$ is given, then there exists $\delta > 0$ such that $\|(x + y)/2\| > 1 - \delta \Rightarrow \|x - y\| < \epsilon$.   A Banach space with this property is said to be *uniformly convex*.   See Taylor [41, p. 231].
6. Give a detailed proof of Theorem C.

53. ORTHOGONAL COMPLEMENTS

Two vectors x and y in a Hilbert space H are said to be *orthogonal* (written $x \perp y$) if $(x,y) = 0$. The symbol $\perp$ is often pronounced "perp." Since $\overline{(x,y)} = (y,x)$, we have $x \perp y \Leftrightarrow y \perp x$. It is also clear that $x \perp 0$ for every x, and $(x,x) = \|x\|^2$ shows that 0 is the only vector orthogonal to itself. One of the simplest geometric facts about orthogonal vectors is the *Pythagorean theorem:*

$$x \perp y \Rightarrow \|x + y\|^2 = \|x - y\|^2 = \|x\|^2 + \|y\|^2.$$

A vector x is said to be orthogonal to a non-empty set S (written $x \perp S$) if $x \perp y$ for every y in S, and the *orthogonal complement* of S—denoted by $S^\perp$—is the set of all vectors orthogonal to S. The following statements are easy consequences of the definition:

$$\{0\}^\perp = H; H^\perp = \{0\};$$
$$S \cap S^\perp \subseteq \{0\};$$
$$S_1 \subseteq S_2 \Rightarrow S_1^\perp \supseteq S_2^\perp;$$
$$S^\perp \text{ is a closed linear subspace of } H.$$

It is customary to write $(S^\perp)^\perp$ in the form $S^{\perp\perp}$. Clearly, $S \subseteq S^{\perp\perp}$.

Let M be a closed linear subspace of H. We know that $M^\perp$ is also a closed linear subspace, and that M and $M^\perp$ are disjoint in the sense that they have only the zero vector in common. Our aim in this section is to prove that $H = M \oplus M^\perp$, and each of our theorems is a step in this direction.

Theorem A. *Let M be a closed linear subspace of a Hilbert space H, let x be a vector not in M, and let d be the distance from x to M. Then there exists a unique vector y_0 in M such that $\|x - y_0\| = d$.*

PROOF. The set $C = x + M$ is a closed convex set, and d is the distance from the origin to C (see Fig. 37). By Theorem 52-B, there exists a unique vector z_0 in C such that $\|z_0\| = d$. The vector $y_0 = x - z_0$ is easily seen to be in M, and $\|x - y_0\| = \|z_0\| = d$. The uniqueness of y_0 follows from the fact that if y_1 is a vector in M such that $y_1 \neq y_0$ and $\|x - y_1\| = d$, then $z_1 = x - y_1$ is a vector in C such that $z_1 \neq z_0$ and $\|z_1\| = d$, which contradicts the uniqueness of z_0.

We use this result to prove

Theorem B. *If M is a proper closed linear subspace of a Hilbert space H, then there exists a non-zero vector z_0 in H such that $z_0 \perp M$.*

PROOF. Let x be a vector not in M, and let d be the distance from x to M. By Theorem A, there exists a vector y_0 in M such that $\|x - y_0\| = d$.

We define z_0 by $z_0 = x - y_0$ (see Fig. 37), and we observe that since $d > 0$, z_0 is a non-zero vector. We conclude the proof by showing that if y is an arbitrary vector in M, then $z_0 \perp y$. For any scalar α, we have

$$\|z_0 - \alpha y\| = \|x - (y_0 + \alpha y)\| \geq d = \|z_0\|,$$

so
$$\|z_0 - \alpha y\|^2 - \|z_0\|^2 \geq 0$$

and
$$-\bar{\alpha}(z_0,y) - \alpha\overline{(z_0,y)} + |\alpha|^2\|y\|^2 \geq 0. \tag{1}$$

If we put $\alpha = \beta(z_0,y)$ for an arbitrary real number β, then (1) becomes

$$-2\beta|(z_0,y)|^2 + \beta^2|(z_0,y)|^2\|y\|^2 \geq 0.$$

If we now put $a = |(z_0,y)|^2$ and $b = \|y\|^2$, we obtain

$$-2\beta a + \beta^2 ab \geq 0,$$

so
$$\beta a(\beta b - 2) \geq 0 \tag{2}$$

for all real β. However, if $a > 0$, then (2) is obviously false for all sufficiently small positive β. We see from this that $a = 0$, which means that $z_0 \perp y$.

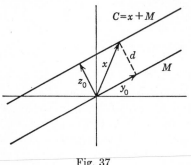

Fig. 37

This proof of Theorem B may strike the reader as being excessively dependent on ingenious computations. If so, he will be pleased to learn that the ideas developed in the next section can be used to provide another proof which is free of computation. In order to state our next theorem, we need the following additional concept. Two non-empty subsets S_1 and S_2 of a Hilbert space are said to be orthogonal (written $S_1 \perp S_2$) if $x \perp y$ for all x in S_1 and y in S_2.

Theorem C. *If M and N are closed linear subspaces of a Hilbert space H such that $M \perp N$, then the linear subspace $M + N$ is also closed.*

PROOF. Let z be a limit point of $M + N$. It suffices to show that z is in $M + N$. There certainly exists a sequence $\{z_n\}$ in $M + N$ such that $z_n \to z$. By the assumption that $M \perp N$, we see that M and N are disjoint, so each z_n can be written uniquely in the form $z_n = x_n + y_n$, where x_n is in M and y_n is in N. The Pythagorean theorem shows that $\|z_m - z_n\|^2 = \|x_m - x_n\|^2 + \|y_m - y_n\|^2$, so $\{x_n\}$ and $\{y_n\}$ are Cauchy sequences in M and N. M and N are closed, and therefore complete, so there exist vectors x and y in M and N such that $x_n \to x$ and $y_n \to y$. Since $x + y$ is in $M + N$, our conclusion follows from the fact that $z = \lim z_n = \lim (x_n + y_n) = \lim x_n + \lim y_n = x + y$.

The way is now clear for the proof of our principal theorem.

Theorem D. *If M is a closed linear subspace of a Hilbert space H, then $H = M \oplus M^\perp$.*

PROOF. Since M and $M^\perp$ are orthogonal closed linear subspaces of H, Theorem C shows that $M + M^\perp$ is also a closed linear subspace of H. We prove that $M + M^\perp$ equals H. If this is not so, then by Theorem B there exists a vector $z_0 \neq 0$ such that $z_0 \perp (M + M^\perp)$. This non-zero vector must evidently lie in $M^\perp \cap M^{\perp\perp}$; and since this is impossible, we infer that $H = M + M^\perp$. To conclude the proof, it suffices to observe that since M and $M^\perp$ are disjoint, the statement that $H = M + M^\perp$ can be strengthened to $H = M \oplus M^\perp$.

The main effect of this theorem is to guarantee that a Hilbert space is always rich in projections. In fact, if M is an arbitrary closed linear subspace of a Hilbert space H, then Theorem 50-D shows that there exists a projection defined on H whose range is M and whose null space is $M^\perp$. This satisfactory state of affairs is to be contrasted with the situation in a general Banach space, as explained in the remarks following Theorem 50-D.

Problems

1. If S is a non-empty subset of a Hilbert space, show that $S^\perp = S^{\perp\perp\perp}$.
2. If M is a linear subspace of a Hilbert space, show that M is closed $\Leftrightarrow M = M^{\perp\perp}$.
3. If S is a non-empty subset of a Hilbert space H, show that the set of all linear combinations of vectors in S is dense in $H \Leftrightarrow S^\perp = \{0\}$.
4. If S is a non-empty subset of a Hilbert space H, show that $S^{\perp\perp}$ is the closure of the set of all linear combinations of vectors in S. This is usually expressed by saying that $S^{\perp\perp}$ is the smallest closed linear subspace of H which contains S.

54. ORTHONORMAL SETS

An *orthonormal set* in a Hilbert space H is a non-empty subset of H which consists of mutually orthogonal unit vectors; that is, it is a non-empty subset $\{e_i\}$ of H with the following properties:

(1) $i \neq j \Rightarrow e_i \perp e_j$;
(2) $\|e_i\| = 1$ for every i.

If H contains only the zero vector, then it has no orthonormal sets. If H contains a non-zero vector x, and if we normalize x by considering $e = x/\|x\|$, then the single-element set $\{e\}$ is clearly an orthonormal set. More generally, if $\{x_i\}$ is a non-empty set of mutually orthogonal non-zero vectors in H, and if the x_i's are normalized by replacing each of them by $e_i = x_i/\|x_i\|$, then the resulting set $\{e_i\}$ is an orthonormal set.

Example 1.   The subset $\{e_1, e_2, \ldots, e_n\}$ of l_2^n, where e_i is the n-tuple with 1 in the ith place and 0's elsewhere, is evidently an orthonormal set in this space.

Example 2.   Similarly, if e_n is the sequence with 1 in the nth place and 0's elsewhere, then $\{e_1, e_2, \ldots, e_n, \ldots\}$ is an orthonormal set in l_2.

At the end of this section, we give some additional examples taken from the field of analysis.

Every aspect of the theory of orthonormal sets depends in one way or another on our first theorem.

Theorem A.   *Let $\{e_1, e_2, \ldots, e_n\}$ be a finite orthonormal set in a Hilbert space H.   If x is any vector in H, then*

$$\sum_{i=1}^{n} |(x,e_i)|^2 \leq \|x\|^2; \tag{1}$$

further,
$$x - \sum_{i=1}^{n} (x,e_i)e_i \perp e_j \tag{2}$$

for each j.

PROOF.   The inequality (1) follows from a computation similar to that used in proving Schwarz's inequality:

$$0 \leq \| x - \sum_{i=1}^{n} (x,e_i)e_i \|^2$$

$$= \Big(x - \sum_{i=1}^{n} (x,e_i)e_i, \; x - \sum_{j=1}^{n} (x,e_j)e_j\Big)$$

$$= (x,x) - \sum_{i=1}^{n} (x,e_i)\overline{(x,e_i)} - \sum_{j=1}^{n} (x,e_j)\overline{(x,e_j)} + \sum_{i=1}^{n} \sum_{j=1}^{n} (x,e_i)\overline{(x,e_j)}(e_i,e_j)$$

$$= \|x\|^2 - \sum_{i=1}^{n} |(x,e_i)|^2.$$

To conclude the proof, we observe that

$$\Big(x - \sum_{i=1}^{n} (x,e_i)e_i, e_j\Big) = (x,e_j) - \sum_{i=1}^{n} (x,e_i)(e_i,e_j) = (x,e_j) - (x,e_j) = 0,$$

from which statement (2) follows at once.

The reader should note that the inequality (1) can be given the following loose but illuminating geometric interpretation: the sum of the squares of the components of a vector in various perpendicular directions does not exceed the square of the length of the vector itself. This is usually called *Bessel's inequality*, though, as we shall see below, it is only a special case of a more general inequality with the same name. In a similar vein, relation (2) says that if we subtract from a vector its components in several perpendicular directions, then the result has no component left in any of these directions.

Our next task is to prove that both parts of Theorem A generalize to the case of an arbitrary orthonormal set. The main problem here is to show that the sums in (1) and (2) can be defined in a reasonable way when no restriction is placed on the number of e_i's under consideration. The key to this problem lies in the following theorem.

Theorem B. *If $\{e_i\}$ is an orthonormal set in a Hilbert space H, and if x is any vector in H, then the set $S = \{e_i : (x,e_i) \neq 0\}$ is either empty or countable.*

PROOF. For each positive integer n, consider the set

$$S_n = \{e_i : |(x,e_i)|^2 > \|x\|^2/n\}.$$

By Bessel's inequality, S_n contains at most $n - 1$ vectors. The conclusion now follows from the fact that $S = \bigcup_{n=1}^{\infty} S_n$.

As our first application of this result, we prove the general form of Bessel's inequality.

Theorem C (Bessel's Inequality). *If $\{e_i\}$ is an orthonormal set in a Hilbert space H, then*

$$\Sigma |(x,e_i)|^2 \leq \|x\|^2 \tag{3}$$

for every vector x in H.

PROOF. Our basic obligation here is to explain what is meant by the sum on the left of (3). Once this is clearly understood, the proof is easy. As in the preceding theorem, we write $S = \{e_i : (x,e_i) \neq 0\}$. If S is empty, we define $\Sigma |(x,e_i)|^2$ to be the number 0; and in this case, (3) is obviously true. We now assume that S is non-empty, and we see from Theorem B that it must be finite or countably infinite. If S is finite, it can be written in the form $S = \{e_1, e_2, \ldots, e_n\}$ for some positive integer n. In this case, we define $\Sigma |(x,e_i)|^2$ to be $\Sigma_{i=1}^{n} |(x,e_i)|^2$, which is clearly independent of the order in which the elements of S are arranged. The inequality (3) now reduces to (1), which has already been proved. All that remains is to consider the case in which S is countably infinite. Let the vectors in S be arranged in a definite order:

$$S = \{e_1, e_2, \ldots, e_n, \ldots\}.$$

By the theory of absolutely convergent series, if $\Sigma_{n=1}^{\infty} |(x,e_n)|^2$ converges, then every series obtained from this by rearranging its terms also converges, and all such series have the same sum. We therefore define $\Sigma|(x,e_i)|^2$ to be $\Sigma_{n=1}^{\infty}|(x,e_n)|^2$, and it follows from the above remark that $\Sigma|(x,e_i)|^2$ is a non-negative extended real number which depends only on S, and not on the arrangement of its vectors. We conclude the proof by observing that in this case, (3) reduces to the assertion that

$$\sum_{n=1}^{\infty}|(x,e_n)|^2 \leq \|x\|^2; \tag{4}$$

and since it follows from (1) that no partial sum of the series on the left of (4) can exceed $\|x\|^2$, it is clear that (4) itself is true.

The second part of Theorem A is generalized in essentially the same way.

Theorem D. *If $\{e_i\}$ is an orthonormal set in a Hilbert space H, and if x is an arbitrary vector in H, then*

$$x - \Sigma(x,e_i)e_i \perp e_j \tag{5}$$

for each j.

PROOF. As in the above proof, we define $\Sigma(x,e_i)e_i$ for each of the various cases, and we prove (5) as we go along. We again write

$$S = \{e_i : (x,e_i) \neq 0\}.$$

When S is empty, we define $\Sigma(x,e_i)e_i$ to be the vector 0, and we observe that (5) reduces to the statement that $x - 0 = x$ is orthogonal to each e_j, which is precisely what is meant by saying that S is empty. When S is non-empty and finite, and can be written in the form

$$S = \{e_1, e_2, \ldots, e_n\},$$

we define $\Sigma(x,e_i)e_i$ to be $\Sigma_{i=1}^{n}(x,e_i)e_i$; and in this case, (5) reduces to (2), which has already been proved.

By Theorem B, we may assume for the remainder of the proof that S is countably infinite. Let the vectors in S be listed in a definite order: $S = \{e_1, e_2, \ldots, e_n, \ldots\}$. We put $s_n = \Sigma_{i=1}^{n}(x,e_i)e_i$, and we note that for $m > n$ we have

$$\|s_m - s_n\|^2 = \|\sum_{i=n+1}^{m}(x,e_i)e_i\|^2 = \sum_{i=n+1}^{m}|(x,e_i)|^2.$$

Bessel's inequality shows that the series $\Sigma_{n=1}^{\infty}|(x,e_n)|^2$ converges, so $\{s_n\}$ is a Cauchy sequence in H; and since H is complete, this sequence converges to a vector s, which we write in the form $s = \Sigma_{n=1}^{\infty}(x,e_n)e_n$. We now define $\Sigma(x,e_i)e_i$ to be $\Sigma_{n=1}^{\infty}(x,e_n)e_n$, and—deferring for a moment the question of what happens when the vectors in S are rearranged—we observe that (5) follows from (2) and the continuity of the inner product:

$$(x - \Sigma(x,e_i)e_i, e_j) = (x - s, e_j) = (x,e_j) - (s,e_j) = (x,e_j) - (\lim s_n, e_j)$$
$$= (x,e_j) - \lim (s_n,e_j) = (x,e_j) - (x,e_j) = 0.$$

All that remains is to show that this definition of $\Sigma(x,e_i)e_i$ is valid, in the sense that it does not depend on the arrangement of the vectors in S. Let the vectors in S be rearranged in any manner:

$$S = \{f_1, f_2, \ldots, f_n, \ldots\}.$$

We put $s_n' = \Sigma_{i=1}^n (x,f_i)f_i$, and we see—as above—that the sequence $\{s_n'\}$ converges to a limit s', which we write in the form $s' = \Sigma_{n=1}^\infty (x,f_n)f_n$. We conclude the proof by showing that s' equals s. Let $\epsilon > 0$ be given, and let n_0 be a positive integer so large that if $n \geq n_0$, then $\|s_n - s\| < \epsilon$, $\|s_n' - s'\| < \epsilon$, and $\Sigma_{i=n_0+1}^\infty |(x,e_i)|^2 < \epsilon^2$. For some positive integer $m_0 > n_0$, all terms of s_{n_0} occur among those of s_{m_0}', so $s_{m_0}' - s_{n_0}$ is a finite sum of terms of the form $(x,e_i)e_i$ for $i = n_0 + 1$, $n_0 + 2$, $\ldots$. This yields $\|s_{m_0}' - s_{n_0}\|^2 \leq \Sigma_{i=n_0+1}^\infty |(x,e_i)|^2 < \epsilon^2$, so $\|s_{m_0}' - s_{n_0}\| < \epsilon$ and

$$\|s' - s\| \leq \|s' - s_{m_0}'\| + \|s_{m_0}' - s_{n_0}\| + \|s_{n_0} - s\| < \epsilon + \epsilon + \epsilon = 3\epsilon.$$

Since ϵ is arbitrary, this shows that $s' = s$.

Let H be a non-zero Hilbert space, so that the class of all its orthonormal sets is non-empty. This class is clearly a partially ordered set with respect to set inclusion. An orthonormal set $\{e_i\}$ in H is said to be *complete* if it is maximal in this partially ordered set, that is, if it is impossible to adjoin a vector e to $\{e_i\}$ in such a way that $\{e_i,e\}$ is an orthonormal set which properly contains $\{e_i\}$.

Theorem E. *Every non-zero Hilbert space contains a complete orthonormal set.*

PROOF. The statement follows at once from Zorn's lemma, since the union of any chain of orthonormal sets is clearly an upper bound for the chain in the partially ordered set of all orthonormal sets.

Orthonormal sets are truly interesting only when they are complete. The reasons for this are presented in our next theorem.

Theorem F. *Let H be a Hilbert space, and let $\{e_i\}$ be an orthonormal set in H. Then the following conditions are all equivalent to one another:*
(1) *$\{e_i\}$ is complete;*
(2) *$x \perp \{e_i\} \Rightarrow x = 0$;*
(3) *if x is an arbitrary vector in H, then $x = \Sigma(x,e_i)e_i$;*
(4) *if x is an arbitrary vector in H, then $\|x\|^2 = \Sigma|(x,e_i)|^2$.*

PROOF. We prove that each of the conditions (1), (2), and (3) implies the one following it and that (4) implies (1).

(1) $\Rightarrow$ (2).   If (2) is not true, there exists a vector $x \neq 0$ such that $x \perp \{e_i\}$. We now define e by $e = x/\|x\|$, and we observe that $\{e_i, e\}$ is an orthonormal set which properly contains $\{e_i\}$. This contradicts the completeness of $\{e_i\}$.

(2) $\Rightarrow$ (3).   By Theorem D, $x - \Sigma(x, e_i)e_i$ is orthogonal to $\{e_i\}$, so (2) implies that $x - \Sigma(x, e_i)e_i = 0$, or equivalently, that $x = \Sigma(x, e_i)e_i$.

(3) $\Rightarrow$ (4).   By the joint continuity of the inner product, the expression in (3) yields

$$\|x\|^2 = (x, x) = (\Sigma(x, e_i)e_i, \Sigma(x, e_j)e_j) = \Sigma(x, e_i)\overline{(x, e_i)} = \Sigma|(x, e_i)|^2.$$

(4) $\Rightarrow$ (1).   If $\{e_i\}$ is not complete, it is a proper subset of an orthonormal set $\{e_i, e\}$. Since e is orthogonal to all the e_i's, (4) yields $\|e\|^2 = \Sigma|(e, e_i)|^2 = 0$, and this contradicts the fact that e is a unit vector.

There is some standard terminology which is often used in connection with this theorem.   Let $\{e_i\}$ be a complete orthonormal set in a Hilbert space H, and let x be an arbitrary vector in H.   The numbers (x, e_i) are called the *Fourier coefficients* of x, the expression $x = \Sigma(x, e_i)e_i$ is called the *Fourier expansion* of x, and the equation $\|x\|^2 = \Sigma|(x, e_i)|^2$ is called *Parseval's equation*—all with respect to the particular complete orthonormal set $\{e_i\}$ under consideration.   These terms come from the classical theory of Fourier series, as indicated in our next example.

Example 3.   Consider the Hilbert space L_2 associated with the measure space $[0, 2\pi]$, where measure is Lebesgue measure and integrals are Lebesgue integrals.[1]   This space essentially consists of all complex functions f defined on $[0, 2\pi]$ which are Lebesgue measurable and square-integrable, in the sense that

$$\int_0^{2\pi} |f(x)|^2 \, dx < \infty.$$

Its norm and inner product are defined by

$$\|f\| = \left(\int_0^{2\pi} |f(x)|^2 \, dx\right)^{1/2}$$

and
$$(f, g) = \int_0^{2\pi} f(x)\overline{g(x)} \, dx.$$

A simple computation shows that the functions e^{inx}, for

$$n = 0, \pm 1, \pm 2, \ldots,$$

---

[1] In order to understand this and the next example, the reader should have some knowledge of the modern theory of measure and integration.   We wish to emphasize once again that these examples are in no way essential to the structure of the book, and may be skipped by any reader without the necessary background.   We advise such a reader to ignore these examples and to proceed at once to the discussion of the Gram-Schmidt process.

are mutually orthogonal in L_2:

$$\int_0^{2\pi} e^{imx}e^{-inx}\,dx = \begin{cases} 0 & m \neq n \\ 2\pi & m = n. \end{cases}$$

It follows from this that the functions e_n $(n = 0, \pm 1, \pm 2, \ldots)$ defined by $e_n(x) = e^{inx}/\sqrt{2\pi}$ form an orthonormal set in L_2. For any function f in L_2, the numbers

$$c_n = (f,e_n) = \frac{1}{\sqrt{2\pi}} \int_0^{2\pi} f(x)e^{-inx}\,dx \tag{6}$$

are its classical *Fourier coefficients*, and *Bessel's inequality* takes the form

$$\sum_{n=-\infty}^{\infty} |c_n|^2 \leq \int_0^{2\pi} |f(x)|^2\,dx.$$

It is a fact of very great importance in the theory of Fourier series that the orthonormal set $\{e_n\}$ is complete in L_2. As we have seen in Theorem F, the completeness of $\{e_n\}$ is equivalent to the assertion that for every f in L_2, Bessel's inequality can be strengthened to *Parseval's equation:*

$$\sum_{n=-\infty}^{\infty} |c_n|^2 = \int_0^{2\pi} |f(x)|^2\,dx.$$

Theorem F also tells us that the completeness of $\{e_n\}$ is equivalent to the statement that each f in L_2 has a *Fourier expansion:*

$$f(x) = \frac{1}{\sqrt{2\pi}} \sum_{n=-\infty}^{\infty} c_n e^{inx}. \tag{7}$$

It must be emphasized that this expansion is not to be interpreted as saying that the series converges pointwise to the function. The meaning of (7) is that the partial sums of the series, that is, the vectors f_n in L_2 defined by

$$f_n(x) = \frac{1}{\sqrt{2\pi}} \sum_{k=-n}^{n} c_k e^{ikx}, \tag{8}$$

converge to the vector f in the sense of L_2:

$$\|f_n - f\| \to 0.$$

This situation is often expressed by saying that f is the *limit in the mean* of the f_n's. We add one final remark to our description of this portion of the theory of Fourier series. If f is an arbitrary function in L_2 with Fourier coefficients c_n defined by (6), then Bessel's inequality tells us that the series $\sum_{n=-\infty}^{\infty} |c_n|^2$ converges. The celebrated *Riesz-Fischer theorem* asserts the converse: if c_n $(n = 0, \pm 1, \pm 2, \ldots)$ are given complex numbers for which $\sum_{n=-\infty}^{\infty} |c_n|^2$ converges, then there exists a function

f in L_2 whose Fourier coefficients are the c_n's. If we grant the completeness of L_2 as a metric space, this is very easy to prove. All that is necessary is to use the c_n's to define a sequence of f_n's in accordance with (8). The functions $e^{inx}/\sqrt{2\pi}$ form an orthonormal set, so for $m > n$ we have

$$\|f_m - f_n\|^2 = \sum_{|k|=n+1}^{m} |c_k|^2. \tag{9}$$

By the convergence of $\Sigma_{n=-\infty}^{\infty} |c_n|^2$, the sum on the right of (9) can be made as small as we please for all sufficiently large n and all $m > n$. This tells us that the f_n's form a Cauchy sequence in L_2; and since L_2 is complete, there exists a function f in L_2 such that $f_n \to f$. This function f is given by (7), and the c_n's are clearly its Fourier coefficients. It is apparent from these remarks that the essence of the Riesz-Fischer theorem lies in the completeness of L_2 as a metric space.

We shall have use for one further item in the general theory of orthonormal sets, namely, the *Gram-Schmidt orthogonalization process*. Suppose that $\{x_1, x_2, \ldots, x_n, \ldots\}$ is a linearly independent set in a Hilbert space H. The problem is to exhibit a constructive procedure for converting this set into a corresponding orthonormal set $\{e_1, e_2, \ldots, e_n, \ldots\}$ with the property that for each n the linear subspace of H spanned by $\{e_1, e_2, \ldots, e_n\}$ is the same as that spanned by $\{x_1, x_2, \ldots, x_n\}$. Our first step is to normalize x_1—which is necessarily nonzero—by putting

$$e_1 = \frac{x_1}{\|x_1\|}.$$

The next step is to subtract from x_2 its component in the direction of e_1 to obtain the vector $x_2 - (x_2, e_1)e_1$ orthogonal to e_1, and then to normalize this by putting

$$e_2 = \frac{x_2 - (x_2, e_1)e_1}{\|x_2 - (x_2, e_1)e_1\|}.$$

We observe that since x_2 is not a scalar multiple of x_1, the vector $x_2 - (x_2, e_1)e_1$ is not zero, so the definition of e_2 is valid. Also, it is clear that e_2 is a linear combination of x_1 and x_2, and that x_2 is a linear combination of e_1 and e_2. The next step is to subtract from x_3 its components in the directions of e_1 and e_2 to obtain a vector orthogonal to e_1 and e_2, and then to normalize this by putting

$$e_3 = \frac{x_3 - (x_3, e_1)e_1 - (x_3, e_2)e_2}{\|x_3 - (x_3, e_1)e_1 - (x_3, e_2)e_2\|}.$$

If this process is continued in the same way, it clearly produces an orthonormal set $\{e_1, e_2, \ldots, e_n, \ldots\}$ with the required property.

Example 4. Many orthonormal sets of great interest and importance in analysis can be obtained conveniently by applying the Gram-Schmidt process to sequences of simple functions.

(a) In the space L_2 associated with the interval $[-1,1]$, the functions x^n ($n = 0, 1, 2, \ldots$) are linearly independent. If we take these functions to be the x_n's in the Gram-Schmidt process, then the e_n's are the *normalized Legendre polynomials*.

(b) Consider the space L_2 over the entire real line. If the x_n's here are taken to be the functions $x^n e^{-x^2/2}$ ($n = 0, 1, 2, \ldots$), then the corresponding e_n's are the *normalized Hermite functions*.

(c) Consider the space L_2 associated with the interval $[0, +\infty)$. If the x_n's are the functions $x^n e^{-x}$ ($n = 0, 1, 2, \ldots$), then the e_n's are the *normalized Laguerre functions*.

Each of the orthonormal sets described in the above example can be shown to be complete in its corresponding Hilbert space. The analysis involved in a detailed study of these matters is quite complicated and has no proper place in the present book. The reader should recognize, however—and this is our only reason for mentioning the material in Examples 3 and 4—that the theory of Hilbert spaces does have significant contacts with many solid topics in analysis.

Problems

1. Let $\{e_1, e_2, \ldots, e_n\}$ be a finite orthonormal set in a Hilbert space H, and let x be a vector in H. If $\alpha_1, \alpha_2, \ldots, \alpha_n$ are arbitrary scalars, show that $\|x - \Sigma_{i=1}^n \alpha_i e_i\|$ attains its minimum value $\Leftrightarrow$

$$\alpha_i = (x, e_i)$$

for each i. (*Hint:* expand $\|x - \Sigma_{i=1}^n \alpha_i e_i\|^2$, add and subtract $\Sigma_{i=1}^n |(x, e_i)|^2$, and obtain an expression of the form $\Sigma_{i=1}^n |(x, e_i) - \alpha_i|^2$ in the result.)

2. Show that the orthonormal sets described in Examples 1 and 2 are complete.

3. Show that every orthonormal set in a Hilbert space is contained in some complete orthonormal set, and use this fact to give an alternative proof of Theorem 53-B.

4. Prove that a Hilbert space H is separable $\Leftrightarrow$ every orthonormal set in H is countable.

5. Show that an orthonormal set in a Hilbert space is linearly independent, and use this to prove that a Hilbert space is finite-dimensional $\Leftrightarrow$ every complete orthonormal set is a basis.

6. Prove that any two complete orthonormal sets in a Hilbert space H have the same cardinal number. This cardinal number is called the

orthogonal dimension of H (if H has no complete orthonormal sets, its orthogonal dimension is said to be 0).

7. If H and H' are Hilbert spaces, prove that H is isometrically isomorphic to $H' \Leftrightarrow$ they have the same orthogonal dimension. (*Hint:* by Eq. 52-(2), an isometric isomorphism T preserves inner products, in the sense that $(T(x), T(y)) = (x, y)$.)

8. Let S be a non-empty set, and let $l_2(S)$ be the set of all complex functions f defined on S with the following two properties:
   (1) $\{s : f(s) \neq 0\}$ is empty or countable;
   (2) $\Sigma |f(s)|^2 < \infty$.
   These functions clearly form a complex linear space with respect to pointwise addition and scalar multiplication. Show that $l_2(S)$ becomes a Hilbert space if the norm and inner product are defined by $\|f\| = (\Sigma |f(s)|^2)^{1/2}$ and $(f, g) = \Sigma f(s)\overline{g(s)}$. Show also that the set of all functions defined on S which have the value 1 at a single point and are 0 elsewhere is a complete orthonormal set in $l_2(S)$. We shall see in the next problem that Hilbert spaces of the type described here are universal models for all non-zero Hilbert spaces.

9. Let $S = \{e_i\}$ be a complete orthonormal set in a Hilbert space H. Each vector x in H determines a function f defined on S by

$$f(e_i) = (x, e_i),$$

and Theorems B and C tell us that f is in $l_2(S)$. Show that the mapping $x \to f$ is an isometric isomorphism of H onto $l_2(S)$.

55. THE CONJUGATE SPACE H^*

We pointed out in the introduction to this chapter that one of the fundamental properties of a Hilbert space H is the fact that there is a natural correspondence between the vectors in H and the functionals in H^*. Our purpose in this section is to develop the features of this correspondence which are relevant to our work with operators in the rest of the chapter.

Let y be a fixed vector in H, and consider the function f_y defined on H by $f_y(x) = (x, y)$. It is easy to see that f_y is linear, for

$$f_y(x_1 + x_2) = (x_1 + x_2, y)$$
$$= (x_1, y) + (x_2, y)$$
$$= f_y(x_1) + f_y(x_2)$$

and

$$f_y(\alpha x) = (\alpha x, y)$$
$$= \alpha(x, y)$$
$$= \alpha f_y(x).$$

Further, f_y is continuous and is therefore a functional, for Schwarz's inequality gives

$$|f_y(x)| = |(x,y)|$$
$$\leq \|x\| \|y\|,$$

which shows that $\|f_y\| \leq \|y\|$. Even more, equality is attained here, that is, $\|f_y\| = \|y\|$. This is clear if $y = 0$; and if $y \neq 0$, it follows from

$$\|f_y\| = \sup \{|f_y(x)| : \|x\| = 1\}$$
$$\geq \left| f_y \left(\frac{y}{\|y\|} \right) \right|$$
$$= \left| \left(\frac{y}{\|y\|}, y \right) \right| = \|y\|.$$

To summarize, we have seen that $y \to f_y$ is a norm-preserving mapping of H into H^*. This observation would be of no more than passing interest if it were not for the fact that *every* functional in H^* arises in just this way.

Theorem A. *Let H be a Hilbert space, and let f be an arbitrary functional in H^*. Then there exists a unique vector y in H such that*

$$f(x) = (x,y) \tag{1}$$

for every x in H.

PROOF. It is easy to see that if such a y exists, then it is necessarily unique. For if we also have $f(x) = (x,y')$ for all x, then $(x,y') = (x,y)$ and $(x, y' - y) = 0$ for all x; and since 0 is the only vector orthogonal to every vector, this implies that $y' - y = 0$ or $y' = y$.

We now turn to the problem of showing that y does exist. If $f = 0$, then it clearly suffices to choose $y = 0$. We may therefore assume that $f \neq 0$. The null space M of f is thus a proper closed linear subspace of H, and by Theorem 53-B, there exists a non-zero vector y_0 which is orthogonal to M. We show that if α is a suitably chosen scalar, then the vector $y = \alpha y_0$ meets our requirements. We first observe that no matter what α may be, (1) is true for every x in M; for $f(x) = 0$ for such an x, and since x is orthogonal to y_0, we also have $(x,y) = 0$. This allows us to focus our attention on choosing α in such a way that (1) is true for $x = y_0$. The condition this imposes on α is that

$$f(y_0) = (y_0, \alpha y_0) = \bar{\alpha} \|y_0\|^2.$$

We therefore choose α to be $\overline{f(y_0)}/\|y_0\|^2$, and it follows that (1) is true for every x in M and for $x = y_0$. It is easily seen that each x in H can be written in the form $x = m + \beta y_0$ with m in M: all that is necessary is to choose β in such a way that $f(x - \beta y_0) = f(x) - \beta f(y_0) = 0$, and this is accomplished by putting $\beta = f(x)/f(y_0)$. Our conclusion that (1) is true for every x in H now follows at once from

$$f(x) = f(m + \beta y_0) = f(m) + \beta f(y_0) = (m,y) + \beta(y_0,y)$$
$$= (m + \beta y_0, y) = (x,y).$$

This result tells us that the norm-preserving mapping of H into H^* defined by

$$y \to f_y, \text{ where } f_y(x) = (x,y), \tag{2}$$

is actually a mapping of H *onto* H^*. It would be pleasant if (2) were also a linear mapping. This is not quite true, however, for

$$f_{y_1+y_2} = f_{y_1} + f_{y_2} \quad \text{and} \quad f_{\alpha y} = \bar{\alpha} f_y. \tag{3}$$

It is an easy consequence of (3) that the mapping (2) is an isometry, for $\|f_x - f_y\| = \|f_{x-y}\| = \|x - y\|$. We state several interesting additional facts about this mapping (and what it enables us to do) in the problems, and we leave their verification to the reader. It should be remembered, however, that the real significance of this entire circle of ideas lies in its influence on the theory of the operators on H. We begin the treatment of these matters in the next section.

Problems

1. Verify relations (3).
2. Let H be a Hilbert space, and show that H^* is also a Hilbert space with respect to the inner product defined by $(f_x,f_y) = (y,x)$. In just the same way, the fact that H^* is a Hilbert space implies that H^{**} is a Hilbert space whose inner product is given by $(F_f,F_g) = (g,f)$.
3. Let H be a Hilbert space. We have two natural mappings of H into H^{**}, the second of which is onto: the Banach space natural imbedding $x \to F_x$, where $F_x(f) = f(x)$, and the product mapping $x \to f_x \to F_{f_x}$, where $f_x(y) = (y,x)$ and $F_{f_x}(f) = (f,f_x)$. Show that these mappings are equal, and conclude that H is reflexive. Show also that $(F_x,F_y) = (x,y)$.

56. THE ADJOINT OF AN OPERATOR

Throughout the rest of this chapter, we focus our attention on a fixed but arbitrary Hilbert space H, and unless we specifically state otherwise, it is to be understood that H is the context for all our discussions and theorems.

Let T be an operator on H. We saw in Sec. 51 that T gives rise to an operator T^* (its conjugate) on H^*, where T^* is defined by

$$(T^*f)x = f(Tx).^1$$

We also saw that the mapping $T \to T^*$ is an isometric isomorphism of $\mathfrak{B}(H)$ into $\mathfrak{B}(H^*)$ which reverses products and preserves the identity transformation. In the same way, T^* gives rise to an operator T^{**} on H^{**}; and since H is reflexive, it follows that $T^{**} = T$ when H^{**} is identified with H by means of the natural imbedding.

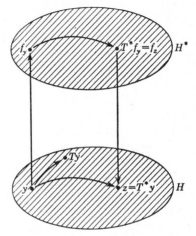

These statements depend only on the fact that H is a reflexive Banach space. We now bring its Hilbert space character into the picture, and we use the natural correspondence between H and H^* discussed in the previous section to pull T^* down to H. The details of this procedure are as follows (see Fig. 38). Let y be a vector in H, and f_y its corresponding functional in H^*; operate with T^* on f_y to obtain a functional $f_z = T^*f_y$; and return to its corresponding vector z in H. There are three mappings under

Fig. 38. The conjugate and the adjoint of T.

consideration here, and we are forming their product:

$$y \to f_y \to T^*f_y = f_z \to z. \tag{1}$$

We write $z = T^*y$, and we call this new mapping T^* of H into itself the *adjoint* of T. The same symbol is used for the adjoint of T as for its conjugate because these two mappings are actually the same if H and H^* are identified by means of the natural correspondence. It is easy to keep track of whether T^* signifies the conjugate or the adjoint of T by noticing whether it operates on functionals or on vectors. The action of the adjoint can be linked more closely to the structure of H by observing that for every vector x we have $(T^*f_y)x = f_y(Tx) = (Tx,y)$ and $(T^*f_y)x = f_z(x) = (x,z) = (x,T^*y)$, so that

$$(Tx,y) = (x,T^*y) \tag{2}$$

for all x and y. Equation (2) is much more than merely a property of the

---

[1] In working with operators, it is common practice to omit parentheses whenever it seems convenient. There is evidently no impairment of clarity in writing $(T^*f)x = f(Tx)$ instead of $[T^*(f)](x) = f(T(x))$, and there will be a considerable gain when we consider operators and inner products together, as we do below.

adjoint of T, for it uniquely determines this adjoint. The proof is simple: if T' is any mapping of H into itself such that $(Tx,y) = (x,T'y)$ for all x and y, then $(x,T'y) = (x,T^*y)$ for all x, so $T'y = T^*y$;[1] and since the latter is true for all y, $T' = T^*$.

Our remarks in the above paragraph have shown that to each operator T on H there corresponds a unique mapping T^* of H into itself (called the adjoint of T) which satisfies relation (2) for all x and y. There is a more direct but less natural approach to these ideas, one which avoids any reference to the conjugate of T. If y is fixed, it is clear that the expression (Tx,y) is a scalar-valued continuous linear function of x. By Theorem 55-A, there exists a unique vector z such that $(Tx,y) = (x,z)$ for all x. We now write $z = T^*y$, and since y is arbitrary, we again have relation (2) for all x and y. The fact that T^* is uniquely determined by (2) follows just as before.

The principal value of our approach to the definition of the adjoint (as opposed to that just mentioned) lies in the motivation it provides for considering adjoints at all. We can express this by emphasizing that an operator on a Banach space always has a conjugate which operates on the conjugate space; and when the Banach space happens to be a Hilbert space, then, as we have seen, the natural correspondence discussed in the previous section makes it almost inevitable that we regard the conjugate as an operator on the space itself. Once the definition of the adjoint is fully understood, however, there is no further need to mention conjugates. All our future work with adjoints will be based on Eq. (2), and from this point on, the symbol T^* will always signify the adjoint of T (and never its conjugate).

As our first step in exploring the properties of adjoints, we verify that T^* actually is an operator on H (all we know so far is that it maps H into itself). For any y and z, and for all x, we have

$$(x, T^*(y + z)) = (Tx, y + z) = (Tx,y) + (Tx,z)$$
$$= (x,T^*y) + (x,T^*z) = (x, T^*y + T^*z),$$

so

$$T^*(y + z) = T^*y + T^*z.$$

The relation
$$T^*(\alpha y) = \alpha T^*y$$

is proved similarly, so T^* is linear. It remains to be seen that T^* is continuous; and to prove this, we note that

$$\|T^*y\|^2 = (T^*y,T^*y) = (TT^*y,y) \le \|TT^*y\| \, \|y\| \le \|T\| \, \|T^*y\| \, \|y\|$$

---

[1] The reasoning here depends on the fact that if y_1 and y_2 are vectors such that $(x,y_1) = (x,y_2)$ for all x, then $(x, y_1 - y_2) = 0$ for all x, so $y_1 - y_2 = 0$ or $y_1 = y_2$.

implies that $\|T^*y\| \leq \|T\| \, \|y\|$ for all y, so

$$\|T^*\| \leq \|T\|.$$

These facts tell us that $T \to T^*$ is a mapping of $\mathcal{B}(H)$ into itself. This mapping is called the *adjoint operation* on $\mathcal{B}(H)$.

Theorem A. *The adjoint operation* $T \to T^*$ *on* $\mathcal{B}(H)$ *has the following properties:*

(1) $(T_1 + T_2)^* = T_1{}^* + T_2{}^*$;
(2) $(\alpha T)^* = \bar{\alpha} T^*$;
(3) $(T_1 T_2)^* = T_2{}^* T_1{}^*$;
(4) $T^{**} = T$;
(5) $\|T^*\| = \|T\|$;
(6) $\|T^*T\| = \|T\|^2$.

PROOF. The arguments used in proving (1) to (4) are all essentially the same. As an illustration of the method, we observe that (3) follows from the fact that for all x and y we have

$$(x,(T_1 T_2)^* y) = (T_1 T_2 x, y) = (T_2 x, T_1{}^* y) = (x, T_2{}^* T_1{}^* y).$$

To prove (5), we note that we already have $\|T^*\| \leq \|T\|$; and if we apply this to T^* instead of T and use (4), we obtain $\|T\| = \|T^{**}\| \leq \|T^*\|$. Half of (6) follows from (5) and the inequality 47-(5), for

$$\|T^*T\| \leq \|T^*\| \, \|T\| = \|T\| \, \|T\| = \|T\|^2;$$

and the fact that $\|T\|^2 \leq \|T^*T\|$ is an immediate consequence of

$$\|Tx\|^2 = (Tx,Tx) = (T^*Tx,x) \leq \|T^*Tx\| \, \|x\| \leq \|T^*T\| \, \|x\|^2.$$

The presence of the adjoint operation is what distinguishes the theory of the operators on H from the more general theory of the operators on a reflexive Banach space.[1] In the next three sections, we use this operation as a tool by means of which we single out for special study certain types of operators on H whose theory is particularly complete and satisfying.

Problems

1. Prove parts (1), (2), and (4) of Theorem A.
2. Show that the adjoint operation is one-to-one onto as a mapping of $\mathcal{B}(H)$ into itself.

[1] See Kakutani and Mackey [23].

3. Show that $0^* = 0$ and $I^* = I$.   Use the latter to show that if T is non-singular, then T^* is also non-singular, and that in this case $(T^*)^{-1} = (T^{-1})^*$.
4. Show that $\|TT^*\| = \|T\|^2$.

57. SELF-ADJOINT OPERATORS

There is an interesting analogy between the set $\mathcal{B}(H)$ of all operators on our Hilbert space H and the set C of all complex numbers.   This can be summarized by observing that each is a complex algebra together with a mapping of the algebra onto itself $(T \to T^*$ and $z \to \bar{z})$ and that these mappings have similar properties.   We shall see that this analogy is quite useful as an intuitive guide to the study of the operators on H. The most significant difference between these systems is that multiplication in the algebra $\mathcal{B}(H)$ is in general non-commutative, and it will become clear as we proceed that this is the primary source of the much greater structural complexity of $\mathcal{B}(H)$.

The most important subsystem of the complex plane is the real line, which is characterized by the relation $z = \bar{z}$.   By analogy, we consider those operators A on H which equal their adjoints, that is, which satisfy the condition $A = A^*$.   Such an operator is said to be *self-adjoint*.   The self-adjoint operators on H are evidently those which are related in the simplest possible way to their adjoints.

We know that $0^* = 0$ and $I^* = I$, so 0 and I are self-adjoint.   If A_1 and A_2 are self-adjoint, and if α and β are real numbers, then

$$(\alpha A_1 + \beta A_2)^* = \bar{\alpha} A_1{}^* + \bar{\beta} A_2{}^* = \alpha A_1 + \beta A_2$$

shows that $\alpha A_1 + \beta A_2$ is also self-adjoint.   Further, if $\{A_n\}$ is a sequence of self-adjoint operators which converges to an operator A, then it is easy to see that A is also self-adjoint; for

$$\|A - A^*\| \leq \|A - A_n\| + \|A_n - A_n{}^*\| + \|A_n{}^* - A^*\| = \|A - A_n\|$$
$$+ \|(A_n - A)^*\| = \|A - A_n\| + \|A_n - A\| = 2\|A_n - A\| \to 0$$

shows that $A - A^* = 0$, so $A = A^*$.   These remarks yield our first theorem.

Theorem A.   *The self-adjoint operators in $\mathcal{B}(H)$ form a closed real linear subspace of $\mathcal{B}(H)$—and therefore a real Banach space—which contains the identity transformation.*

The reader will notice that we have said nothing here about the product of two self-adjoint operators.   Very little is known about such products, and the following simple result represents almost the extent of our information.

Theorem B. *If A_1 and A_2 are self-adjoint operators on H, then their product A_1A_2 is self-adjoint $\Leftrightarrow A_1A_2 = A_2A_1$.*
PROOF. This is an obvious consequence of

$$(A_1A_2)^* = A_2{}^*A_1{}^* = A_2A_1.$$

The order properties of self-adjoint operators are more interesting, and we devote the remainder of the section to establishing some of the simpler facts in this direction.

If T is an arbitrary operator on H, it is easy to see that

$$T = 0 \Leftrightarrow (Tx,y) = 0$$

for all x and y. It is also clear that $T = 0 \Rightarrow (Tx,x) = 0$ for all x. We shall need the converse of this implication.

Theorem C. *If T is an operator on H for which $(Tx,x) = 0$ for all x, then $T = 0$.*
PROOF. It suffices to show that $(Tx,y) = 0$ for any x and y, and the proof of this depends on the following easily verified identity:

$$(T(\alpha x + \beta y), \alpha x + \beta y) - |\alpha|^2(Tx,x) - |\beta|^2(Ty,y)$$
$$= \alpha\bar{\beta}(Tx,y) + \bar{\alpha}\beta(Ty,x). \quad (1)$$

We first observe that by our hypothesis, the left side of (1)—and therefore the right side as well—equals 0 for all α and β. If we put $\alpha = 1$ and $\beta = 1$, then (1) becomes

$$(Tx,y) + (Ty,x) = 0; \quad (2)$$

and if we put $\alpha = i$ and $\beta = 1$, we get

$$i(Tx,y) - i(Ty,x) = 0. \quad (3)$$

Dividing (3) by i and adding the result to (2) yields $2(Tx,y) = 0$, so $(Tx,y) = 0$ and the proof is complete.

It is worth emphasizing that this proof makes essential use of the fact that the scalars are the complex numbers (and not merely the real numbers).

We now apply this result to proving our next theorem, which indicates that self-adjoint operators are linked to real numbers by stronger ties than might be suspected from the loose analogy that led to their definition.

Theorem D.   *An operator T on H is self-adjoint $\Leftrightarrow$ (Tx,x) is real for all x.*
PROOF.   If T is self-adjoint, then

$$\overline{(Tx,x)} = (x,Tx) = (x,T^*x) = (Tx,x)$$

shows that (Tx,x) is real for all x.   On the other hand, if (Tx,x) is real for all x, then $(Tx,x) = \overline{(Tx,x)} = \overline{(x,T^*x)} = (T^*x,x)$ or

$$([T - T^*]x, x) = 0$$

for all x.   By Theorem C, this implies that $T - T^* = 0$, so $T = T^*$.

This theorem enables us to define a respectable and useful order relation on the set of all self-adjoint operators.   If A_1 and A_2 are self-adjoint, we write $A_1 \leq A_2$ if $(A_1x,x) \leq (A_2x,x)$ for all x.   The main elementary facts about this relation are summarized in

Theorem E.   *The real Banach space of all self-adjoint operators on H is a partially ordered set whose linear structure and order structure are related by the following properties:*
   (1)   *if $A_1 \leq A_2$, then $A_1 + A \leq A_2 + A$ for every A;*
   (2)   *if $A_1 \leq A_2$ and $\alpha \geq 0$, then $\alpha A_1 \leq \alpha A_2$.*
PROOF.   The relation in question is obviously reflexive and transitive (see Sec. 8).   To show that it is also antisymmetric, we assume that $A_1 \leq A_2$ and $A_2 \leq A_1$.   This implies at once that $([A_1 - A_2]x, x) = 0$ for all x, so by Theorem C, $A_1 - A_2 = 0$ and $A_1 = A_2$.   The proofs of properties (1) and (2) are easy.   For instance, if $A_1 \leq A_2$, so that $(A_1x,x) \leq (A_2x,x)$ for all x, then $(A_1x,x) + (Ax,x) \leq (A_2x,x) + (Ax,x)$ or $([A_1 + A]x, x) \leq ([A_2 + A]x, x)$ for all x, so $A_1 + A \leq A_2 + A$.   The proof of (2) is similar.

A self-adjoint operator A is said to be *positive* if $A \geq 0$, that is, if $(Ax,x) \geq 0$ for all x.   It is clear that 0 and I are positive, as are T^*T and TT^* for an arbitrary operator T.

Theorem F.   *If A is a positive operator on H, then $I + A$ is non-singular. In particular, $I + T^*T$ and $I + TT^*$ are non-singular for an arbitrary operator T on H.*
PROOF.   We must show that $I + A$ is one-to-one onto as a mapping of H into itself.   First, it is one-to-one, for

$$(I + A)x = 0 \Rightarrow Ax = -x \Rightarrow (Ax,x) = (-x,x) = -\|x\|^2 \geq 0 \Rightarrow x = 0.$$

We next show that the range M of $I + A$ is closed.   It follows from $\|(I + A)x\|^2 = \|x\|^2 + \|Ax\|^2 + 2(Ax,x)$—and the assumption that A is positive—that $\|x\| \leq \|(I + A)x\|$.   By this inequality and the completeness of H, M is complete and therefore closed.   We conclude the proof by observing that $M = H$; for otherwise there would exist a non-zero vector x_0 orthogonal to M, and this would contradict the fact that $(x_0, [I + A]x_0) = 0 \Rightarrow \|x_0\|^2 = -(Ax_0, x_0) \leq 0 \Rightarrow x_0 = 0$.

If the reader wonders why we fail to show that the partially ordered set of all self-adjoint operators is a lattice, the reason is simple: it isn't true. As a matter of fact, this system is about as far from being a lattice as a partially ordered set can be, for it can be shown that two operators in the set have a greatest lower bound $\Leftrightarrow$ they are comparable. This whole situation is intimately related to questions of commutativity for algebras of operators and is too complicated for us to explore here. For further details, see Kadison [22].

Problems

1. Define a new operation of "multiplication" for self-adjoint operators by $A_1 \circ A_2 = (A_1 A_2 + A_2 A_1)/2$, and note that $A_1 \circ A_2$ is always self-adjoint and that it equals $A_1 A_2$ whenever A_1 and A_2 commute. Show that this operation has the following properties:

$$A_1 \circ A_2 = A_2 \circ A_1,$$
$$A_1 \circ (A_2 + A_3) = A_1 \circ A_2 + A_1 \circ A_3,$$
$$\alpha(A_1 \circ A_2) = (\alpha A_1) \circ A_2 = A_1 \circ (\alpha A_2),$$

and $A \circ I = I \circ A = A$. Show also that $A_1 \circ (A_2 \circ A_3) = (A_1 \circ A_2) \circ A_3$ whenever A_1 and A_3 commute.

2. If T is any operator on H, it is clear that $|(Tx,x)| \leq \|Tx\| \|x\| \leq \|T\| \|x\|^2$; so if $H \neq \{0\}$, we have $\sup \{|(Tx,x)|/\|x\|^2 : x \neq 0\} \leq \|T\|$. Prove that if T is self-adjoint, then equality holds here. (*Hint:* write $a = \sup \{|(Tx,x)|/\|x\|^2 : x \neq 0\} = \sup \{|(Tx,x)| : \|x\| = 1\}$, and show that $\|Tx\| \leq a$ whenever $\|x\| = 1$ by putting $b = \|Tx\|^{1/2}$—if $Tx \neq 0$—and considering

$$4\|Tx\|^2 = (T(bx + b^{-1}Tx), bx + b^{-1}Tx)$$
$$- (T(bx - b^{-1}Tx), bx - b^{-1}Tx) \leq a[\|bx + b^{-1}Tx\|^2$$
$$+ \|bx - b^{-1}Tx\|^2] = 4a\|Tx\|.)$$

58. NORMAL AND UNITARY OPERATORS

An operator N on H is said to be *normal* if it commutes with its adjoint, that is, if $NN^* = N^*N$. The reason for the importance of normal operators will not become clear until the next chapter. We shall see that they are the most general operators on H for which a simple and revealing structure theory is possible. Our purpose in this section is to present a few of their more elementary properties which are necessary for our later work.

It is obvious that every self-adjoint operator is normal, and that if N is normal and α is any scalar, then αN is also normal. Further, the limit N of any convergent sequence $\{N_k\}$ of normal operators is normal; for we know that $N_k^* \to N^*$, so

$$\|NN^* - N^*N\| \leq \|NN^* - N_kN_k^*\| + \|N_kN_k^* - N_k^*N_k\|$$
$$+ \|N_k^*N_k - N^*N\| = \|NN^* - N_kN_k^*\| + \|N_k^*N_k - N^*N\| \to 0,$$

which implies that $NN^* - N^*N = 0$. These remarks prove

Theorem A. *The set of all normal operators on H is a closed subset of $\mathcal{B}(H)$ which contains the set of all self-adjoint operators and is closed under scalar multiplication.*

It is natural to wonder whether the sum and product of two normal operators are necessarily normal. They are not, but nevertheless, we can say a little in this direction.

Theorem B. *If N_1 and N_2 are normal operators on H with the property that either commutes with the adjoint of the other, then $N_1 + N_2$ and N_1N_2 are normal.*

PROOF. It is clear by taking adjoints that

$$N_1N_2^* = N_2^*N_1 \Leftrightarrow N_2N_1^* = N_1^*N_2,$$

so the assumption implies that each commutes with the adjoint of the other. To show that $N_1 + N_2$ is normal under the stated conditions, we have only to compare the results of the following computations:

$$(N_1 + N_2)(N_1 + N_2)^* = (N_1 + N_2)(N_1^* + N_2^*)$$
$$= N_1N_1^* + N_1N_2^* + N_2N_1^* + N_2N_2^*$$

and $(N_1 + N_2)^*(N_1 + N_2) = (N_1^* + N_2^*)(N_1 + N_2)$
$$= N_1^*N_1 + N_1^*N_2 + N_2^*N_1 + N_2^*N_2.$$

The fact that N_1N_2 is normal follows similarly from

$$N_1N_2(N_1N_2)^* = N_1N_2N_2^*N_1^* = N_1N_2^*N_2N_1^* = N_2^*N_1N_1^*N_2$$
$$= N_2^*N_1^*N_1N_2 = (N_1N_2)^*N_1N_2.$$

By definition, a self-adjoint operator A is one which satisfies the identity $A^*x = Ax$. Many properties of self-adjoint operators do not depend on this, but only on the weaker identity $\|A^*x\| = \|Ax\|$. Our next theorem shows that all such properties are shared by normal operators.

Theorem C. *An operator T on H is normal* $\Leftrightarrow \|T^*x\| = \|Tx\|$ *for every x.*

PROOF. In view of Theorem 57-C, this is implied by the fact that

$$\|T^*x\| = \|Tx\| \Leftrightarrow \|T^*x\|^2 = \|Tx\|^2 \Leftrightarrow (T^*x, T^*x)$$
$$= (Tx, Tx) \Leftrightarrow (TT^*x, x) = (T^*Tx, x) \Leftrightarrow ([TT^* - T^*T]x, x) = 0.$$

The following consequence of this result will be useful in our later work.

Theorem D. *If N is a normal operator on H, then* $\|N^2\| = \|N\|^2$.

PROOF. The preceding theorem shows that

$$\|N^2x\| = \|NNx\| = \|N^*Nx\|$$

for every x, and this implies that $\|N^2\| = \|N^*N\|$. By Theorem 56-A, we have $\|N^*N\| = \|N\|^2$, so the proof is complete.

We know that any complex number z can be expressed uniquely in the form $z = a + ib$ where a and b are real numbers, and that these real numbers are called the real and imaginary parts of z and are given by $a = (z + \bar{z})/2$ and $b = (z - \bar{z})/2i$. The analogy between general operators and complex numbers, and between self-adjoint operators and real numbers, suggests that for an arbitrary operator T on H we form $A_1 = (T + T^*)/2$ and $A_2 = (T - T^*)/2i$. A_1 and A_2 are clearly self-adjoint, and they have the property that $T = A_1 + iA_2$. The uniqueness of this expression for T follows at once from the fact that

$$T^* = A_1 - iA_2.$$

The self-adjoint operators A_1 and A_2 are called the *real part* and the *imaginary part* of T.

We emphasized earlier that the complicated structure of $\mathfrak{B}(H)$ is due in large part to the fact that operator multiplication is in general non-commutative. Since our future work will be focused mainly on normal operators, it is of interest to see—as the following theorem shows—that the existence of non-normal operators can be traced directly to the non-commutativity of self-adjoint operators.

Theorem E. *If T is an operator on H, then T is normal* $\Leftrightarrow$ *its real and imaginary parts commute.*

PROOF. If A_1 and A_2 are the real and imaginary parts of T, so that $T = A_1 + iA_2$ and $T^* = A_1 - iA_2$, then

$$TT^* = (A_1 + iA_2)(A_1 - iA_2) = A_1^2 + A_2^2 + i(A_2A_1 - A_1A_2)$$

and

$$T^*T = (A_1 - iA_2)(A_1 + iA_2) = A_1^2 + A_2^2 + i(A_1A_2 - A_2A_1).$$

It is clear that if $A_1A_2 = A_2A_1$, then $TT^* = T^*T$. Conversely, if $TT^* = T^*T$, then $A_1A_2 - A_2A_1 = A_2A_1 - A_1A_2$, so $2A_1A_2 = 2A_2A_1$ and $A_1A_2 = A_2A_1$.

Perhaps the most important subsystem of the complex plane after the real line is the unit circle, which is characterized by either of the equivalent identities $|z| = 1$ or $z\bar{z} = \bar{z}z = 1$. An operator U on H which satisfies the equation $UU^* = U^*U = I$ is said to be *unitary*. Unitary operators—which are obviously normal— are thus the natural analogues of complex numbers of absolute value 1. It is clear from the definition that the unitary operators on H are precisely the non-singular operators whose inverses equal their adjoints. The geometric significance of these operators is best understood in the light of our next theorem.

Theorem F. *If T is an operator on H, then the following conditions are all equivalent to one another:*

(1) $T^*T = I$;

(2) $(Tx,Ty) = (x,y)$ for all x and y;

(3) $\|Tx\| = \|x\|$ for all x.

PROOF. If (1) is true, then $(T^*Tx,y) = (x,y)$ or $(Tx,Ty) = (x,y)$ for all x and y, so (2) is true; and if (2) is true, then by taking $y = x$ we obtain $(Tx,Tx) = (x,x)$ or $\|Tx\|^2 = \|x\|^2$ for all x, so (3) is true. The fact that (3) implies (1) is a consequence of Theorem 57-C and the following chain of implications:

$$\|Tx\| = \|x\| \Rightarrow \|Tx\|^2 = \|x\|^2 \Rightarrow (Tx,Tx) = (x,x) \Rightarrow (T^*Tx,x)$$
$$= (x,x) \Rightarrow ([T^*T - I]x,x) = 0.$$

An operator on H with property (3) of this theorem is simply an isometric isomorphism of H into itself. That an operator of this kind need not be unitary is easily seen by considering the operator on l_2 defined by

$$T\{x_1, x_2, \ldots\} = \{0, x_1, x_2, \ldots\},$$

which preserves norms but has no inverse. These ideas lead at once to

Theorem G. *An operator T on H is unitary $\Leftrightarrow$ it is an isometric isomorphism of H onto itself.*

PROOF. If T is unitary, then we know from the definition that it is onto; and since by Theorem F it preserves norms, it is an isometric isomorphism of H onto itself. Conversely, if T is an isometric isomorphism of H onto itself, then T^{-1} exists, and by Theorem F we have $T^*T = I$. It now follows that $(T^*T)T^{-1} = IT^{-1}$, so $T^* = T^{-1}$ and $TT^* = T^*T = I$, which shows that T is unitary.

This theorem makes quite clear the nature of unitary operators: they are precisely those one-to-one mappings of H onto itself which preserve all structure—the linear operations, the norm, and the inner product.

Problems

1. If T is an arbitrary operator on H, and if α and β are scalars such that $|\alpha| = |\beta|$, show that $\alpha T + \beta T^*$ is normal.
2. If H is finite-dimensional, show that every isometric isomorphism of H into itself is unitary.
3. Show that an operator T on H is unitary $\Leftrightarrow T(\{e_i\})$ is a complete orthonormal set whenever $\{e_i\}$ is.
4. Show that the unitary operators on H form a group.

59. PROJECTIONS

According to the definition given in Sec. 50, a projection on a Banach space B is an idempotent operator on B, that is, an operator P with the property that $P^2 = P$. It was proved in that section that each projection P determines a pair of closed linear subspaces M and N—the range and null space of P—such that $B = M \oplus N$, and also, conversely, that each such pair of closed linear subspaces M and N determines a projection P with range M and null space N. In this way, there is established a one-to-one correspondence between projections on B and pairs of closed linear subspaces of B which span the whole space and have only the zero vector in common.

The context of our present work, however, is the Hilbert space H, and not a general Banach space, and the structure which H enjoys in addition to being a Banach space enables us to single out for special attention those projections whose range and null space are orthogonal. Our first theorem gives a convenient characterization of these projections.

Theorem A. *If P is a projection on H with range M and null space N, then $M \perp N \Leftrightarrow P$ is self-adjoint; and in this case, $N = M^\perp$.*

PROOF. Each vector z in H can be written uniquely in the form $z = x + y$ with x and y in M and N. If $M \perp N$, so that $x \perp y$, then $P^* = P$ will follow by Theorem 57-C from $(P^*z,z) = (Pz,z)$; and this is a consequence of

$$(P^*z,z) = (z,Pz) = (z,x) = (x + y,\ x) = (x,x) + (y,x) = (x,x)$$

and $(Pz,z) = (x,z) = (x, x + y) = (x,x) + (x,y) = (x,x)$. If, conversely, $P^* = P$, then the conclusion that $M \perp N$ follows from the fact that for any x and y in M and N we have

$$(x,y) = (Px,y) = (x,P^*y) = (x,Py) = (x,0) = 0.$$

All that remains is to see that if $M \perp N$, then $N = M^\perp$. It is clear that $N \subseteq M^\perp$; and if N is a proper subset of $M^\perp$, and therefore a proper closed linear subspace of the Hilbert space $M^\perp$, then Theorem 53-B implies that there exists a non-zero vector z_0 in $M^\perp$ such that $z_0 \perp N$. Since $z_0 \perp M$ and $z_0 \perp N$, and since $H = M \oplus N$, it follows that $z_0 \perp H$. This is impossible, so we conclude that $N = M^\perp$.

A projection on H whose range and null space are orthogonal is sometimes called a *perpendicular projection*. The only projections considered in the theory of Hilbert spaces are those which are perpendicular, so it is customary to omit the adjective and to refer to them simply as projections. In the light of this agreement and Theorem A, a *projection on H* can be defined as an operator P which satisfies the conditions $P^2 = P$ and $P^* = P$. The operators 0 and I are projections, and they are distinct $\Leftrightarrow H \neq \{0\}$.

The great importance of the projections on H rests mainly on Theorem 53-D, which allows us to set up a natural one-to-one correspondence between projections and closed linear subspaces. To each projection P there corresponds its range $M = \{Px : x \in H\}$, which is a closed linear subspace; and conversely, to each closed linear subspace M there corresponds the projection P with range M defined by $P(x + y) = x$, where x and y are in M and $M^\perp$. Either way, we speak of P as *the projection on M*.

It is clear that P is the projection on $M \Leftrightarrow I - P$ is the projection on $M^\perp$. Also, if P is the projection on M, then

$$x \in M \Leftrightarrow Px = x \Leftrightarrow \|Px\| = \|x\|.$$

The first equivalence here was proved in Problem 44-11; and since for every x in H we have

$$\|x\|^2 = \|Px + (I - P)x\|^2 = \|Px\|^2 + \|(I - P)x\|^2, \tag{1}$$

the non-trivial part of the second is given by the following chain of implications:

$$\|Px\| = \|x\| \Rightarrow \|Px\|^2 = \|x\|^2 \Rightarrow \|(I - P)x\|^2 = 0 \Rightarrow Px = x.$$

Relation (1) also shows that $\|Px\| \leq \|x\|$ for every x, so $\|P\| \leq 1$. If x is an arbitrary vector in H, it is easy to see that

$$(Px,x) = (PPx,x) = (Px,P^*x) = (Px,Px) = \|Px\|^2 \geq 0, \tag{2}$$

so P is a positive operator $(0 \leq P)$ in the sense of Sec. 57. Since $I - P$ is also a projection, we also have $0 \leq I - P$ or $P \leq I$, so $0 \leq P \leq I$.

Let T be an operator on H. A closed linear subspace M of H is said to be *invariant* under T if $T(M) \subseteq M$. When this happens, the restriction of T to M can be regarded as an operator on M alone, and the action of T on vectors outside of M can be ignored. If both M and $M^\perp$ are invariant under T, we say that M *reduces* T, or that T is *reduced* by M. This situation is much more interesting, for it allows us to replace the study of T as a whole by the study of its restrictions to M and $M^\perp$, and it invites the hope that these restrictions will turn out to be operators of some particularly simple type. In the following four theorems, we translate these concepts into relations between T and the projection on M.

Theorem B. *A closed linear subspace M of H is invariant under an operator $T \Leftrightarrow M^\perp$ is invariant under T^*.*

PROOF. Since $M^{\perp\perp} = M$ and $T^{**} = T$, it suffices by symmetry to prove that if M is invariant under T, then $M^\perp$ is invariant under T^*. If y is a vector in $M^\perp$, our conclusion will follow from $(x, T^*y) = 0$ for all x in M. But this is an easy consequence of $(x, T^*y) = (Tx, y)$, for the invariance of M under T implies that $(Tx, y) = 0$.

Theorem C. *A closed linear subspace M of H reduces an operator $T \Leftrightarrow M$ is invariant under both T and T^*.*

PROOF. This is obvious from the definitions and the preceding theorem.

Theorem D. *If P is the projection on a closed linear subspace M of H, then M is invariant under an operator $T \Leftrightarrow TP = PTP$.*

PROOF. If M is invariant under T and x is an arbitrary vector in H, then TPx is in M, so $PTPx = TPx$ and $PTP = TP$. Conversely, if $TP = PTP$ and x is a vector in M, then $Tx = TPx = PTPx$ is also in M, so M is invariant under T.

Theorem E. *If P is the projection on a closed linear subspace M of H, then M reduces an operator $T \Leftrightarrow TP = PT$.*

PROOF. M reduces $T \Leftrightarrow M$ is invariant under T and $T^* \Leftrightarrow TP = PTP$ and $T^*P = PT^*P \Leftrightarrow TP = PTP$ and $PT = PTP$. The last statement in this chain clearly implies that $TP = PT$; it also follows from it, as we see by multiplying $TP = PT$ on the right and left by P.

Our next theorem shows how projections can be used to express the statement that two closed linear subspaces of H are orthogonal.

Theorem F. *If P and Q are the projections on closed linear subspaces M and N of H, then $M \perp N \Leftrightarrow PQ = 0 \Leftrightarrow QP = 0$.*

PROOF. We first remark that the equivalence of $PQ = 0$ and $QP = 0$ is clear by taking adjoints. If $M \perp N$, so that $N \subseteq M^\perp$, then the fact that Qx is in N for every x implies that $PQx = 0$, so $PQ = 0$. If, conversely, $PQ = 0$, then for every x in N we have $Px = PQx = 0$, so $N \subseteq M^\perp$ and $M \perp N$.

Motivated by this result, we say that two projections P and Q are *orthogonal* if $PQ = 0$.

Our final theorem describes the circumstances under which a sum of projections is also a projection.

Theorem G. *If $P_1, P_2, \ldots, P_n$ are the projections on closed linear subspaces $M_1, M_2, \ldots, M_n$ of H, then $P = P_1 + P_2 + \cdots + P_n$ is a projection $\Leftrightarrow$ the P_i's are pairwise orthogonal (in the sense that $P_iP_j = 0$ whenever $i \neq j$); and in this case, P is the projection on*

$$M = M_1 + M_2 + \cdots + M_n.$$

PROOF. Since P is clearly self-adjoint, it is a projection $\Leftrightarrow$ it is idempotent. If the P_i's are pairwise orthogonal, then a simple computation shows at once that P is idempotent. To prove the converse, we assume that P is idempotent. Let x be a vector in the range of P_i, so that $x = P_ix$. Then

$$\|x\|^2 = \|P_ix\|^2 \leq \sum_{j=1}^{n} \|P_jx\|^2 = \sum_{j=1}^{n} (P_jx,x) = (Px,x) = \|Px\|^2 \leq \|x\|^2.$$

We conclude that equality must hold all along the line here, so

$$\sum_{j=1}^{n} \|P_jx\|^2 = \|P_ix\|^2$$

and $$\|P_jx\| = 0 \quad \text{for } j \neq i.$$

Thus the range of P_i is contained in the null space of P_j, that is, $M_i \subseteq M_j^\perp$, for every $j \neq i$. This means that $M_i \perp M_j$ whenever $i \neq j$, and our conclusion that the P_i's are pairwise orthogonal now follows from the preceding theorem. We prove the final statement in two steps. First, we observe that since $\|Px\| = \|x\|$ for every x in M_i, each M_i is contained in the range of P, and therefore M is also contained in the range of P. Second, if x is a vector in the range of P, then

$$x = Px = P_1x + P_2x + \cdots + P_nx$$

is evidently in M.

There are many other ways in which the algebraic structure of the set of all projections on H can be related to the geometry of its closed linear subspaces, and several of these are given in the problems below.

The significance of projections in the general theory of operators on H is the theme of the next chapter. As we shall see, the essence of the matter (the spectral theorem) is that every normal operator is made of projections in a way which clearly reveals the geometric nature of its action on the vectors in H.

Problems

1. If P and Q are the projections on closed linear subspaces M and N of H, prove that PQ is a projection $\Leftrightarrow PQ = QP$. In this case, show that PQ is the projection on $M \cap N$.

2. If P and Q are the projections on closed linear subspaces M and N of H, prove that the following statements are all equivalent to one another:
   (a) $P \leq Q$;
   (b) $\|Px\| \leq \|Qx\|$ for every x;
   (c) $M \subseteq N$;
   (d) $PQ = P$;
   (e) $QP = P$.
   (*Hint:* the equivalence of (a) and (b) is easy to prove, as is that of (c), (d), and (e); prove that (d) implies (a) by using

   $$(Px,x) = \|Px\|^2 = \|PQx\|^2 \leq \|Qx\|^2 = (Qx,x);$$

   and prove that (b) implies (c) by observing that if x is in M, then $\|x\| = \|Px\| \leq \|Qx\| \leq \|x\|$.)

3. Show that the projections on H form a complete lattice with respect to their natural ordering as self-adjoint operators. (Compare this situation with that described in the last paragraph of Sec. 57.)

4. If P and Q are the projections on closed linear subspaces M and N of H, prove that $Q - P$ is a projection $\Leftrightarrow P \leq Q$. In this case, show that $Q - P$ is the projection on $N \cap M^{\perp}$.

CHAPTER ELEVEN

Finite-dimensional Spectral Theory

If T is an operator on a Hilbert space H, then the simplest thing T can do to a vector x is to transform it into a scalar multiple of itself:

$$Tx = \lambda x. \tag{1}$$

A non-zero vector x such that Eq. (1) is true for some scalar λ is called an *eigenvector* of T, and a scalar λ such that (1) holds for some non-zero x is called an *eigenvalue* of T.[1] Each eigenvalue has one or more eigenvectors associated with it, and to each eigenvector there corresponds precisely one eigenvalue. If H has no non-zero vectors at all, then T certainly has no eigenvectors. In this case the whole theory collapses into triviality, so we assume throughout the present chapter that $H \neq \{0\}$.

Let λ be an eigenvalue of T, and consider the set M of all its corresponding eigenvectors together with the vector 0 (note that 0 is not an eigenvector). M is thus the set of all vectors x which satisfy the equation

$$(T - \lambda I)x = 0,$$

and it is clearly a non-zero closed linear subspace of H. We call M the *eigenspace* of T corresponding to λ. It is evident that M is invariant under T and that the restriction of T to M is a very simple operator, namely, scalar multiplication by λ.

In order to place the ideas of this chapter in their proper framework,

---

[1] The equivalent terms *characteristic vector* and *characteristic value*, and *proper vector* and *proper value*, are used by many writers.

278

we lay down several rather sweeping hypotheses, whose validity we examine later:

(a) T actually has eigenvalues, and there are finitely many of them, say $\lambda_1, \lambda_2, \ldots, \lambda_m$—which are understood to be distinct— with corresponding eigenspaces $M_1, M_2, \ldots, M_m$;

(b)   the M_i's are pairwise orthogonal, that is, $i \neq j \Rightarrow M_i \perp M_j$;

(c)   the M_i's span H.

Putting aside for a moment the question of whether these statements are true or not, we investigate their implications. By (b) and (c), every vector x in H can be expressed uniquely in the form

$$x = x_1 + x_2 + \cdots + x_m, \tag{2}$$

where x_i is in M_i for each i and the x_i's are pairwise orthogonal. It now follows from (a) that

$$\begin{aligned} Tx &= Tx_1 + Tx_2 + \cdots + Tx_m \\ &= \lambda_1 x_1 + \lambda_2 x_2 + \cdots + \lambda_m x_m. \end{aligned} \tag{3}$$

This relation exhibits the action of T over all of H in a manner which renders its structure perfectly clear from the geometric point of view. It will be convenient to express this result in terms of the projections P_i on the eigenspaces M_i. By Theorem 59-F, (b) is equivalent to the following statement:

$$\text{the } P_i\text{'s are pairwise orthogonal.} \tag{4}$$

Also, since for each i and for every $j \neq i$ we have $M_j \subseteq M_i^\perp$, Eq. (2) yields

$$P_i x = x_i;$$

and it follows at once from this that

$$\begin{aligned} Ix = x &= x_1 + x_2 + \cdots + x_m \\ &= P_1 x + P_2 x + \cdots + P_m x \\ &= (P_1 + P_2 + \cdots + P_m)x \end{aligned}$$

for every x in H, so

$$I = P_1 + P_2 + \cdots + P_m. \tag{5}$$

Relation (3) now tells us that

$$\begin{aligned} Tx &= \lambda_1 x_1 + \lambda_2 x_2 + \cdots + \lambda_m x_m \\ &= \lambda_1 P_1 x + \lambda_2 P_2 x + \cdots + \lambda_m P_m x \\ &= (\lambda_1 P_1 + \lambda_2 P_2 + \cdots + \lambda_m P_m)x \end{aligned}$$

for every x, so

$$T = \lambda_1 P_1 + \lambda_2 P_2 + \cdots + \lambda_m P_m. \tag{6}$$

The expression for T given by (6)—when it exists—is called the *spectral resolution* of T. Whenever this term is used, it is to be understood that the λ_i's are distinct and that the P_i's are non-zero projections which satisfy conditions (4) and (5). We shall see later that the spectral resolution of T is unique when it exists.

All our inferences from (a), (b), and (c) are perfectly rigorous, but the status of these three hypotheses remains entirely up in the air. First of all, with reference to (a), does an arbitrary operator T on H necessarily have an eigenvalue? The answer to this is no, as the reader will easily verify by considering the operator T on l_2 defined by

$$T\{x_1, x_2, \ldots\} = \{0, x_1, x_2, \ldots\}.$$

On the other hand, if H is finite-dimensional, then we shall see in Sec. 61 that every operator has an eigenvalue. For this reason, we assume for the remainder of the chapter—unless we specifically state otherwise—that H is finite-dimensional with dimension n.

We have seen that if T satisfies conditions (a), (b), and (c), then it has the spectral resolution (6). It is too much to hope that every operator on H meets these requirements, so the question arises as to what restrictions they impose on T. This question is easy to answer: T must be normal. For it follows from (6) that

$$T^* = \overline{\lambda_1}P_1 + \overline{\lambda_2}P_2 + \cdots + \overline{\lambda_m}P_m,$$

and by using (4) we readily obtain

$$TT^* = (\lambda_1 P_1 + \lambda_2 P_2 + \cdots + \lambda_m P_m)(\overline{\lambda_1}P_1 + \overline{\lambda_2}P_2 + \cdots + \overline{\lambda_m}P_m)$$
$$= |\lambda_1|^2 P_1 + |\lambda_2|^2 P_2 + \cdots + |\lambda_m|^2 P_m$$

and, similarly,

$$T^*T = |\lambda_1|^2 P_1 + |\lambda_2|^2 P_2 + \cdots + |\lambda_m|^2 P_m.$$

This entire circle of ideas will be completed in the neatest possible way if we can show that every normal operator on H satisfies conditions (a), (b), and (c), and therefore has a spectral resolution. Our aim in the present chapter is to prove this assertion, which is known as the *spectral theorem*, and the machinery treated in the following sections is directed exclusively toward this end. We emphasize once again that H is understood to be finite-dimensional with dimension $n > 0$.

60. MATRICES

Our first goal is to prove that every operator on H has an eigenvalue, and in pursuing this we make use of certain elementary portions of the theory of matrices. We adopt the view that the reader is probably familiar with this theory to some degree and that it suffices here to give a brief sketch of its basic ideas. Our discussion in this section is entirely independent of the Hilbert space character of H and applies equally well to any non-trivial finite-dimensional linear space.

Let $B = \{e_1, e_2, \ldots, e_n\}$ be an ordered basis for H, so that each vector in H is uniquely expressible as a linear combination of the e_i's. If T is an operator on H, then for each e_j we have

$$Te_j = \sum_{i=1}^{n} \alpha_{ij}e_i. \tag{1}$$

The n^2 scalars α_{ij} which are determined in this way by T form the *matrix* of T relative to the ordered basis B. We symbolize this matrix by $[T]$, or if it seems desirable to indicate the ordered basis under consideration, by $[T]_B$. It is customary to write out a matrix as a square array:

$$[T] = \begin{bmatrix} \alpha_{11} & \alpha_{12} & \cdots & \alpha_{1n} \\ \alpha_{21} & \alpha_{22} & \cdots & \alpha_{2n} \\ \cdots\cdots\cdots\cdots\cdots \\ \alpha_{n1} & \alpha_{n2} & \cdots & \alpha_{nn} \end{bmatrix}. \tag{2}$$

The array of scalars $(\alpha_{i1}, \alpha_{i2}, \ldots, \alpha_{in})$ is the *ith row* of the matrix $[T]$, and $(\alpha_{1j}, \alpha_{2j}, \ldots, \alpha_{nj})$ is its *jth column*. As this terminology shows, the first subscript on the entry α_{ij} always indicates the row to which it belongs, and the second the column. In our work, we generally write (2) more concisely in the form

$$[T] = [\alpha_{ij}]. \tag{3}$$

The reader should make sure that he has a perfectly clear understanding of the rule according to which the matrix of T is constructed: write Te_j as a linear combination of $e_1, e_2, \ldots, e_n$, and use the resulting coefficients to form the jth column of $[T]$.

We offer several comments on the above paragraph. First, the term *matrix* has not been defined at all, but only "the matrix of an operator relative to an ordered basis." A matrix—defined simply as a square array of scalars—is sometimes regarded as an object worthy of interest in its own right. For the most part, however, we shall consider a matrix to be associated with a definite operator relative to a particular ordered basis, and we shall regard matrices as little more than computational devices which are occasionally useful in handling operators. Next, the matrices we work with are all square matrices. Rectangular matrices occur in connection with linear transformations of one linear space into another and are of no interest to us here. Finally, we took B to be an ordered basis rather than merely a basis, because the appearance of the array (2) clearly depends on the arrangement of the e_i's as well as on the e_i's themselves. In most theoretical considerations, however, the order of the rows and columns of a matrix is as irrelevant as the order of the vectors in a basis. For this reason, we usually omit the adjective and speak of "the matrix of an operator relative to a basis."

By using the fixed basis $B = \{e_i\}$, we have assigned a matrix $[T] = [\alpha_{ij}]$ to each operator T on H, and the mapping $T \rightarrow [T]$ from operators to matrices is described by $Te_j = \Sigma_{i=1}^n \alpha_{ij}e_i$. The importance of matrices is based primarily on two facts: $T \rightarrow [T]$ is a one-to-one mapping of the set of all operators on H onto the set of all matrices; and algebraic operations can be defined on the set of all matrices in such a manner that the mapping $T \rightarrow [T]$ preserves the algebraic structure of $\mathfrak{B}(H)$.

The first of these statements is easy to prove. If we know that $[\alpha_{ij}]$ is the matrix of T, then this information fully determines Tx for every x; for if $x = \Sigma_{j=1}^n \beta_j e_j$, then

$$
\begin{aligned}
Tx &= \sum_{j=1}^n \beta_j Te_j \\
&= \sum_{j=1}^n \beta_j \left(\sum_{i=1}^n \alpha_{ij}e_i \right) \\
&= \sum_{i=1}^n \left(\sum_{j=1}^n \alpha_{ij}\beta_j \right) e_i.
\end{aligned}
$$

This shows that $T \rightarrow [T]$ is one-to-one. We see that this mapping is onto by means of the following reasoning: if $[\alpha_{ij}]$ is any matrix, then $Te_j = \Sigma_{i=1}^n \alpha_{ij}e_i$ defines T for the vectors in B, and when T is extended by linearity to all of H, it is clear that the resulting operator has $[\alpha_{ij}]$ as its matrix.

To establish the second statement, it suffices to discover how to add and multiply two matrices and how to multiply a matrix by a scalar, in such a way that the following matrix equations are true for all operators T_1 and T_2 on H: $[T_1 + T_2] = [T_1] + [T_2]$, $[\alpha T_1] = \alpha[T_1]$, and

$$
[T_1 T_2] = [T_1][T_2].
$$

Let $[\alpha_{ij}]$ and $[\beta_{ij}]$ be the matrices of T_1 and T_2. The computation

$$
\begin{aligned}
(T_1 + T_2)e_j &= T_1 e_j + T_2 e_j \\
&= \sum_{i=1}^n \alpha_{ij}e_i + \sum_{i=1}^n \beta_{ij}e_i \\
&= \sum_{i=1}^n (\alpha_{ij} + \beta_{ij})e_i
\end{aligned}
$$

shows that if we define addition for matrices by

$$[\alpha_{ij}] + [\beta_{ij}] = [\alpha_{ij} + \beta_{ij}], \tag{4}$$

then we obtain $[T_1 + T_2] = [T_1] + [T_2].$

Similarly, if we multiply a matrix by a scalar in accordance with

$$\alpha[\alpha_{ij}] = [\alpha\alpha_{ij}], \tag{5}$$

then $[\alpha T_1] = \alpha[T_1].$

Finally, the computation

$$(T_1 T_2)e_j = T_1(T_2 e_j) = T_1 \left(\sum_{k=1}^{n} \beta_{kj} e_k \right)$$

$$= \sum_{k=1}^{n} \beta_{kj} T_1 e_k$$

$$= \sum_{k=1}^{n} \beta_{kj} \left(\sum_{i=1}^{n} \alpha_{ik} e_i \right)$$

$$= \sum_{i=1}^{n} \left(\sum_{k=1}^{n} \alpha_{ik} \beta_{kj} \right) e_i$$

shows that if we define multiplication for matrices by

$$[\alpha_{ij}][\beta_{ij}] = \left[\sum_{k=1}^{n} \alpha_{ik} \beta_{kj} \right], \tag{6}$$

then we get $[T_1 T_2] = [T_1][T_2].$

The operations defined by (4), (5), and (6) are the standard algebraic operations for matrices. In words, we add two matrices by adding corresponding entries, and we multiply a matrix by a scalar by multiplying each of its entries by that scalar. The verbal description of (6) is more complicated, and is often called the *row-by-column* rule: to find the entry in the ith row and jth column of the product $[\alpha_{ij}][\beta_{ij}]$, take the ith row $(\alpha_{i1}, \alpha_{i2}, \ldots, \alpha_{in})$ of the first factor and the jth column $(\beta_{1j}, \beta_{2j}, \ldots, \beta_{nj})$ of the second, multiply corresponding entries, and add:

$$\sum_{k=1}^{n} \alpha_{ik} \beta_{kj} = \alpha_{i1} \beta_{1j} + \alpha_{i2} \beta_{2j} + \cdots + \alpha_{in} \beta_{nj}.$$

It is worth noting that the image of the zero operator under the mapping $T \to [T]$ is the *zero matrix*, all of whose entries are 0. Further, it is equally clear that the image of the identity operator is the *identity matrix*, which has 1's down the main diagonal (where $i = j$) and 0's elsewhere. If we introduce the standard *Kronecker delta*, which is defined by

$$\delta_{ij} = \begin{cases} 0 & \text{if } i \neq j \\ 1 & \text{if } i = j, \end{cases}$$

then the identity matrix can be written $[\delta_{ij}]$.

We now reverse our point of view for a moment (but only a moment) and consider the set A_n of all $n \times n$ matrices as an algebraic system in its own right, with addition, scalar multiplication, and multiplication defined by (4), (5), and (6). It can be verified directly from these definitions that A_n is a complex algebra with identity (the identity matrix), called the *total matrix algebra* of degree n. If we ignore the ideas leading to (4), (5), and (6), then the structure of A_n is defined, and can be studied, without any reference to its origin as a representing system for the operators on H. This approach would make very little sense, however, because the primary reason for considering matrices in the first place is that they provide a computational tool which is useful in treating certain aspects of the theory of these operators.

Let us return to our original position and observe two facts: that $\mathscr{B}(H)$ is an algebra; and that the structure of A_n is defined in just such a way as to guarantee that the one-to-one mapping $T \to [T]$ of $\mathscr{B}(H)$ onto A_n preserves addition, scalar multiplication, and multiplication. It now follows at once that A_n is an algebra, and that $T \to [T]$ is an isomorphism (see Problem 45-4) of $\mathscr{B}(H)$ onto A_n.

We give the following formal summary of our work so far.

Theorem A. *If $B = \{e_i\}$ is a basis for H, then the mapping $T \to [T]$, which assigns to each operator T its matrix relative to B, is an isomorphism of the algebra $\mathscr{B}(H)$ onto the total matrix algebra A_n.*

If T is a non-singular operator whose matrix relative to B is $[\alpha_{ij}]$, then T^{-1} clearly has a matrix whose entries are determined in some way by the α_{ij}'s. The formulas involved here are rather clumsy and complicated, and since they have no importance for us, we shall say nothing further about them.

It is necessary, however, to know what is meant by the inverse of a matrix, when it is considered purely as an element of A_n and without reference to any operator which it may represent. We first remark that the identity matrix $[\delta_{ij}]$ is easily seen by direct matrix multiplication to be an identity element for the algebra A_n, in the sense that we have

$$[\alpha_{ij}][\delta_{ij}] = [\delta_{ij}][\alpha_{ij}] = [\alpha_{ij}]$$

for every matrix $[\alpha_{ij}]$; and by the theory of rings, this identity is unique. A matrix $[\alpha_{ij}]$ is said to be *non-singular* if there exists a matrix $[\beta_{ij}]$ such that

$$[\alpha_{ij}][\beta_{ij}] = [\beta_{ij}][\alpha_{ij}] = [\delta_{ij}];$$

and, again by the theory of rings, if such a matrix exists, then it is unique, it is denoted by $[\alpha_{ij}]^{-1}$, and it is called the *inverse* of $[\alpha_{ij}]$.

These ideas are connected with operators by the following considerations. Suppose that $[\alpha_{ij}]$ is the matrix of an operator T relative to B.

We know that the non-singularity of T is equivalent to the existence of an operator T^{-1} such that

$$TT^{-1} = T^{-1}T = I.$$

The isomorphism of Theorem A transforms this operator equation into the matrix equation

$$[T][T^{-1}] = [T^{-1}][T] = [I],$$

which is equivalent to

$$[\alpha_{ij}][T^{-1}] = [T^{-1}][\alpha_{ij}] = [\delta_{ij}].$$

We therefore have

Theorem B. *Let B be a basis for H, and T an operator whose matrix relative to B is $[\alpha_{ij}]$. Then T is non-singular $\Leftrightarrow [\alpha_{ij}]$ is non-singular, and in this case $[\alpha_{ij}]^{-1} = [T^{-1}]$.*

There is one further issue which requires discussion. If T is a fixed operator on H, then its matrix $[T]_B$ relative to B obviously depends on the choice of B. If B changes, how does $[T]_B$ change? More specifically, if $B' = \{f_1, f_2, \ldots, f_n\}$ is also a basis for H, what is the relation between $[T]_B$ and $[T]_{B'}$? The answer to this question is best given in terms of the non-singular operator A defined by $Ae_i = f_i$. Let $[\alpha_{ij}]$ and $[\beta_{ij}]$ be the matrices of T relative to B and B', so that

$$Te_j = \sum_{i=1}^{n} \alpha_{ij}e_i$$

and $Tf_j = \sum_{i=1}^{n} \beta_{ij}f_i$. Let $[\gamma_{ij}]$ be the matrix of A relative to B, so that $Ae_j = \sum_{i=1}^{n} \gamma_{ij}e_i$. By Theorem B, $[\gamma_{ij}]$ is non-singular. We now compute Tf_j in two different ways:

$$
\begin{aligned}
Tf_j &= \sum_{k=1}^{n} \beta_{kj}f_k = \sum_{k=1}^{n} \beta_{kj}Ae_k \\
&= \sum_{k=1}^{n} \beta_{kj} \left(\sum_{i=1}^{n} \gamma_{ik}e_i \right) \\
&= \sum_{i=1}^{n} \left(\sum_{k=1}^{n} \gamma_{ik}\beta_{kj} \right) e_i;
\end{aligned}
$$

and

$$
\begin{aligned}
Tf_j = TAe_j &= T \left(\sum_{k=1}^{n} \gamma_{kj}e_k \right) \\
&= \sum_{k=1}^{n} \gamma_{kj}Te_k \\
&= \sum_{k=1}^{n} \gamma_{kj} \left(\sum_{i=1}^{n} \alpha_{ik}e_i \right) \\
&= \sum_{i=1}^{n} \left(\sum_{k=1}^{n} \alpha_{ik}\gamma_{kj} \right) e_i.
\end{aligned}
$$

A comparison of these results shows that

$$\sum_{k=1}^{n} \gamma_{ik}\beta_{kj} = \sum_{k=1}^{n} \alpha_{ik}\gamma_{kj}$$

for all i and j, so

$$[\gamma_{ij}][\beta_{ij}] = [\alpha_{ij}][\gamma_{ij}]$$

or
$$[\beta_{ij}] = [\gamma_{ij}]^{-1}[\alpha_{ij}][\gamma_{ij}]. \tag{7}$$

If we now write this in the form

$$[T]_{B'} = [A]_B^{-1}[T]_B[A]_B,$$

then it becomes quite clear how the matrix of T changes when B is replaced by B'.

Two matrices $[\alpha_{ij}]$ and $[\beta_{ij}]$ are said to be *similar* if there exists a non-singular matrix $[\gamma_{ij}]$ such that (7) is true. The analysis given above proves half of the following theorem (we leave the proof of the other half to the reader).

Theorem C. *Two matrices in A_n are similar $\Leftrightarrow$ they are the matrices of a single operator on H relative to (possibly) different bases.*

We are now in a position to formulate the fundamental problem of the classical theory of matrices. A given operator on H may have many different matrices relative to different bases, and Theorem C shows in purely matrix terms how these matrices are related to one another. The question arises as to whether it is possible to find, for each operator (or for each operator of a special kind), a basis relative to which its matrix assumes some particularly simple form. This is the *canonical form problem* of matrix theory, and the most important theorem in this direction is the spectral theorem, which we state in the language of matrices in Sec. 62. In the classical approach to these ideas, it was customary to work exclusively with matrices. However, the great advances in the understanding of algebra which have taken place in recent years have made it plain that problems of this kind are best treated intrinsically, that is, directly in terms of the linear spaces and linear transformations involved. As matters now stand, it is possible—and preferable—to state the main canonical form theorems of matrix theory without mentioning matrices at all. Nevertheless, matrices remain useful for some purposes, notably (from our point of view) in the problem of proving that an arbitrary operator on H has an eigenvalue.

Problems

1. Show that the dimension of $\mathfrak{B}(H)$ is n^2.
2. A *scalar matrix* in A_n is one which has the same scalar in every position on the main diagonal and 0's elsewhere. Show that a scalar

matrix commutes with every matrix, and that a matrix which commutes with every matrix is necessarily scalar. What does this imply about $\mathfrak{B}(H)$? (See Problem 45-3.)

3. A *diagonal matrix* in A_n is one which has arbitrary scalars on the main diagonal and 0's elsewhere. Show that all diagonal matrices commute with one another, and that a matrix is necessarily diagonal if it commutes with all diagonal matrices.

4. Complete the proof of Theorem C.

5. Let θ be a fixed real number, and show that the following two matrices in A_2 are similar:

$$\begin{bmatrix} \cos\theta & -\sin\theta \\ \sin\theta & \cos\theta \end{bmatrix} \quad \text{and} \quad \begin{bmatrix} e^{i\theta} & 0 \\ 0 & e^{-i\theta} \end{bmatrix}.$$

(*Hint:* let T be the operator on l_2^2 whose matrix relative to the basis $B = \{e_1, e_2\}$—where $e_1 = (1,0)$ and $e_2 = (0,1)$—is the first of those given, and find another basis $B' = \{f_1, f_2\}$ such that $Tf_1 = e^{i\theta}f_1$ and $Tf_2 = e^{-i\theta}f_2$.)

6. Let T_1 and T_2 be operators on H, and show that there exist bases B and B' such that $[T_1]_B = [T_2]_{B'} \Leftrightarrow$ there exists a non-singular operator A such that $T_2 = AT_1A^{-1}$. (*Hint:* if $[T_1]_B = [T_2]_{B'}$, let A be the operator which carries B onto B'; and if $T_2 = AT_1A^{-1}$, let B be any basis and B' its image under A.)

61. DETERMINANTS AND THE SPECTRUM OF AN OPERATOR

Determinants are often advertised to students of elementary mathematics as a computational device of great value and efficiency for solving numerical problems involving systems of linear equations. This is somewhat misleading, for their value in problems of this kind is very limited. On the other hand, they do have definite importance as a theoretical tool. Briefly, they provide a numerical means of distinguishing between singular and non-singular matrices (and operators).

This is not the place for developing the theory of determinants in any detail. Instead, we assume that the reader already knows something about them, and we confine ourselves to listing a few of their simpler properties which are relevant to our present interests.

Let $[\alpha_{ij}]$ be an $n \times n$ matrix. The *determinant* of this matrix, which we denote by $\det([\alpha_{ij}])$, is a scalar associated with it in such a way that

(1) $\det([\delta_{ij}]) = 1$;

(2) $\det([\alpha_{ij}][\beta_{ij}]) = \det([\alpha_{ij}]) \det([\beta_{ij}])$;

(3) $\det([\alpha_{ij}]) \neq 0 \Leftrightarrow [\alpha_{ij}]$ is non-singular; and

(4) $\det([\alpha_{ij}] - \lambda[\delta_{ij}])$ is a polynomial, with complex coefficients, of degree n in the variable λ.

The determinant function *det* is thus a scalar-valued function of matrices which has certain properties. In elementary work, the determinant of a matrix is usually written out with vertical bars, as follows,

$$\det([\alpha_{ij}]) = \begin{vmatrix} \alpha_{11} & \alpha_{12} & \cdots & \alpha_{1n} \\ \alpha_{21} & \alpha_{22} & \cdots & \alpha_{2n} \\ \cdots\cdots\cdots\cdots\cdots \\ \alpha_{n1} & \alpha_{n2} & \cdots & \alpha_{nn} \end{vmatrix},$$

and is evaluated by complicated procedures which are of no concern to us here.

We now consider an operator T on H. If B and B' are bases for H, then the matrices $[\alpha_{ij}]$ and $[\beta_{ij}]$ of T relative to B and B' may be entirely different, but nevertheless they have the same determinant. For we know from the previous section that there exists a non-singular matrix $[\gamma_{ij}]$ such that

$$[\beta_{ij}] = [\gamma_{ij}]^{-1}[\alpha_{ij}][\gamma_{ij}];$$

and therefore, by properties (1), (2), and (3), we have

$$\begin{aligned} \det([\beta_{ij}]) &= \det([\gamma_{ij}]^{-1}[\alpha_{ij}][\gamma_{ij}]) \\ &= \det([\gamma_{ij}]^{-1})\det([\alpha_{ij}])\det([\gamma_{ij}]) \\ &= \det([\gamma_{ij}]^{-1})\det([\gamma_{ij}])\det([\alpha_{ij}]) \\ &= \det([\gamma_{ij}]^{-1}[\gamma_{ij}])\det([\alpha_{ij}]) \\ &= \det([\delta_{ij}])\det([\alpha_{ij}]) \\ &= \det([\alpha_{ij}]). \end{aligned}$$

This result allows us to speak of *the determinant of the operator T*, meaning, of course, the determinant of its matrix relative to any basis; and from this point on, we shall regard the determinant function primarily as a scalar-valued function of the operators on H. We at once obtain the following four properties for this function, which are simply translations of those stated above:

(1′) $\det(I) = 1$;

(2′) $\det(T_1 T_2) = \det(T_1)\det(T_2)$;

(3′) $\det(T) \neq 0 \Leftrightarrow T$ is non-singular; and

(4′) $\det(T - \lambda I)$ is a polynomial, with complex coefficients, of degree n in the variable λ.

We are now in a position to take up once again, and to settle, the problem of the existence of eigenvalues.

Let T be an operator on H. If we recall Problem 44-6, it is clear that a scalar λ is an eigenvalue of $T \Leftrightarrow$ there exists a non-zero vector x such that $(T - \lambda I)x = 0 \Leftrightarrow T - \lambda I$ is singular $\Leftrightarrow \det(T - \lambda I) = 0$. The eigenvalues of T are therefore precisely the distinct roots of the equation

$$\det(T - \lambda I) = 0, \tag{1}$$

which is called the *characteristic equation* of T.   It may illuminate matters somewhat if we choose a basis B for H, find the matrix $[\alpha_{ij}]$ of T relative to B, and write the characteristic equation in the extended form

$$\begin{vmatrix} \alpha_{11} - \lambda & \alpha_{12} & \cdots & \alpha_{1n} \\ \alpha_{21} & \alpha_{22} - \lambda & \cdots & \alpha_{2n} \\ \cdots & \cdots & \cdots & \cdots \\ \alpha_{n1} & \alpha_{n2} & \cdots & \alpha_{nn} - \lambda \end{vmatrix} = 0.$$

Our search for eigenvalues of T is reduced in this way to a search for roots of Eq. (1).   Property (4′) tells us that this is a polynomial equation, with complex coefficients, of degree n in the complex variable λ.   We now appeal to the fundamental theorem of algebra, which guarantees that an equation of this kind always has exactly n complex roots.   Some of these roots may of course be repeated, in which case there are fewer than n distinct roots.   In summary, we have

Theorem A.   *If T is an arbitrary operator on H, then the eigenvalues of T constitute a non-empty finite subset of the complex plane.   Furthermore, the number of points in this set does not exceed the dimension n of the space H.*

The set of eigenvalues of T is called its *spectrum*, and is denoted by $\sigma(T)$.   For future reference, we observe that $\sigma(T)$ is a compact subspace of the complex plane.

It should now be reasonably clear why we required in the definition of a Hilbert space that its scalars be the complex numbers.   The reader will easily convince himself that in the Euclidean plane the operation of rotation about the origin through 90 degrees is an operator on this real Banach space which has no eigenvalues at all, for no non-zero vector is transformed into a real multiple of itself.   The existence of eigenvalues is therefore linked in an essential way to properties of the complex numbers which are not enjoyed by the real numbers, and the most significant of these properties is that stated in the fundamental theorem of algebra.   The mechanism of matrices and determinants turns out to be simply a device for making effective use of this theorem in our basic problem of proving that eigenvalues exist.   We also remark that Theorem A and its proof remain valid in the case of an arbitrary linear transformation on any complex linear space of finite dimension $n > 0$.

Problems

1.   Let T be an operator on H, and prove the following statements:
    (a) T is singular $\Leftrightarrow$ $0 \in \sigma(T)$;
    (b)   if T is non-singular, then $\lambda \in \sigma(T) \Leftrightarrow \lambda^{-1} \in \sigma(T^{-1})$;
    (c)   if A is non-singular, then $\sigma(ATA^{-1}) = \sigma(T)$;

(d)   if $\lambda \, \varepsilon \, \sigma(T)$, and if p is any polynomial, then $p(\lambda) \, \varepsilon \, \sigma(p(T))$;

(e)   if $T^k = 0$ for some positive integer k, then $\sigma(T) = \{0\}$.

2.   Let the dimension n of H be 2, let $B = \{e_1, e_2\}$ be a basis for H, and assume that the determinant of a 2×2 matrix $[\alpha_{ij}]$ is given by $\alpha_{11}\alpha_{22} - \alpha_{12}\alpha_{21}$.

(a)   Find the spectrum of the operator T on H defined by $Te_1 = e_2$ and $Te_2 = -e_1$.

(b)   If T is an arbitrary operator on H whose matrix relative to B is $[\alpha_{ij}]$, show that $T^2 - (\alpha_{11} + \alpha_{22})T + (\alpha_{11}\alpha_{22} - \alpha_{12}\alpha_{21})I = 0$. Give a verbal statement of this result.

62. THE SPECTRAL THEOREM

We now return to the central purpose of this chapter, namely, the statement and proof of the spectral theorem.

Let T be an arbitrary operator on H. We know by Theorem 61-A that the distinct eigenvalues of T form a non-empty finite set of complex numbers. Let $\lambda_1, \lambda_2, \ldots, \lambda_m$ be these eigenvalues; let $M_1, M_2, \ldots, M_m$ be their corresponding eigenspaces; and let $P_1, P_2, \ldots, P_m$ be the projections on these eigenspaces. We consider the following three statements.

I. The M_i's are pairwise orthogonal and span H.

II. The P_i's are pairwise orthogonal, $I = \Sigma_{i=1}^{m} P_i$, and $T = \Sigma_{i=1}^{m} \lambda_i P_i$.

III. T is normal.

We take the *spectral theorem* to be the assertion that these statements are all equivalent to one another. It was proved in the introduction to this chapter that I $\Rightarrow$ II $\Rightarrow$ III. We now complete the cycle by showing that III $\Rightarrow$ I.

The hypothesis that T is normal plays its most critical role in our first theorem.

Theorem A.   *If T is normal, then x is an eigenvector of T with eigenvalue $\lambda \Leftrightarrow x$ is an eigenvector of T^* with eigenvalue $\bar{\lambda}$.*

PROOF.   Since T is normal, it is easy to see that the operator $T - \lambda I$ (whose adjoint is $T^* - \bar{\lambda}I$) is also normal for any scalar λ. By Theorem 58-C, we have

$$\|Tx - \lambda x\| = \|T^*x - \bar{\lambda}x\|$$

for every vector x, and the statements of the theorem follow at once from this.

The way is now clear for

Theorem B.   *If T is normal, then the M_i's are pairwise orthogonal.*

PROOF. Let x_i and x_j be vectors in M_i and M_j for $i \neq j$, so that $Tx_i = \lambda_i x_i$ and $Tx_j = \lambda_j x_j$. The preceding theorem shows that

$$\lambda_i(x_i,x_j) = (\lambda_i x_i,x_j) = (Tx_i,x_j) = (x_i,T^*x_j)$$
$$= (x_i,\overline{\lambda_j}x_j) = \lambda_j(x_i,x_j);$$

and since $\lambda_i \neq \lambda_j$, it is clear that we must have $(x_i,x_j) = 0$.

Our next step is to prove that the M_i's span H when T is normal, and for this we need the following preliminary fact.

Theorem C. *If T is normal, then each M_i reduces T.*

PROOF. It is obvious that each M_i is invariant under T, so it suffices, by Theorem 59-C, to show that each M_i is also invariant under T^*. This is an immediate consequence of Theorem A, for if x_i is a vector in M_i, so that $Tx_i = \lambda_i x_i$, then $T^*x_i = \overline{\lambda_i}x_i$ is also in M_i.

Finally, we have

Theorem D. *If T is normal, then the M_i's span H.*

PROOF. The fact that the M_i's are pairwise orthogonal implies, by Theorems 59-F and 59-G, that $M = M_1 + M_2 + \cdots + M_m$ is a closed linear subspace of H, and that its associated projection is

$$P = P_1 + P_2 + \cdots + P_m.$$

Since each M_i reduces T, we see by Theorem 59-E that $TP_i = P_iT$ for each P_i. It follows from this that $TP = PT$, so M also reduces T, and consequently $M^\perp$ is invariant under T. If $M^\perp \neq \{0\}$, then, since all the eigenvectors of T are contained in M, the restriction of T to $M^\perp$ is an operator on a non-trivial finite-dimensional Hilbert space which has no eigenvectors, and hence no eigenvalues. Theorem 61-A shows that this is impossible. We therefore conclude that $M^\perp = \{0\}$, so $M = H$ and the M_i's span H.

This completes the proof of the spectral theorem and, in particular, of the fact that if T is normal, then it has a *spectral resolution*

$$T = \lambda_1 P_1 + \lambda_2 P_2 + \cdots + \lambda_m P_m. \tag{1}$$

We now make several observations which will be useful in carrying out our promise to show that this expression for T is unique. Since the P_i's are pairwise orthogonal, if we square both sides of (1) we obtain

$$T^2 = \sum_{i=1}^{m} \lambda_i{}^2 P_i.$$

More generally, if n is any positive integer, then

$$T^n = \sum_{i=1}^{m} \lambda_i{}^n P_i. \tag{2}$$

If we make the customary agreement that $T^0 = I$, then the fact that $I = \Sigma_{i=1}^m P_i$ shows that (2) is also valid for the case $n = 0$. Next, let $p(z)$ be any polynomial, with complex coefficients, in the complex variable z. By taking linear combinations, (2) can evidently be extended to

$$p(T) = \sum_{i=1}^m p(\lambda_i)P_i. \tag{3}$$

We would like to find a polynomial p such that the right side of (3) collapses to a specified one of the P_i's, say P_j. What is needed is a polynomial p_j with the property that $p_j(\lambda_i) = 0$ if $i \neq j$ and $p_j(\lambda_j) = 1$. We define p_j as follows:

$$p_j(z) = \frac{(z - \lambda_1) \cdots (z - \lambda_{j-1})(z - \lambda_{j+1}) \cdots (z - \lambda_m)}{(\lambda_j - \lambda_1) \cdots (\lambda_j - \lambda_{j-1})(\lambda_j - \lambda_{j+1}) \cdots (\lambda_j - \lambda_m)}.$$

Since p_j is a polynomial, and since $p_j(\lambda_i) = \delta_{ij}$, (3) yields

$$P_j = p_j(T). \tag{4}$$

In order to interpret these remarks to our advantage, we point out that only three facts about (1) have been used in obtaining (4): the λ_i's are distinct complex numbers; the P_i's are pairwise orthogonal projections; and $I = \Sigma_{i=1}^m P_i$. By using these properties of (1), and these alone, we have shown that the P_i's are uniquely determined as specific polynomials in T.

We now assume that we have another expression for T similar to (1),

$$T = \alpha_1 Q_1 + \alpha_2 Q_2 + \cdots + \alpha_k Q_k, \tag{5}$$

and that this is also a spectral resolution of T, in the sense that the α_i's are distinct complex numbers, the Q_i's are non-zero pairwise orthogonal projections, and $I = \Sigma_{i=1}^k Q_i$. We wish to show that (5) is actually identical with (1), except for notation and order of terms. We begin by proving, in two steps, that the α_i's are precisely the eigenvalues of T. First, since $Q_i \neq 0$, there exists a non-zero vector x in the range of Q_i; and since $Q_i x = x$ and $Q_j x = 0$ for $j \neq i$, we see from (5) that $Tx = \alpha_i x$, so each α_i is an eigenvalue of T. Next, if λ is an eigenvalue of T, so that $Tx = \lambda x$ for some non-zero x, then

$$Tx = \lambda x = \lambda I x = \lambda \sum_{i=1}^k Q_i x = \sum_{i=1}^k \lambda Q_i x$$

and

$$Tx = \sum_{i=1}^k \alpha_i Q_i x,$$

so

$$\sum_{i=1}^k (\lambda - \alpha_i) Q_i x = 0.$$

Since the $Q_i x$'s are pairwise orthogonal, the non-zero vectors among them—there is at least one, for $x \neq 0$—are linearly independent, and this implies that $\lambda = \alpha_i$ for some i. These arguments show that the set of α_i's equals the set of λ_i's, and therefore, by changing notation if necessary, we can write (5) in the form

$$T = \lambda_1 Q_1 + \lambda_2 Q_2 + \cdots + \lambda_m Q_m. \tag{6}$$

The discussion in the preceding paragraph now applies to (6) and gives

$$Q_j = p_j(T) \tag{7}$$

for every j. On comparing (7) with (4), we see that the Q_j's equal the P_j's. This shows that (5) is exactly the same as (1)—except for notation and the order of terms—and completes our proof of the fact that the spectral resolution of T is unique.

We conclude with a brief look at the matrix interpretation of statements I and II at the beginning of this section. Assume that I is true, that is, that the eigenspaces $M_1, M_2, \ldots, M_m$ of T are pairwise orthogonal and span H. For each M_i, choose a basis which consists of mutually orthogonal unit vectors. This can always be done, for a basis of this kind—called an *orthonormal basis*—is precisely a complete orthonormal set for M_i. It is easy to see that the union of these little bases is an orthonormal basis for all of H; and relative to this, the matrix of T has the following diagonal form (all entries off the main diagonal are understood to be 0):

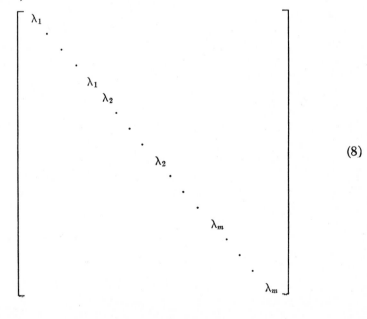

$$\tag{8}$$

We next assume that H has an orthonormal basis relative to which the matrix of T is diagonal.   If we rearrange the basis vectors in such a way that equal matrix entries adjoin one another on the main diagonal, then the matrix of T relative to this new orthonormal basis will have the form (8).   It is easy to see from this that T can be written in the form

$$T = \sum_{i=1}^{m} \lambda_i P_i,$$

where the λ_i's are distinct complex numbers, the P_i's are non-zero pairwise orthogonal projections, and $I = \Sigma_{i=1}^{m} P_i$.   The uniqueness of the spectral resolution now guarantees that the λ_i's are the distinct eigenvalues of T and that the P_i's are the projections on the corresponding eigenspaces. The spectral theorem tells us that statements I, II, and III are equivalent to one another.   The above remarks carry us a bit further, for they constitute a proof of the fact that these statements are also equivalent to

IV. There exists an orthonormal basis for H relative to which the matrix of T is diagonal.

It is interesting to realize that the implication III $\Rightarrow$ IV, which we proved by showing that III $\Rightarrow$ I and I $\Rightarrow$ IV, can be made to depend more directly on matrix computations.   This proof is outlined in the last three problems below.

Problems

1. Show that an operator T on H is normal $\Leftrightarrow$ its adjoint T^* is a polynomial in T.

2. Let T be an arbitrary operator on H, and N a normal operator. Show that if T commutes with N, then T also commutes with N^*.

3. Let T be a normal operator on H with spectrum $\{\lambda_1, \lambda_2, \ldots, \lambda_m\}$, and use the spectral resolution of T to prove the following statements: (a) T is self-adjoint $\Leftrightarrow$ each λ_i is real; (b) T is positive $\Leftrightarrow \lambda_i \geq 0$ for each i; (c) T is unitary $\Leftrightarrow |\lambda_i| = 1$ for each i.

4. Show that a positive operator T on H has a unique positive square root; that is, show that there exists a unique positive operator A on H such that $A^2 = T$.

5. Let $B = \{e_1, e_2, \ldots, e_n\}$ be an orthonormal basis for H.   If T is an operator on H whose matrix relative to B is $[\alpha_{ij}]$, show that the matrix of T^* relative to B is $[\beta_{ij}]$, where $\beta_{ij} = \overline{\alpha_{ji}}$. $[\beta_{ij}]$ is often called the *conjugate transpose* of $[\alpha_{ij}]$.

6. Let T be an arbitrary operator on H, and prove that there exist n closed linear subspaces $M_1, M_2, \ldots, M_n$ such that

$$\{0\} \subset M_1 \subset M_2 \subset \cdots \subset M_n = H,$$

the dimension of each M_i is i, and each M_i is invariant under T

(*Hint:* if $n = 1$, the statement is clear; and if $n > 1$, assume it for all Hilbert spaces of dimension $n - 1$, and prove it for H by using Theorem 59-B and the fact that T^* has an eigenvector.)

7. Let T be an arbitrary operator on H, and use the previous problem to show that there exists a basis B relative to which the matrix $[\alpha_{ij}]$ of T is *triangular*, in the sense that $i > j \Rightarrow \alpha_{ij} = 0$. If T is normal, show that there exists an orthonormal basis B' relative to which the matrix of T is diagonal. (*Hint:* generate B' by applying the Gram-Schmidt process to B, observe that the matrix of T relative to B' is still triangular, and use Problem 5 to show that this matrix is actually diagonal.)

63. A SURVEY OF THE SITUATION

The spectral theorem is often stated in a somewhat more restricted form than that given in the previous section. The usual version is that each normal operator N on H has a spectral resolution, that is, that there exist distinct complex numbers $\lambda_1, \lambda_2, \ldots, \lambda_m$ and non-zero pairwise orthogonal projections $P_1, P_2, \ldots, P_m$ such that $\Sigma_{i=1}^m P_i = I$, with the property that

$$N = \sum_{i=1}^m \lambda_i P_i. \tag{1}$$

In our version, we attempted to give equal emphasis to both the geometric and the algebraic sides of the matter. Most writers, however, confine their statement of the theorem to that given above, and for a very good reason: it is (1) that generalizes to the infinite-dimensional case.

There are two ways of carrying out this generalization, and we give a brief description of each.

First, there is the analytic approach. For the sake of simplicity, we consider a self-adjoint operator A, and we write (1) in the form

$$A = \sum_{i=1}^m \lambda_i P_i. \tag{2}$$

Our reason for making this assumption is that the eigenvalues of A are real numbers and are therefore ordered in a natural way. We further assume that the notation in (2) is chosen so that $\lambda_1 < \lambda_2 < \cdots < \lambda_m$, and we use the P_i's to define new projections:

$$E_{\lambda_0} = 0;$$
$$E_{\lambda_1} = P_1;$$
$$E_{\lambda_2} = P_1 + P_2;$$
$$\cdots \cdots \cdots \cdots \cdots \cdots \cdots \cdots$$
$$E_{\lambda_m} = P_1 + P_2 + \cdots + P_m.$$

(The subscript λ_0 is introduced solely for notational convenience and has no significance beyond this.)   The E_{λ_i}'s enable us to rewrite (2) as follows:

$$
\begin{aligned}
A &= \lambda_1 P_1 + \lambda_2 P_2 + \cdots + \lambda_m P_m \\
&= \lambda_1 (E_{\lambda_1} - E_{\lambda_0}) + \lambda_2 (E_{\lambda_2} - E_{\lambda_1}) + \cdots + \lambda_m (E_{\lambda_m} - E_{\lambda_{m-1}}) \\
&= \sum_{i=1}^{m} \lambda_i (E_{\lambda_i} - E_{\lambda_{i-1}}).
\end{aligned}
$$

If we denote $E_{\lambda_i} - E_{\lambda_{i-1}}$ by ΔE_{λ_i}, then we can compress this to

$$
A = \sum_{i=1}^{m} \lambda_i \, \Delta E_{\lambda_i},
$$

which suggests an integral representation

$$
A = \int \lambda \, dE_\lambda. \tag{3}
$$

In this form, the spectral resolution remains valid for self-adjoint operators on infinite-dimensional Hilbert spaces.   A similar result holds for normal operators,

$$
N = \int \lambda \, dE_\lambda. \tag{4}
$$

There are many difficulties to be surmounted in reaching the level of (3) and (4).   We have already met one of these, namely, the fact that an operator T on an arbitrary Hilbert space $H \neq \{0\}$ need not have any eigenvalues at all.   In this general case, the *spectrum* of T is defined by

$$
\sigma(T) = \{\lambda : T - \lambda I \text{ is singular}\}.
$$

When H is finite-dimensional, we have seen that $\sigma(T)$ consists entirely of eigenvalues.   This made our work in the present chapter relatively easy, but it is not true in general.   What is true is that $\sigma(T)$ is always non-empty, closed, and bounded, and is thus a compact subspace of the complex plane.   Once this difficulty is dealt with, there remain substantial problems in giving meaning to integrals like those in (3) and (4) and in proving the validity of these relations.[1]

The second approach to generalizing (1) is essentially algebraic and topological in nature.   Its starting point is the observation made in the previous section that the spectral resolution

$$
N = \sum_{i=1}^{m} \lambda_i P_i
$$

[1] For a general discussion of the spectral theorem from this analytic point of view, see Lorch [28].   A full treatment can be found in Riesz and Sz.-Nagy [35, chap. 7].

leads to, and is actually part of, the fact that

$$p(N) = \sum_{i=1}^{m} p(\lambda_i)P_i \qquad (5)$$

for any polynomial p. The set of all polynomials in N is evidently an algebra of operators, a subalgebra of $\mathcal{B}(H)$. Let us now consider the corresponding algebra of all polynomial functions defined on the set of λ_i's. We have seen that this algebra contains polynomials p_j such that $p_j(\lambda_i) = \delta_{ij}$, and it therefore consists of all complex functions defined on the set of λ_i's. If we denote the latter set by X for a moment and think of it as a compact subspace of the complex plane, then, since X is finite, the algebra in question is precisely $\mathcal{C}(X)$, the algebra of all continuous complex functions defined on X. The mapping

$$p(N) \rightarrow p, \qquad (6)$$

which makes correspond to each $p(N)$ the function p in $\mathcal{C}(X)$, is easily seen by the properties of (5) to preserve all algebraic operations. We know from the previous section that the eigenvalues of $p(N)$ are—with possible repetitions—the $p(\lambda_i)$'s on the right of (5); and we shall see later, as an unexpected dividend, that the norm of a normal operator always equals the maximum of the absolute values of its eigenvalues. It follows from these remarks that the mapping (6) is an isometric isomorphism of the algebra of all $p(N)$'s onto $\mathcal{C}(X)$. These ideas constitute an extended version of (5) and are thus, in a sense, a generalization of the spectral resolution of N. They apply virtually without change to the case of a normal operator on an infinite-dimensional Hilbert space, and one of our aims in the next three chapters is to treat them in detail.

PART THREE

Algebras of Operators

CHAPTER TWELVE

General

Preliminaries on Banach Algebras

Our work in Part 1 of this book was primarily concerned with topological spaces and the continuous functions carried by them.

The ideas of Part 2, on the other hand, were essentially algebraic in nature. The function spaces we encountered earlier led us to begin a study of Banach spaces for their own sake, and as we proceeded, we found our attention focusing more and more closely on the properties of their operators. Except for a few elementary notions about metric spaces, we used very little genuine topology in Part 2. As a matter of fact, our treatment of spectral theory in Chap. 11 was completely independent of topology, for in the finite-dimensional case, all linear transformations are continuous.

In the following three chapters, these two apparently diverse trains of thought—the topological and the algebraic—are united by a single elegant concept: that of a Banach algebra. Our remarks in the last section of the previous chapter suggested that there may be important links between algebras of operators on Hilbert spaces and algebras of the type $\mathcal{C}(X)$, where X is a compact Hausdorff space. Banach algebras are the systems which enable us to establish these connections on a firm footing. They are also interesting in that they constitute a field of study in which a wide variety of mathematical ideas meet in significant contact.

Our main task in the present chapter is to provide a number of miscellaneous tools which are necessary for the structure theory developed later.

64. THE DEFINITION AND SOME EXAMPLES

A *Banach algebra* is a complex Banach space which is also an algebra with identity 1, and in which the multiplicative structure is related to the norm by the following requirements:

(1) $\|xy\| \leq \|x\| \, \|y\|$;

(2) $\|1\| = 1$.

It follows from (1) that multiplication is jointly continuous in any Banach algebra, that is, that if $x_n \to x$ and $y_n \to y$, then $x_n y_n \to xy$ (proof:

$$\|x_n y_n - xy\| = \|x_n(y_n - y) + (x_n - x)y\| \leq \|x_n\| \, \|y_n - y\| + \|x_n - x\| \, \|y\|).$$

A *Banach subalgebra* of a Banach algebra A is a closed subalgebra of A which contains 1. The Banach subalgebras of A are precisely those subsets of A which are themselves Banach algebras with respect to the same algebraic operations, the same identity, and the same norm.

The definition of a Banach algebra is sometimes given without the restriction that the scalars are the complex numbers. The complex case, however, is the only one that concerns us, and by framing the definition as we do, we avoid the necessity of treating the additional complications which arise in the real case. We have further assumed, for the sake of simplicity, that every Banach algebra has an identity. It is possible, at a considerable sacrifice of clarity, to develop most of the important ideas without this assumption, and this is done whenever the primary purpose of the theory is the study of group algebras of locally compact but not discrete groups. Since our attention will be directed chiefly to the structure of operator algebras, there is no need for us to strain for the added generality obtained by not requiring the presence of an identity.

The Banach algebras of principal interest to us are described in the following examples. The reader will notice that they all consist of functions or operators and that the linear operations in all of them are defined pointwise. They can be classified in a general way into *function algebras*, *operator algebras*, or *group algebras*, according as multiplication is defined pointwise, by composition, or by convolution.

Example 1. (*a*) One of the most important Banach algebras is the set $\mathcal{C}(X)$ of all bounded continuous complex functions defined on a topological space X. The case in which X is a compact Hausdorff space will have particular significance for our later work. If X has only one point, then $\mathcal{C}(X)$ can be identified with the simplest of all Banach algebras, the algebra of complex numbers.

(*b*) Consider the closed unit disc $D = \{z : |z| \leq 1\}$ in the complex plane. The subset of $\mathcal{C}(D)$ which consists of all functions analytic in the interior of D is obviously a subalgebra which contains the identity. A simple application of Morera's theorem from complex analysis shows that it is closed and is therefore a Banach subalgebra of $\mathcal{C}(D)$. This Banach algebra is called the *disc algebra*. It has a number of interesting properties, which are, of course, intimately related to the special character of its functions.

Example 2. (*a*) If B is a non-trivial complex Banach space, then the set $\mathcal{B}(B)$ of all operators on B is a Banach algebra. We assume that B is non-trivial in order to guarantee that the identity operator is an identity in the algebraic sense.

(*b*) If we consider a non-trivial Hilbert space H, then $\mathcal{B}(H)$ is a Banach algebra. This is a special case of $\mathcal{B}(B)$, and it is important to observe that additional structure is present here, namely, the adjoint operation $T \rightarrow T^*$.

(*c*) A subalgebra of $\mathcal{B}(H)$ is said to be *self-adjoint* if it contains the adjoint of each of its operators. Banach subalgebras of $\mathcal{B}(H)$'s which are self-adjoint are called *C*-algebras*. We shall return to the subject of commutative C^*-algebras in Chap. 14.

(*d*) The *weak operator topology* on $\mathcal{B}(H)$ is the weak topology generated by all functions of the form $T \rightarrow (Tx,y)$; that is, it is the weakest topology with respect to which all these functions are continuous. It is easy to see from the inequality $|(Tx,y) - (T_0 x,y)| \leq \|T - T_0\| \, \|x\| \, \|y\|$ that this topology is weaker than the usual norm topology, so that its closed sets are also closed in the usual sense. A C^*-algebra with the further property of being closed in the weak operator topology is called a *W*-algebra*. Algebras of this kind are also called *rings of operators*, or *von Neumann algebras*. They are among the most interesting of all Banach algebras, but their theory is quite beyond the scope of this book.[1]

Example 3. (*a*) If $G = \{g_1, g_2, \ldots, g_n\}$ is a finite group, then its *group algebra* $L_1(G)$ is the set of all complex functions defined on G. Addition and scalar multiplication are defined pointwise, and the norm by $\|f\| = \Sigma_{i=1}^n |f(g_i)|$. In order to see what underlies the definition of multiplication, it is convenient to regard a typical element f of $L_1(G)$ as a formal sum $\Sigma_{i=1}^n \alpha_i g_i$, where α_i is the value of f at g_i. With this interpretation, we use the given multiplication in G to define multiplication in $L_1(G)$, as follows:

$$\left(\sum_{i=1}^n \alpha_i g_i \right) \left(\sum_{j=1}^n \beta_j g_j \right) = \sum_{k=1}^n \gamma_k g_k, \tag{1}$$

where
$$\gamma_k = \sum_{g_i g_j = g_k} \alpha_i \beta_j. \tag{2}$$

The meaning of the sum in (2) is that the summation is to be extended

---

[1] See Dixmier [7].

over all subscripts i and j such that $g_i g_j = g_k$. In effect, therefore, we formally multiply out the sums on the left of (1), and we then gather together all the resulting terms which contain the same element of G. With these ideas as an intuitive guide, we revert to our first point of view, in which the elements of $L_1(G)$ are functions, and we see that our definition of multiplication can be expressed in the following way. If two functions f and g in $L_1(G)$ are given, then their product, which is denoted by $f * g$ and called their *convolution*, is that function whose value at g_k is

$$
\begin{aligned}
(f * g)(g_k) &= \sum_{g_i g_j = g_k} f(g_i) g(g_j) \\
&= \sum_{j=1}^{n} f(g_k g_j^{-1}) g(g_j).
\end{aligned}
\tag{3}
$$

We note that if each element of G is identified with the function whose value is 1 at that element and 0 elsewhere, then G becomes a subset of $L_1(G)$. Further, multiplication in G agrees with convolution in $L_1(G)$, and the element of $L_1(G)$ which corresponds to the identity in G is an identity for $L_1(G)$. We conclude this description by observing that every element of G has norm 1, so that $\|1\| = 1$, and that the basic norm inequality for a Banach algebra is satisfied:

$$
\begin{aligned}
\|f * g\| &= \sum_{k=1}^{n} |(f * g)(g_k)| \\
&= \sum_{k=1}^{n} \left| \sum_{j=1}^{n} f(g_k g_j^{-1}) g(g_j) \right| \\
&\leq \sum_{k=1}^{n} \sum_{j=1}^{n} |f(g_k g_j^{-1})| \, |g(g_j)| \\
&= \sum_{j=1}^{n} \sum_{k=1}^{n} |f(g_k g_j^{-1})| \, |g(g_j)| \\
&= \sum_{j=1}^{n} |g(g_j)| \sum_{k=1}^{n} |f(g_k g_j^{-1})| \\
&= \sum_{j=1}^{n} |g(g_j)| \, \|f\| \\
&= \|f\| \sum_{j=1}^{n} |g(g_j)| \\
&= \|f\| \, \|g\|.
\end{aligned}
$$

(b) Let $G = \{\ldots, -2, -1, 0, 1, 2, \ldots\}$ be the additive group of integers. Its group algebra $L_1(G)$ is the set of all complex functions f defined on G for which $\sum_{n=-\infty}^{\infty} |f(n)|$ converges. The linear operations are defined pointwise, the norm by $\|f\| = \sum_{n=-\infty}^{\infty} |f(n)|$, and the convolution of f and g—see Eq. (3)—by

$$(f * g)(n) = \sum_{m=-\infty}^{\infty} f(n - m)g(m).$$

Just as in (a), G is contained in $L_1(G)$ in a natural way, and $L_1(G)$ is a Banach algebra. Any attempt to discuss the group algebra of a nondiscrete topological group like the real line must clearly be based on an adequate theory of integration. It should also have available a theory of Banach algebras in which no identity is assumed to be present. These ideas constitute a rich and beautiful field of modern analysis. They are, however, outside the scope of this work.[1]

The Banach algebras described above are many and diverse, and there are yet others which we have not mentioned. Our attention in the following chapters will be centered on $\mathcal{C}(X)$'s and commutative C^*-algebras, but the general theory we develop is equally applicable to all. It is worthy of notice that an arbitrary Banach algebra A can be regarded as a Banach subalgebra of $\mathcal{B}(A)$. In a sense, therefore, Example $2a$ and its Banach subalgebras include all possible Banach algebras. To see this, we recall from Problem 45-4 that $a \rightarrow M_a$, where $M_a(x) = ax$, is an isomorphism of A into $\mathcal{B}(A)$. It is easy to see that M_1 is the identity operator on A, so all that remains is to observe that $\|a\| = \|M_a\|$ for every a (proof: $\|M_a(x)\| = \|ax\| \leq \|a\|\,\|x\|$ shows that $\|M_a\| \leq \|a\|$, and the fact that $\|a\| \leq \|M_a\|$ follows from

$$\|M_a\| = \sup\,\{\|M_a(x)\| : \|x\| \leq 1\} \geq \|M_a(1)\| = \|a\|).$$

The mapping $a \rightarrow M_a$ is thus an isometric isomorphism of A onto a Banach subalgebra of $\mathcal{B}(A)$, and it allows us to identify the abstract Banach algebra A with a concrete Banach algebra of operators on A.

65. REGULAR AND SINGULAR ELEMENTS

Let A be a Banach algebra. We denote the set of regular elements in A by G, and its complement, the set of singular elements, by S. It is clear that G contains 1 and is a group, and that S contains 0. Several important issues depend on the character of G and S. Our first result along these lines is

---

[1] Loomis [27] is the standard reference in this subject. For a general exposition of the main ideas, see Mackey [30]. A brief treatment of the classical analysis which underlies the modern theory can be found in Goldberg [14].

Theorem A. *Every element x for which $\|x - 1\| < 1$ is regular, and the inverse of such an element is given by the formula $x^{-1} = 1 + \Sigma_{n=1}^{\infty} (1 - x)^n$.*
PROOF. If we put $r = \|x - 1\|$, so that $r < 1$, then

$$\|(1 - x)^n\| \leq \|1 - x\|^n = r^n$$

shows that the partial sums of the series $\Sigma_{n=1}^{\infty} (1 - x)^n$ form a Cauchy sequence in A. Since A is complete, these partial sums converge to an element of A, which we denote by $\Sigma_{n=1}^{\infty} (1 - x)^n$. If we define y by $y = 1 + \Sigma_{n=1}^{\infty}(1 - x)^n$, then the joint continuity of multiplication in A implies that

$$y - xy = (1 - x)y = (1 - x) + \sum_{n=2}^{\infty} (1 - x)^n = \sum_{n=1}^{\infty} (1 - x)^n = y - 1,$$

so $xy = 1$. Similarly, $yx = 1$.

We now use this as a tool to prove

Theorem B. *G is an open set, and therefore S is a closed set.*
PROOF. Let x_0 be an element in G, and let x be any element in A such that $\|x - x_0\| < 1/\|x_0^{-1}\|$. It is clear that

$$\|x_0^{-1}x - 1\| = \|x_0^{-1}(x - x_0)\| \leq \|x_0^{-1}\| \, \|x - x_0\| < 1,$$

so we see by Theorem A that $x_0^{-1}x$ is in G. Since $x = x_0(x_0^{-1}x)$, it follows that x is also in G, so G is open.

It was shown in Problem 32-5 that every Banach space is locally connected, so A is also locally connected. A direct application of Theorem 34-A yields the fact that the components of G are themselves open sets.

As our final result, we have

Theorem C. *The mapping $x \to x^{-1}$ of G into G is continuous and is therefore a homeomorphism of G onto itself.*
PROOF. Let x_0 be an element of G, and x another element of G such that $\|x - x_0\| < 1/(2\|x_0^{-1}\|)$. Since

$$\|x_0^{-1}x - 1\| = \|x_0^{-1}(x - x_0)\| \leq \|x_0^{-1}\| \, \|x - x_0\| < \tfrac{1}{2},$$

we see by Theorem A that $x_0^{-1}x$ is in G and

$$x^{-1}x_0 = (x_0^{-1}x)^{-1} = 1 + \sum_{n=1}^{\infty} (1 - x_0^{-1}x)^n.$$

Our conclusion now follows from

$$\|x^{-1} - x_0^{-1}\| = \|(x^{-1}x_0 - 1)x_0^{-1}\| \leq \|x_0^{-1}\| \; \|x^{-1}x_0 - 1\| = \|x_0^{-1}\|$$

$$\|\sum_{n=1}^{\infty} (1 - x_0^{-1}x)^n\| \leq \|x_0^{-1}\| \sum_{n=1}^{\infty} \|1 - x_0^{-1}x\|^n$$

$$= \|x_0^{-1}\| \; \|1 - x_0^{-1}x\| \sum_{n=0}^{\infty} \|1 - x_0^{-1}x\|^n$$

$$= \frac{\|x_0^{-1}\| \; \|1 - x_0^{-1}x\|}{1 - \|1 - x_0^{-1}x\|}$$

$$< 2\|x_0^{-1}\| \; \|1 - x_0^{-1}x\| \leq 2\|x_0^{-1}\|^2\|x - x_0\|.$$

If x is an element in A, it should always be kept in mind that the regularity or singularity of x depends on A as well as on x itself. If x is regular in A, and if we pass to a Banach subalgebra A' of A which also contains x, then x may lose its inverse and become singular in A'. By the same token, if x is singular in A, and if A is regarded as a Banach subalgebra of a larger Banach algebra A'', then x may acquire an inverse and become regular in A''. In the next section, we study certain elements in A which are singular and remain singular with respect to all possible enlargements of A.

66. TOPOLOGICAL DIVISORS OF ZERO

An element z in our Banach algebra A is called a *topological divisor of zero* if there exists a sequence $\{z_n\}$ in A such that $\|z_n\| = 1$ and either $zz_n \to 0$ or $z_n z \to 0$. It is clear that every divisor of zero is also a topological divisor of zero. We denote the set of all topological divisors of zero by Z.

Theorem A. *Z is a subset of S.*

PROOF. Let z be an element of Z and $\{z_n\}$ a sequence such that $\|z_n\| = 1$ and (say) $zz_n \to 0$. If z were in G, then by the joint continuity of multiplication we would have $z^{-1}(zz_n) = z_n \to 0$, contrary to $\|z_n\| = 1$.

Our next theorem relates to the manner in which Z is distributed within S.

Theorem B. *The boundary of S is a subset of Z.*

PROOF. Since S is closed, its boundary consists of all points in S which are limits of convergent sequences in G. We show that if z is such a point, that is, if z is in S and there exists a sequence $\{r_n\}$ in G such that $r_n \to z$, then z is in Z. First, we see from $r_n^{-1}z - 1 = r_n^{-1}(z - r_n)$ that the sequence $\{r_n^{-1}\}$ is unbounded; for otherwise, we would have

$$\|r_n^{-1}z - 1\| < 1$$

for some n, so that $r_n^{-1}z$, and therefore $z = r_n(r_n^{-1}z)$, would be regular. Since $\{r_n^{-1}\}$ is unbounded, we may assume that $\|r_n^{-1}\| \to \infty$. If z_n is now defined by $z_n = r_n^{-1}/\|r_n^{-1}\|$, then our conclusion follows from the observations that $\|z_n\| = 1$ and

$$zz_n = \frac{zr_n^{-1}}{\|r_n^{-1}\|} = \frac{1 + (z - r_n)r_n^{-1}}{\|r_n^{-1}\|} = \frac{1}{\|r_n^{-1}\|} + (z - r_n)z_n \to 0.$$

In order to understand the significance of these facts, let us suppose that A is imbedded as a Banach subalgebra in a larger Banach algebra A'. As we remarked in the previous section, an element which is singular in A may cease to be so in A'. However, if it is a topological divisor of zero in A, then it is in A' as well, so it is singular in A'. The topological divisors of zero in A are thus "permanently singular," in the sense that they are singular and remain so with respect to every possible enlargement of the containing Banach algebra. Theorem B tells us that no matter what happens to S as a whole in such a process, its boundary is "permanent" in this sense.

67. THE SPECTRUM

Let T be an operator on a non-trivial Hilbert space. In the previous chapter, we defined the spectrum of T to be the set

$$\sigma(T) = \{\lambda : T - \lambda I \text{ is singular}\},$$

and we devoted a good deal of attention to the geometric ideas leading to this concept. We found—at least in the finite-dimensional case—that a number in $\sigma(T)$ is a *value assumed by* T, in the sense that T acts on some non-zero vector as if it were scalar multiplication by that number. We shall see later that this formulation of the meaning of the spectrum has a much wider significance than we might at first suspect.

Let us now consider an element x in our general Banach algebra A. By analogy with the above, we define the *spectrum* of x to be the following subset of the complex plane:

$$\sigma(x) = \{\lambda : x - \lambda 1 \text{ is singular}\}.$$

Whenever it is desirable to express the fact that the spectrum of x depends on A as well as x, we use the notation $\sigma_A(x)$. It is easy to see that $x - \lambda 1$ is a continuous function of λ with values in A; and since the set of singular elements in A is closed, it follows at once that $\sigma(x)$ is closed. We further observe that $\sigma(x)$ is a subset of the closed disc $\{z : |z| \leq \|x\|\}$, for if λ is a complex number such that $|\lambda| > \|x\|$, then $\|x/\lambda\| < 1$, $\|1 - (1 - x/\lambda)\|$ < 1, $1 - x/\lambda$ is regular, and therefore $x - \lambda 1$ is regular.

Our first task is to establish the fact that $\sigma(x)$ is always non-empty, and for this we need a few preliminary notions. The *resolvent set* of x, denoted by $\rho(x)$, is the complement of $\sigma(x)$; it is clearly an open subset of the complex plane which contains $\{z : |z| > \|x\|\}$. The *resolvent* of x is the function with values in A defined on $\rho(x)$ by

$$x(\lambda) = (x - \lambda 1)^{-1}.$$

Theorem 65-C tells us that $x(\lambda)$ is a continuous function of λ; and the fact that $x(\lambda) = \lambda^{-1}(x/\lambda - 1)^{-1}$ for $\lambda \neq 0$ implies that $x(\lambda) \to 0$ as $\lambda \to \infty$. If λ and μ are both in $\rho(x)$, then

$$\begin{aligned}
x(\lambda) &= x(\lambda)[x - \mu 1]x(\mu) \\
&= x(\lambda)[x - \lambda 1 + (\lambda - \mu)1]x(\mu) \\
&= [1 + (\lambda - \mu)x(\lambda)]x(\mu) \\
&= x(\mu) + (\lambda - \mu)x(\lambda)x(\mu),
\end{aligned}$$

so

$$x(\lambda) - x(\mu) = (\lambda - \mu)x(\lambda)x(\mu).$$

This relation is called the *resolvent equation*.

Theorem A. $\sigma(x)$ *is non-empty*.

PROOF.   Let f be a functional on A—that is, an element of the conjugate space A^*—and define $f(\lambda)$ by $f(\lambda) = f(x(\lambda))$. It is clear that $f(\lambda)$ is a complex function which is defined and continuous on the resolvent set $\rho(x)$. The resolvent equation shows that

$$\frac{f(\lambda) - f(\mu)}{\lambda - \mu} = f(x(\lambda)x(\mu)),$$

and it follows from this that

$$\lim_{\lambda \to \mu} \frac{f(\lambda) - f(\mu)}{\lambda - \mu} = f(x(\mu)^2),$$

so $f(\lambda)$ has a derivative at each point of $\rho(x)$. Further,

$$|f(\lambda)| \leq \|f\| \, \|x(\lambda)\|,$$

so $f(\lambda) \to 0$ as $\lambda \to \infty$. We now assume that $\sigma(x)$ is empty, so that $\rho(x)$ is the entire complex plane. Liouville's theorem from complex analysis allows us to conclude that $f(\lambda) = 0$ for all λ. Since f is an arbitrary functional on A, Theorem 48-B implies that $x(\lambda) = 0$ for all λ. This is impossible, for no inverse can equal 0, and therefore it cannot be true that $\sigma(x)$ is empty.

If the reader is surprised by the appearance of Liouville's theorem in such a context, he should recall two facts. First, our proof of Theorem 61-A, which is a special case of the above result, required the use of the fundamental theorem of algebra. And second, the fundamental theorem of algebra is most commonly proved as a simple consequence of Liouville's theorem. It is therefore only to be expected that some tool from analysis comparable in depth with Liouville's theorem should be necessary for the proof of Theorem A.

Now that we know that $\sigma(x)$ is non-empty, we also know that it is a compact subspace of the complex plane. The number $r(x)$ defined by

$$r(x) = \sup \{|\lambda| : \lambda \in \sigma(x)\}$$

is called the *spectral radius* of x. It is clear that $0 \leq r(x) \leq \|x\|$. The concept of the spectral radius will be useful in certain parts of our later work.

We recall that a division algebra is an algebra with identity in which each non-zero element is regular. The most important single consequence of Theorem A is

Theorem B. *If A is a division algebra, then it equals the set of all scalar multiples of the identity.*

PROOF. We must show that if x is an element of A, then x equals $\lambda 1$ for some scalar λ. Suppose, on the contrary, that $x \neq \lambda 1$ for every λ. Then $x - \lambda 1 \neq 0$ for every λ, $x - \lambda 1$ is regular for every λ, and therefore $\sigma(x)$ is empty. This contradicts Theorem A and completes the proof.

The mapping $\lambda 1 \to \lambda$ is clearly an isometric isomorphism of the set of all scalar multiples of the identity onto the Banach algebra C of all complex numbers. We may therefore identify this set with C; and in terms of this identification, Theorem B says that *any Banach algebra which is a division algebra equals C.* This fact is the foundation on which we build the structure theory presented in the next chapter.

It is obvious that C itself, which is the simplest of all Banach algebras, is a division algebra, so Theorem B characterizes C as the only Banach algebra with this property. In the next two theorems, we give some other interesting characterizations of C among all possible Banach algebras.

Since 0 is a divisor of zero, it is a topological divisor of zero in every Banach algebra. In the Banach algebra C, 0 is plainly the only topological divisor of zero. Conversely, we have

Theorem C. *If 0 is the only topological divisor of zero in A, then $A = C$.*

PROOF. Let x be an element of A. Its spectrum $\sigma(x)$ is non-empty, so it has a boundary point λ; and $x - \lambda 1$ is easily seen to be a boundary point of the set S of all singular elements. By Theorem 66-B, $x - \lambda 1$ is a topological divisor of zero, so it follows from our hypothesis that $x - \lambda 1 = 0$ or $x = \lambda 1$.

The basic link between multiplication in A and the norm is given by the inequality $\|xy\| \leq \|x\| \, \|y\|$, and when $A = C$, this inequality can be reversed. The following result shows to what extent this reversibility is true in general.

Theorem D. *If the norm in A satisfies the inequality $\|xy\| \geq K\|x\| \, \|y\|$ for some positive constant K, then $A = C$.*

PROOF. In the light of Theorem C, it suffices to observe that the hypothesis here implies that 0 is the only topological divisor of zero.

We next look into the question of what happens to the spectrum of an element x in A when A is enlarged.

Theorem E. *If A is a Banach subalgebra of a Banach algebra A', then the spectra of an element x in A with respect to A and A' are related as follows:* (1) $\sigma_{A'}(x) \subseteq \sigma_A(x)$; (2) *each boundary point of $\sigma_A(x)$ is also a boundary point of $\sigma_{A'}(x)$.*

PROOF. If $x - \lambda 1$ is singular in A', then it is certainly singular in A, so (1) is clear. To prove (2), we let λ be a boundary point of $\sigma_A(x)$. It is easy to see that $x - \lambda 1$ is a boundary point of the set of singular elements in A, so by Theorem 66-B, it is a topological divisor of zero in A. It is therefore a topological divisor of zero in A' as well, so it is singular in A' and λ is in $\sigma_{A'}(x)$. The fact that λ is actually a boundary point of $\sigma_{A'}(x)$ is immediate from (1), so the proof of (2) is complete.

This result shows that in general the spectrum of an element shrinks when its containing Banach algebra is enlarged, and further, that since its boundary points cannot be lost in this process, it must shrink by "hollowing out." An illuminating example of this phenomenon is provided by the disc algebra A of all complex functions which are defined and continuous on $D = \{z : |z| \leq 1\}$ and analytic in the interior of this set. If f is a function in A, then the maximum modulus theorem from complex analysis implies that

$$\|f\| = \sup \{|f(z)| : |z| \leq 1\}$$
$$= \sup \{|f(z)| : |z| = 1\}.$$

This allows us to identify A with the Banach algebra of all the restrictions of its functions to the boundary of D, which is a Banach subalgebra of $A' = C(\{z : |z| = 1\})$. If we now consider the element f in A defined by $f(z) = z$, then it is easy to see that $\sigma_A(f)$ equals D and that $\sigma_{A'}(f)$ equals the boundary of D.

68. THE FORMULA FOR THE SPECTRAL RADIUS

Let x be an element in our general Banach algebra A, and consider its spectral radius $r(x)$, which is defined by

$$r(x) = \sup \{|\lambda| : \lambda \in \sigma_A(x)\}.$$

Now let A' be the Banach subalgebra of A generated by x, that is, the closure of the set of all polynomials in x. Theorem 67-E shows that $r(x)$ has the same value if it is computed with respect to A':

$$r(x) = \sup \{|\lambda| : \lambda \in \sigma_{A'}(x)\}.$$

This suggests quite strongly that $r(x)$ depends only on the sequence of powers of x. The formula for $r(x)$ is given in Theorem A below, and our purpose in this section is to prove it. It is convenient to begin with the following preliminary result.

Lemma. $\sigma(x^n) = \sigma(x)^n$.

PROOF. Let λ be a non-zero complex number and $\lambda_1, \lambda_2, \ldots, \lambda_n$ its distinct nth roots, so that

$$x^n - \lambda 1 = (x - \lambda_1 1)(x - \lambda_2 1) \cdots (x - \lambda_n 1).$$

The statement of the lemma follows easily from the fact that $x^n - \lambda 1$ is singular $\Leftrightarrow x - \lambda_i 1$ is singular for at least one i.

Theorem A. $r(x) = \lim \|x^n\|^{1/n}$.

PROOF. Our lemma shows that $r(x^n) = r(x)^n$, and since $r(x^n) \leq \|x^n\|$, we have $r(x)^n \leq \|x^n\|$ or $r(x) \leq \|x^n\|^{1/n}$ for every n. To conclude the proof, it suffices to show that if a is any real number such that $r(x) < a$, then $\|x^n\|^{1/n} \leq a$ for all but a finite number of n's, and this we now do.

It follows from Theorem 65-A and our work in Sec. 67 that if $|\lambda| > \|x\|$, then

$$
\begin{aligned}
x(\lambda) = (x - \lambda 1)^{-1} &= \lambda^{-1} \left(\frac{x}{\lambda} - 1 \right)^{-1} \\
&= -\lambda^{-1} \left(1 - \frac{x}{\lambda} \right)^{-1} \\
&= -\lambda^{-1} \left[1 + \sum_{n=1}^{\infty} \frac{x^n}{\lambda^n} \right].
\end{aligned}
\tag{1}
$$

If f is any functional on A, then (1) yields

$$f(x(\lambda)) = -\lambda^{-1}\left[f(1) + \sum_{n=1}^{\infty} f\left(\frac{x^n}{\lambda^n}\right)\right]$$

$$= -\lambda^{-1}\left[f(1) + \sum_{n=1}^{\infty} f(x^n)\lambda^{-n}\right] \tag{2}$$

for all $|\lambda| > \|x\|$. We saw in the proof of Theorem 67-A that $f(x(\lambda))$ is an analytic function in the region $|\lambda| > r(x)$; and since (2) is its Laurent expansion for $|\lambda| > \|x\|$, we know from complex analysis that this expansion is valid for $|\lambda| > r(x)$. If we now let α be any real number such that $r(x) < \alpha < a$, then it follows from the preceding remark that the series $\sum_{n=1}^{\infty} f(x^n/\alpha^n)$ converges, so its terms form a bounded sequence. Since this is true for every f in A^*, an application of Theorem 51-B shows that the elements x^n/α^n form a bounded sequence in A. Thus

$$\|x^n/\alpha^n\| \leq K$$

or $\|x^n\|^{1/n} \leq K^{1/n}\alpha$ for some positive constant K and every n. Since $K^{1/n}\alpha \leq a$ for every sufficiently large n, we have $\|x^n\|^{1/n} \leq a$ for all but a finite number of n's, and the proof is complete.

The applications we make of this formula will appear in the next chapter.

69. THE RADICAL AND SEMI-SIMPLICITY

Our final preliminary task is to reach a clear understanding of what is meant by the statement that our Banach algebra A is semi-simple. For this, it is necessary to give an adequate definition of the radical of A, and this in turn depends on a detailed analysis of its ideals.

We recall that an *ideal* in A was defined in Sec. 45 to be a subset I with the following three properties:

(1) I is a linear subspace of A;
(2) $i \in I \Rightarrow xi \in I$ for every element $x \in A$;
(3) $i \in I \Rightarrow ix \in I$ for every element $x \in A$.

If I is assumed only to satisfy conditions (1) and (2) [or conditions (1) and (3)], it is called a *left ideal* (or a *right ideal*). For the sake of clarity and emphasis, an ideal in our previous sense—one which satisfies all three of these conditions—is often called a *two-sided ideal*. In the commutative case, of course, these three concepts coincide with one another.

The properties of the ideals in A are closely related to the properties of its regular and singular elements. In our work so far, the statement that an element x in A is *regular* has meant that there exists an element y such that $xy = yx = 1$. For our present purposes, it is useful to refine this notion slightly, as follows. We say that x is *left regular* if there exists an element y such that $yx = 1$; and if x is not left regular, it is called *left singular*. The terms *right regular* and *right singular* are defined similarly. If x is both left regular and right regular, so that there exist elements y and z such that $yx = 1$ and $xz = 1$, then the relation

$$y = y1 = y(xz) = (yx)z = 1z = z$$

shows that x is regular in the ordinary sense and that $x^{-1} = y = z$.

The concept of maximality for two-sided ideals was introduced in Sec. 41. By analogy, we define a *maximal left ideal* in A to be a proper left ideal which is not properly contained in any other proper left ideal. A straightforward application of Zorn's lemma shows that any proper left ideal can be imbedded in a maximal left ideal; and since the zero ideal $\{0\}$ is a proper left ideal, maximal left ideals certainly exist. We now define the *radical* R of A to be the intersection of all its maximal left ideals. It will be convenient to abbreviate this definition by writing $R = \cap MLI$. R is clearly a proper left ideal.

These ideas can be formulated just as easily for right ideals as for left ideals, and there is no reason for giving preference to either side over the other. The purpose of the following chain of lemmas is to show that R is also the intersection of all the maximal right ideals in A, that is, that $R = \cap MRI$.

Lemma. *If r is an element of R, then $1 - r$ is left regular.*

PROOF. We assume that $1 - r$ is left singular, so that

$$L = A(1 - r) = \{x - xr : x \in A\}$$

is a proper left ideal which contains $1 - r$. We next imbed L in a maximal left ideal M, which of course also contains $1 - r$. Since r is in R, it is also in M, and therefore $1 = (1 - r) + r$ is in M. This implies that $M = A$, which is a contradiction.

Lemma. *If r is an element of R, then $1 - r$ is regular.*

PROOF. By the lemma just proved, there exists an element s such that $s(1 - r) = 1$, so s is right regular and $s = 1 - (-s)r$. The fact that R is a left ideal implies that $(-s)r$ is in R along with r, and another application of the preceding lemma shows that $1 - (-s)r = s$ is left regular. Since s is both left regular and right regular, it is regular with inverse $1 - r$, so $1 - r$ is also regular.

Lemma. *If r is an element of R, then $1 - xr$ is regular for every x.*

PROOF. R is a left ideal, so xr is in R and the statement follows from the lemma just proved.

Lemma. *If r is an element of A with the property that $1 - xr$ is regular for every x, then r is in R.*

PROOF. We assume that r is not in R, so that r is not in some maximal left ideal M. It is easy to see that the set

$$M + Ar = \{m + xr : m \in M \text{ and } x \in A\}$$

is a left ideal which contains both M and r, so $M + Ar = A$ and

$$m + xr = 1$$

for some m and x. It now follows that $1 - xr = m$ is a regular element in M, and this is impossible, for no proper ideal can contain any regular element.

The effect of these lemmas is to establish the equality of two sets:

$$\cap MLI = \{r : 1 - xr \text{ is regular for every } x\}. \tag{1}$$

Precisely the same arguments, when applied to maximal right ideals, show that

$$\cap MRI = \{r : 1 - rx \text{ is regular for every } x\}. \tag{2}$$

We now prove that all four of these sets are the same by showing that the two sets on the right of (1) and (2) are equal to one another. By symmetry, it evidently suffices to prove the

Lemma. *If $1 - xr$ is regular, then $1 - rx$ is also regular.*

PROOF. We assume that $1 - xr$ is regular with inverse

$$s = (1 - xr)^{-1}.$$

This means, of course, that $(1 - xr)s = s(1 - xr) = 1$. We leave it to the reader to show, by a simple computation, that

$$(1 - rx)(1 + rsx) = (1 + rsx)(1 - rx) = 1,$$

so that $1 - rx$ is regular with inverse $1 + rsx$. (The formula for $(1 - rx)^{-1}$ is less mysterious than it looks, as the reader can see by inspecting the meaningless but suggestive expressions

$$s = (1 - xr)^{-1} = 1 + xr + (xr)^2 + \cdots$$

and

$$(1 - rx)^{-1} = 1 + rx + (rx)^2 + (rx)^3 + \cdots = 1 + rx + rxrx$$
$$+ rxrxrx + \cdots = 1 + r(1 + xr + xrxr + \cdots)x = 1 + rsx.)$$

We summarize our results in

Theorem A. *The radical R of A equals each of the four sets in* (1) *and* (2) *and is therefore a proper two-sided ideal.*

A is said to be *semi-simple* if its radical equals the zero ideal $\{0\}$, that is, if each non-zero element of A is outside of some maximal left ideal.

It will be observed that the ideas discussed above are purely algebraic in nature. They can be applied not only to our Banach algebra A, but also to any algebra or ring with identity. Our interest, however, is in A, and we now bring to bear upon these notions the results of Sec. 65, notably, the fact that the set S of all singular elements in A is closed.

We begin by noting that if I is any ideal in A (left, right, or two-sided), then by the joint continuity of the algebraic operations, its closure $\bar{I}$ is an ideal of the same kind. Next, since any proper ideal is contained in the proper closed set S, the closure of any proper ideal is a proper ideal of the same kind. It is an easy step from these facts to

Theorem B. *Every maximal left ideal in A is closed.*

PROOF. If any maximal left ideal L is not closed, then L is a proper subset of the proper left ideal $\bar{L}$; and this cannot happen, for it contradicts the maximality of L.

Taken together, the above two theorems yield

Theorem C. *The radical R of A is a proper closed two-sided ideal.*

We shall also need

Theorem D. *If I is a proper closed two-sided ideal in A, then the quotient algebra A/I is a Banach algebra.*

PROOF. Theorem 46-A tells us that A/I is a non-trivial complex Banach space with respect to the norm defined by

$$\|x + I\| = \inf \{\|x + i\| : i \in I\}.$$

Further, A/I is clearly an algebra with identity $1 + I$, and

$$\|1 + I\| = \inf \{\|1 + i\| : i \in I\} \leq \|1\| = 1.$$

The multiplicative inequality for the norm is easily proved as follows:

$$
\begin{aligned}
\|(x + I)(y + I)\| = \|xy + I\| &= \inf \{\|xy + i\| : i \in I\} \\
&\leq \inf \{\|(x + i_1)(y + i_2)\| : i_1, i_2 \in I\} \\
&\leq \inf \{\|x + i_1\| \, \|y + i_2\| : i_1, i_2 \in I\} \\
&= [\inf \{\|x + i_1\| : i_1 \in I\}][\inf \{\|y + i_2\| : i_2 \in I\}] \\
&= \|x + I\| \, \|y + I\|.
\end{aligned}
$$

All that remains is to show that $\|1 + I\| = 1$; and since we already have $\|1 + I\| \leq 1$, this is an immediate consequence of the fact that $\|1 + I\| = \|(1 + I)^2\| \leq \|1 + I\|^2$ implies $1 \leq \|1 + I\|$.

As a final result, we state

Theorem E.   *A/R is a semi-simple Banach algebra.*

PROOF.   It suffices to observe that the natural homomorphism $x \rightarrow x + R$ of A onto A/R induces a one-to-one correspondence between the maximal left ideals in A and those in A/R.

In the following chapters, we shall be concerned almost exclusively with commutative Banach algebras. An algebra of this kind is of course much easier to handle than one which is not commutative, for all its ideals are two-sided and its radical is simply the intersection of its maximal ideals. Our reason for studying the general case here is that when it becomes necessary to assume commutativity, as it will in the next section, we want the force of this assumption, and the issues that depend on it, to be quite clear.

CHAPTER THIRTEEN

The Structure of
Commutative Banach Algebras

The set $\mathcal{C}(X)$ of all bounded continuous complex functions defined on a topological space X is the simplest of the really interesting Banach algebras. Our purpose in this chapter is to prove the famous *Gelfand-Neumark theorem*, which says that every commutative Banach algebra A of a certain type is essentially identical with $\mathcal{C}(X)$ for a suitable compact Hausdorff space X. More precisely, we shall prove that a compact Hausdorff space X can be built out of the inner structure of A, that X is accompanied by a natural mapping of A into $\mathcal{C}(X)$, and that this mapping is one-to-one onto and preserves all the structure assumed to be present in A.

70. THE GELFAND MAPPING

Let A be an arbitrary commutative Banach algebra. Our first theorem below is the principal source of the structure theory of A, and the remainder of the chapter will be devoted entirely to shaping its consequences into the elegant form of the Gelfand-Neumark theorem.

Theorem A. *If M is a maximal ideal in A, then the Banach algebra A/M is a division algebra, and therefore equals the Banach algebra C of complex numbers. The natural homomorphism $x \to x + M$ of A onto $A/M = C$ assigns to each element x in A a complex number $x(M)$ defined by*

$$x(M) = x + M,$$

and the mapping $x \rightarrow x(M)$ *has the following properties:*

(1) $(x + y)(M) = x(M) + y(M)$;

(2) $(\alpha x)(M) = \alpha x(M)$;

(3) $(xy)(M) = x(M)y(M)$;

(4) $x(M) = 0 \Leftrightarrow x \,\varepsilon\, M$;

(5) $1(M) = 1$;

(6) $|x(M)| \leq \|x\|$.

PROOF. Theorems 69-B and 69-D tell us that A/M is indeed a Banach algebra. Since A contains an identity, M is maximal as a ring ideal (see the comments on this matter in Sec. 45); and therefore, by Theorem 41-C, A/M is a division algebra. We now appeal to Theorem 67-B to conclude that A/M equals C. (Actually, of course, A/M equals the set of all scalar multiples of its own identity, but we identify this set with C in accordance with the remarks following Theorem 67-B.) Finally, properties (1) to (5) are obvious consequences of the nature of the homomorphism under discussion, and (6) follows from

$$|x(M)| = |x + M| = \|x + M\| = \inf \{\|x + m\| : m \,\varepsilon\, M\} \leq \|x\|.$$

It is interesting to observe that this proof depends, either directly or indirectly, on virtually every major theorem in the previous chapter. We also note that the ultimate reason for assuming that A is commutative lies in Theorem 41-A, which is definitely not true in the non-commutative case (see Problem 41-1).

The language of Theorem A is oriented toward the idea that $x(M)$ is a function of x for each fixed M. The notation, however, suggests that we reverse this point of view and that for each fixed x we regard $x(M)$ as a complex function defined on the set $\mathfrak{M}$ of all maximal ideals in A. This is the direction in which we now proceed.

If x is a given element of A, we denote by $\hat{x}$ the function defined on $\mathfrak{M}$ by $\hat{x}(M) = x(M)$, and we put $\hat{A} = \{\hat{x} : x \,\varepsilon\, A\}$. Our next step is to define a topology for $\mathfrak{M}$ in such a manner that every function in $\hat{A}$ is continuous. The most natural way of doing this is to introduce the weak topology generated by $\hat{A}$. It will be recalled that this is the weakest topology on $\mathfrak{M}$ relative to which every function $\hat{x}$ is continuous and that a typical subbasic open set has the form

$$S(\hat{x}, M_0, \epsilon) = \{M : M \,\varepsilon\, \mathfrak{M} \text{ and } |\hat{x}(M) - \hat{x}(M_0)| < \epsilon\}.$$

We call the topological space $\mathfrak{M}$ the *space of maximal ideals*, or the *maximal ideal space*, and the mapping $x \rightarrow \hat{x}$ of A onto $\hat{A}$ will be referred to as the *Gelfand mapping*.

We are now in a position to reformulate Theorem A, and to extend it, in such a way that the Gelfand mapping is displayed as the object of central importance.

Theorem B. *The Gelfand mapping* $x \to \hat{x}$ *is a norm-decreasing (and therefore continuous) homomorphism of* A *into* $\mathcal{C}(\mathfrak{M})$ *with the following properties:*

(1) *the image* $\hat{A}$ *of* A *is a subalgebra of* $\mathcal{C}(\mathfrak{M})$ *which separates the points of* $\mathfrak{M}$ *and contains the identity of* $\mathcal{C}(\mathfrak{M})$;

(2) *the radical* R *of* A *equals the set of all elements* x *for which* $\hat{x} = 0$, *so* $x \to \hat{x}$ *is an isomorphism* $\Leftrightarrow A$ *is semi-simple;*

(3) *an element* x *in* A *is regular* $\Leftrightarrow$ *it does not belong to any maximal ideal* $\Leftrightarrow \hat{x}(M) \neq 0$ *for every* M;

(4) *if* x *is an element of* A, *then its spectrum equals the range of the function* $\hat{x}$ *and its spectral radius equals the norm of* $\hat{x}$, *that is,* $\sigma(x) = \hat{x}(\mathfrak{M})$ *and* $r(x) = \sup |\hat{x}(M)| = \|\hat{x}\|$.

PROOF. The definition of the topology on $\mathfrak{M}$ guarantees that each function $\hat{x}$ is continuous, and part (6) of Theorem A shows that $\hat{x}$ is bounded and that $\|\hat{x}\| = \sup |\hat{x}(M)| \leq \|x\|$, so $x \to \hat{x}$ is a norm-decreasing mapping of A into $\mathcal{C}(\mathfrak{M})$. The fact that this mapping is a homomorphism is immediate from parts (1), (2), and (3) of Theorem A.

Since $x \to \hat{x}$ is a homomorphism, $\hat{A}$ is obviously a subalgebra of $\mathcal{C}(\mathfrak{M})$. The stated properties of $\hat{A}$ follow readily from parts (4) and (5) of Theorem A: if $M_1 \neq M_2$, and if (say) x is in M_1 but not in M_2, then $\hat{x}(M_1) = 0$ and $\hat{x}(M_2) \neq 0$; and $\hat{1}(M) = 1$ for every M.

If we recall that R is the intersection of all the M's, then the proof of (2) is easy: we have only to notice that part (4) of Theorem A tells us that $\hat{x}(M) = 0$ for every $M \Leftrightarrow x$ is in every M.

To prove (3), it suffices—in view of part (4) of Theorem A—to show that x is regular $\Leftrightarrow$ it does not belong to any M. It is elementary that a regular element cannot lie in any proper ideal, so we confine our attention to showing that if x is singular, then it does belong to some M. We prove this by observing that the singularity of x implies that $Ax = \{yx : y \in A\}$ is a proper ideal which contains x and can therefore be imbedded in a maximal ideal M which also contains x.

Finally, we use (3) to prove (4). By the definition of the spectrum of x, we have $\lambda \in \sigma(x) \Leftrightarrow x - \lambda 1$ is singular $\Leftrightarrow \widehat{(x - \lambda 1)}(M) = 0$ for at least one $M \Leftrightarrow (\hat{x} - \lambda 1)(M) = 0$ for at least one $M \Leftrightarrow \hat{x}(M) = \lambda$ for at least one M, so $\sigma(x)$ equals the range of $\hat{x}$. The rest of (4) follows from this statement and the definition of the spectral radius.

We add the final touch to this portion of the theory by showing that $\mathfrak{M}$ is a compact Hausdorff space. The reader will recall that if A^* is the conjugate space of A, then its closed unit sphere

$$S^* = \{f : f \in A^* \text{ and } \|f\| \leq 1\}$$

is a compact Hausdorff space in the weak* topology (see Theorem 49-A).

Our strategy is to identify $\mathfrak{M}$, both as a set and as a topological space, with a closed subspace of S^*.

A *multiplicative functional* on A is a functional f in the ordinary sense —that is, an element of the conjugate space A^*—which is non-zero and satisfies the additional condition $f(xy) = f(x)f(y)$. Theorem A shows that to each M in $\mathfrak{M}$ there corresponds a multiplicative functional f_M defined by $f_M(x) = x(M)$. It is important for us to know that $M \to f_M$ is a one-to-one mapping of $\mathfrak{M}$ onto the set of all multiplicative functionals. It will facilitate our work if we begin by proving the

Lemma. *If f_1 and f_2 are multiplicative functionals on A with the same null space M, then $f_1 = f_2$.*

PROOF. We first show that $f_1 = \alpha f_2$ for some scalar α. Let x_0 be an element of A which is not in M. If x is an arbitrary element of A, it is easy to see that x can be expressed uniquely in the form $x = m + \beta x_0$ with m in M (set $\beta = f_2(x)/f_2(x_0)$), put $m = x - \beta x_0$, and observe that $f_2(m) = 0$). It now follows that

$$f_1(x) = f_1(m) + \beta f_1(x_0) = \beta f_1(x_0) = [f_1(x_0)/f_2(x_0)]f_2(x),$$

so $f_1 = \alpha f_2$ with $\alpha = f_1(x_0)/f_2(x_0)$. We complete the proof by showing that α equals 1. Let x be an element not in M, so that $f_2(x) \neq 0$. Then $\alpha f_2(x)^2 = \alpha f_2(x^2) = f_1(x^2) = f_1(x)^2 = [\alpha f_2(x)]^2 = \alpha^2 f_2(x)^2$ implies that

$$\alpha^2 = \alpha,$$

so $\alpha = 0$ or $\alpha = 1$. Since $f_1 \neq 0$, we conclude that $\alpha = 1$.

We now use this to prove

Theorem C. *$M \to f_M$ is a one-to-one mapping of the set $\mathfrak{M}$ of all maximal ideals in A onto the set of all its multiplicative functionals.*

PROOF. The mapping is easily seen to be one-to-one, for if $M_1 \neq M_2$, and if (say) x is in M_1 and not in M_2, then $f_{M_1}(x) = 0$ and $f_{M_2}(x) \neq 0$. To prove that it is onto, let f be an arbitrary multiplicative functional, and consider its null space $M = \{x : f(x) = 0\}$. It is clear by the assumed properties of f that M is a proper closed ideal in A. Furthermore, M is maximal, for if it were properly contained in a proper ideal I, then $f(I)$ would be a non-trivial ideal in C, contrary to Theorem 41-A. Since f and f_M are multiplicative functionals with the same null space, the lemma just proved implies that $f = f_M$, and our proof is complete.

In some of its more concrete applications, this theorem is used to replace the algebraic problem of determining the maximal ideals in A by the analytic problem of finding its multiplicative functionals. Its importance for our current task of showing that $\mathfrak{M}$ is a compact Hausdorff space is that it enables us to regard $\mathfrak{M}$ as a subset of A^*. We can say even more than this, for parts (5) and (6) of Theorem A tell us that every multiplicative functional f_M has norm 1, so $\mathfrak{M}$ is a subset of the closed unit sphere S^*. We recalled earlier that S^* is a compact Hausdorff space with respect to the weak* topology, which is (see Sec. 49) the weak topology generated by all the functions F_x defined on S^* by $F_x(f) = f(x)$. We now observe that when F_x is restricted to $\mathfrak{M}$, it is precisely $\hat{x}$, for

$$F_x(f_M) = f_M(x) = x(M) = \hat{x}(M).$$

Therefore, by Problem 19-1c, the topology which $\mathfrak{M}$ has as a subspace of S^* is exactly its topology as the space of maximal ideals. These considerations permit us to regard $\mathfrak{M}$ as a *subspace* of S^*.

Theorem D. *The maximal ideal space $\mathfrak{M}$ is a compact Hausdorff space.*
PROOF. In view of the above discussion, it suffices to show that $\mathfrak{M}$ is a closed subspace of S^*. We accomplish this by forming the subspace X of S^* defined by

$$X = \bigcap_{x,y \in A} \{f : f \in S^* \text{ and } f(xy) = f(x)f(y)\}.$$

It is evident that X is simply $\mathfrak{M}$ together with the zero functional; and since we have

$$\begin{aligned}
X &= \bigcap_{x,y \in A} \{f : f \in S^* \text{ and } f(xy) - f(x)f(y) = 0\} \\
&= \bigcap_{x,y \in A} \{f : f \in S^* \text{ and } F_{xy}(f) - F_x(f)F_y(f) = 0\} \\
&= \bigcap_{x,y \in A} \{f : f \in S^* \text{ and } (F_{xy} - F_x F_y)(f) = 0\},
\end{aligned}$$

it is easy to see that X is closed in S^* (note that each of the sets last written has this property). We next remark that F_1 is continuous on X and equals 1 on $\mathfrak{M}$ and 0 at the zero functional. It follows from this that $\mathfrak{M}$ is closed in X and is therefore closed in S^*.

It is worthy of notice that the topology we imposed on $\mathfrak{M}$ is the only one which makes it into a compact Hausdorff space on which all the functions $\hat{x}$ are continuous, for by Theorem 26-E, any stronger compact Hausdorff topology must equal the given one.

When Theorems B and D are taken together, the result is often called the *Gelfand representation theorem*. In essence, this tells us that every commutative semi-simple Banach algebra is isomorphic to an algebra of continuous complex functions on a suitable compact Hausdorff space. In general, the norm is not preserved by this isomorphism and the representing algebra does not exhaust the continuous functions on the underlying space. We shall remove these deficiencies in the following sections by assuming that additional structure is present in the Banach algebra under discussion.

71. APPLICATIONS OF THE FORMULA $r(x) = \lim \|x^n\|^{1/n}$

We continue our study of an arbitrary commutative Banach algebra A and of the Gelfand mapping $x \to \hat{x}$ of A onto the subalgebra $\hat{A}$ of $\mathcal{C}(\mathfrak{M})$. Our first theorem provides a simple way of guaranteeing that this mapping preserves norms.

Theorem A. *The following conditions on A are all equivalent to one another:*
  (1) $\|x^2\| = \|x\|^2$ *for every x;*
  (2) $r(x) = \|x\|$ *for every x;*
  (3) $\|\hat{x}\| = \|x\|$ *for every x.*

PROOF.   It follows from condition (1) that

$$\|x^4\| = \|(x^2)^2\| = \|x^2\|^2 = \|x\|^4$$

and, in general, that $\|x^{2^k}\| = \|x\|^{2^k}$ for every positive integer k.   The formula for the spectral radius now yields

$$r(x) = \lim \|x^n\|^{1/n} = \lim \|x^{2^k}\|^{1/2^k} = \lim \|x\| = \|x\|,$$

so (1) implies (2).   The fact that (2) implies (1) is immediate from $\|x^2\| = r(x^2) = r(x)^2 = \|x\|^2$.   In view of the equation $r(x) = \|\hat{x}\|$ (see Theorem 70-B), the equivalence of (2) and (3) is obvious.

Our next problem is to devise a way of making sure that the representing algebra $\hat{A}$ comes as close as it can to exhausting $\mathcal{C}(\mathfrak{M})$, and we accomplish this by introducing the following property. A is said to be *self-adjoint* if for each x in A there exists an element y in A such that $\hat{y}(M) = \overline{\hat{x}(M)}$ for every M.

Theorem B. *If A is self-adjoint, then $\hat{A}$ is dense in $\mathcal{C}(\mathfrak{M})$.*

PROOF.   By part (1) of Theorem 70-B, we know that $\hat{A}$ is a subalgebra of $\mathcal{C}(\mathfrak{M})$ which separates the points of $\mathfrak{M}$ and contains the identity function.   Problem 20-3 and our hypothesis now tell us that the closure of $\hat{A}$ is a closed subalgebra of $\mathcal{C}(\mathfrak{M})$ which separates points, contains the identity function, and contains the conjugate of each of its functions.   Theorem 36-B (the complex Stone-Weierstrass theorem) shows that this closure equals $\mathcal{C}(\mathfrak{M})$, so $\hat{A}$ itself is dense in $\mathcal{C}(\mathfrak{M})$.

If we put together the results obtained in the above two theorems, we have

Theorem C. *If A is self-adjoint, and if $\|x^2\| = \|x\|^2$ for every x, then the Gelfand mapping $x \to \hat{x}$ is an isometric isomorphism of A onto $\mathcal{C}(\mathfrak{M})$.*

PROOF.   By Theorem A, the mapping $x \to \hat{x}$ preserves norms.   It is therefore an isometric isomorphism of A onto $\hat{A}$, and we see from this that $\hat{A}$ is closed in $\mathcal{C}(\mathfrak{M})$. Since $\hat{A}$ is dense in $\mathcal{C}(\mathfrak{M})$ by Theorem B, it follows that $\hat{A}$ equals $\mathcal{C}(\mathfrak{M})$, and the proof is complete.

This theorem lacks a certain simplicity which it ought to have, for the condition of self-adjointness is rather far removed from the intrinsic structure of A. Our work in the next two sections will remedy this defect and at the same time will establish closer connections with the operator algebras to which we apply our final result.

72. INVOLUTIONS IN BANACH ALGEBRAS

A Banach algebra A is called a *Banach *-algebra* if it has an *involution*, that is, if there exists a mapping $x \to x^*$ of A into itself with the following properties:

(1) $(x + y)^* = x^* + y^*$;

(2) $(\alpha x)^* = \bar{\alpha} x^*$;

(3) $(xy)^* = y^* x^*$;

(4) $x^{**} = x$.

It is an easy consequence of (4) that the involution $x \to x^*$ is actually a one-to-one mapping of A onto itself. We also note that $0^* = 0$ and $1^* = 1$, as we see from $0 + x^* = x^* = (0 + x)^* = 0^* + x^*$ and $1^* = 11^* = 1^{**}1^* = (11^*)^* = (1^*)^* = 1^{**} = 1$. The element x^* is called the *adjoint* of x, and a subalgebra of A is said to be *self-adjoint* if it contains the adjoint of each of its elements. If A' is also a Banach *-algebra, and if f is an isomorphism of A onto A', then f is called a **-isomorphism* if it preserves the involution in the sense that $f(x^*) = f(x)^*$.

We naturally want the involution in a Banach *-algebra to be linked in some useful way to the norm. The property $\|x^*\| = \|x\|$ clearly implies that the involution is continuous; for if $x_n \to x$, then

$$\|x_n^* - x^*\| = \|(x_n - x)^*\| = \|x_n - x\|$$

shows that $x_n^* \to x^*$. A much stronger relation between the involution and the norm is given by the condition

$$\|x^*x\| = \|x\|^2,$$

and any Banach *-algebra which satisfies it is called a *B*-algebra*. It is easy to see that we have $\|x^*\| = \|x\|$ in every B^*-algebra; for

$$\|x\|^2 = \|x^*x\| \leq \|x^*\| \, \|x\|$$

shows that $\|x\| \leq \|x^*\|$ for every x, so $\|x^*\| \leq \|x^{**}\| = \|x\|$, and therefore $\|x^*\| = \|x\|$. It follows from this that the relation $\|x^*x\| = \|x^*\| \, \|x\|$ is also true.

Several of the Banach algebras described in Sec. 64 are also Banach

*-algebras with respect to natural involutions. If X is any topological space, then $\mathcal{C}(X)$ is clearly a commutative B^*-algebra relative to the involution defined by $f^*(x) = \overline{f(x)}$. The disc algebra, however, is not, for if f is the function defined by $f(z) = z$, then $f^*(z) = \bar{z}$ is not analytic at any point. If H is a non-trivial Hilbert space, then $\mathcal{B}(H)$ is a B^*-algebra with the adjoint operation $T \to T^*$ taken as the involution (see Theorem 56-A). Since C^*-algebras are the self-adjoint Banach sub-algebras of $\mathcal{B}(H)$'s, they too are B^*-algebras. Finally, the group algebra $L_1(G)$ of a finite group G is a Banach *-algebra with respect to the involution defined by $f^*(g_i) = \overline{f(g_i^{-1})}$, and it is easy to see that $\|f^*\| = \|f\|$.

It should be reasonably clear that Banach *-algebras (and especially B^*-algebras) are modeled along lines suggested by $\mathcal{B}(H)$. We have already called the element x^* in such an algebra the adjoint of x. By analogy, we say that x is *self-adjoint* if $x = x^*$, *normal* if $xx^* = x^*x$, and a *projection* if $x = x^*$ and $x^2 = x$.

Theorem A. *If x is a normal element in a B^*-algebra, then $\|x^2\| = \|x\|^2$.*
PROOF. It is obvious that $\|x^2\| \leq \|x\|^2$. The inequality in the other direction is a consequence of the following computation:

$$\|x^*\|^2\|x\|^2 = (\|x^*\|\,\|x\|)^2 = \|x^*x\|^2 = \|(x^*x)^*x^*x\| = \|x^*xx^*x\|$$
$$= \|x^*x^*xx\| = \|(x^*)^2x^2\| = \|(x^2)^*x^2\| = \|(x^2)^*\|\,\|x^2\|$$
$$= \|(x^*)^2\|\,\|x^2\| \leq \|x^*\|^2\|x^2\|.$$

This result suggests more strongly than ever that there are close connections between B^*-algebras and algebras of operators on Hilbert spaces (see Theorem 58-D). We describe the true state of affairs in this matter at the end of the next section.

73. THE GELFAND-NEUMARK THEOREM

We are now in a position to give Theorem 71-C its final form.

Theorem A. *If A is a commutative B^*-algebra, then the Gelfand mapping $x \to \hat{x}$ is an isometric *-isomorphism of A onto the commutative B^*-algebra $\mathcal{C}(\mathfrak{M})$.*
PROOF. Since A is commutative, each of its elements is normal, and it follows from Theorem 72-A that $\|x^2\| = \|x\|^2$ for every x. By Theorem 71-C, it now suffices to show that $\widehat{x^*}(M) = \overline{\hat{x}(M)}$ for each x in A and M in $\mathfrak{M}$.

Our first step is to prove that if x is self-adjoint, then $\hat{x}(M)$ is real for every M. We assume the contrary, namely, that there exists an M such that $\hat{x}(M) = \alpha + i\beta$ with $\beta \neq 0$. Since x is self-adjoint,

$$y = (x - \alpha 1)/\beta$$

is also self-adjoint. We further note that $\hat{y}(M) = i$, so $y - i1$ is in M. It is obvious from the properties of the involution in A that

$$M^* = \{m^* : m \in M\}$$

is a maximal ideal; and since it contains $(y - i1)^* = y + i1$, we see that $\hat{y}(M^*) = -i$. If K is any positive number, then

$$(\widehat{y - iK1})(M^*) = -i(1 + K)$$

and $(\widehat{y + iK1})(M) = i(1 + K)$. It follows from this that $1 + K \leq \|\widehat{y - iK1}\| \leq \|y - iK1\|$ and, similarly, that $1 + K \leq \|y + iK1\|$. On multiplying these two inequalities, we obtain

$$
\begin{aligned}
(1 + K)^2 \leq \|y - iK1\|\,\|y + iK1\| &= \|(y + iK1)^*\|\,\|y + iK1\| \\
&= \|(y + iK1)^*(y + iK1)\| = \|(y - iK1)(y + iK1)\| \\
&= \|y^2 + K^2 1\| \leq \|y^2\| + K^2,
\end{aligned}
$$

so $1 + 2K \leq \|y^2\|$. Since K is arbitrary, this is impossible, and this portion of our proof is complete.

We now conclude the proof by showing that if x is any element of A, then $\widehat{x^*}(M) = \overline{\hat{x}(M)}$ for every M. It is clear that $y = (x + x^*)/2$ and $z = (x - x^*)/(2i)$ are self-adjoint, and that $x = y + iz$; and therefore, by the result of the above paragraph, we have

$$
\begin{aligned}
\widehat{x^*}(M) = (\widehat{y - iz})(M) = \hat{y}(M) - i\hat{z}(M) &= \overline{\hat{y}(M)} - i\overline{\hat{z}(M)} \\
&= \overline{\hat{y}(M) + i\hat{z}(M)} = \overline{\hat{x}(M)}.
\end{aligned}
$$

We already know that if X is any compact Hausdorff space, then $\mathcal{C}(X)$ is a commutative B^*-algebra. The theorem just proved—it is called the *Gelfand-Neumark representation theorem*—tells us that commutative B^*-algebras are simply abstract $\mathcal{C}(X)$'s, in the sense that every such algebra is abstractly identical with $\mathcal{C}(X)$ for a suitable compact Hausdorff space X.

There is another Gelfand-Neumark theorem of great interest, which applies to arbitrary B^*-algebras. We observed in the previous section that every C^*-algebra is a B^*-algebra. The converse of this is also true, for if A is a B^*-algebra, then there exists a Hilbert space H with the property that A is isometrically *-isomorphic to a C^*-algebra of operators on H.[1] General B^*-algebras are therefore abstract C^*-algebras. The proof of this theorem evidently requires that a suitable Hilbert space be constructed out of the given structure of A. The details of this construction are beyond the scope of this book, and we content ourselves with merely stating the facts.

[1] See Rickart [34, p. 244].

CHAPTER FOURTEEN

Some Special

Commutative Banach Algebras

Our discussion in Sec. 63 foreshadowed a generalized form of the spectral theorem, and the principal purpose of the present chapter is to formulate and prove this result. We begin with some additional material relating to Banach algebras of continuous functions. In particular, we keep the promise made in Sec. 30 by showing that the Stone-Čech compactification of a completely regular space is essentially unique.

74. IDEALS IN $\mathcal{C}(X)$ AND THE BANACH-STONE THEOREM

Let X be a compact Hausdorff space, and consider the commutative B^*-algebra $\mathcal{C}(X)$. If $\mathfrak{M}$ is the space of maximal ideals in $\mathcal{C}(X)$, then the developments of the previous chapter lead us to expect that $\mathfrak{M}$ can be identified with X and that the Gelfand mapping is the identity mapping of $\mathcal{C}(X)$ onto itself.

In order to substantiate this conjecture, we begin by observing that to each point x in X there corresponds a proper ideal M_x in $\mathcal{C}(X)$, defined by

$$M_x = \{f : f \, \varepsilon \, \mathcal{C}(X) \text{ and } f(x) = 0\}.$$

M_x is easily seen to be maximal and is thus an element of $\mathfrak{M}$, for it is the null space of the multiplicative functional f_{M_x} defined by $f_{M_x}(f) = f(x)$, which assigns to each function in $\mathcal{C}(X)$ its value at x. Since X is compact Hausdorff, and therefore normal, Urysohn's lemma tells us that for each point $y \neq x$ there exists a function f in $\mathcal{C}(X)$ such that $f(x) = 0$ and

$f(y) \neq 0$. This shows that $x \rightarrow M_x$ is a one-to-one mapping of X into $\mathfrak{M}$. Our next step is to prove that this mapping is onto, and for this it clearly suffices to show that if M is any maximal ideal in $\mathcal{C}(X)$, then there exists a point in X at which every function in M vanishes. We assume the contrary, namely, that for each point x in X there exists a function f in M such that $f(x) \neq 0$. Since f is continuous, x has a neighborhood at no point of which f vanishes. We now vary x to obtain an open cover for X, and we use compactness to infer that this open cover has a finite subcover. Let $f_1, f_2, \ldots, f_n$ be the corresponding functions in M. M is an ideal, so the function $g = \Sigma_{i=1}^n f_i \overline{f_i} = \Sigma_{i=1}^n |f_i|^2$ is also in M; and by the manner of its construction, it clearly has the property that $g(x) > 0$ for every x. It follows that g is a regular element of $\mathcal{C}(X)$, and this contradicts the fact that it lies in the proper ideal M. We therefore conclude that $x \rightarrow M_x$ is a one-to-one mapping of X onto $\mathfrak{M}$.

These considerations enable us to identify the set $\mathfrak{M}$ with the set X, and in terms of this identification, we regard X and $\mathfrak{M}$ as two possibly different compact Hausdorff spaces built on the same underlying set of points. By our work in the previous chapter, we know that the Gelfand mapping $f \rightarrow \hat{f}$ is a one-to-one mapping of $\mathcal{C}(X)$ onto $\mathcal{C}(\mathfrak{M})$. If we use the notation established there, then we find that

$$\hat{f}(M_x) = f(M_x) = f_{M_x}(f) = f(x),$$

so $\hat{f} = f$ and $\mathcal{C}(\mathfrak{M}) = \mathcal{C}(X)$. We now recall that any compact Hausdorff topology on a non-empty set is uniquely determined as the weak topology generated by the set of all its continuous complex functions (see Problem 27-3). It follows from this that X and $\mathfrak{M}$ are equal as topological spaces.

We summarize the results of this discussion in

Theorem A. *Let X be a compact Hausdorff space and $\mathfrak{M}$ the space of maximal ideals in the commutative B*-algebra $\mathcal{C}(X)$. Then to each point x in X there corresponds a maximal ideal M_x defined by*

$$M_x = \{f : f \in \mathcal{C}(X) \text{ and } f(x) = 0\},$$

and $x \rightarrow M_x$ is a one-to-one mapping of X onto $\mathfrak{M}$. If this mapping is used to identify $\mathfrak{M}$ with X, then $\mathfrak{M}$ and X are equal as topological spaces, $\mathcal{C}(\mathfrak{M})$ equals $\mathcal{C}(X)$, and the Gelfand mapping $f \rightarrow \hat{f}$ is the identity mapping of $\mathcal{C}(X)$ onto itself.

The main idea of this theorem is that the maximal ideals in $\mathcal{C}(X)$ correspond in a natural way to the points of X. Our next step is to extend this idea and to obtain a similar characterization of the proper closed ideals in $\mathcal{C}(X)$.

We again consider a compact Hausdorff space X, and we begin our discussion with the observation that to each non-empty closed subset F

of X there corresponds a proper closed ideal $I(F)$ in $\mathcal{C}(X)$, defined by

$$I(F) = \{f : f \in \mathcal{C}(X) \text{ and } f(F) = 0\}.$$

If x is any point not in F, then it follows from the complete regularity of X that there exists a function f in $\mathcal{C}(X)$ such that $f(x) \neq 0$ and $f(F) = 0$. This shows that $F \to I(F)$ is a one-to-one mapping of the class of all non-empty closed subsets of X into the set of all proper closed ideals in $\mathcal{C}(X)$. We shall prove that this mapping is onto, that is, that every proper closed ideal in $\mathcal{C}(X)$ arises in this way from some F.

Let I be a proper closed ideal in $\mathcal{C}(X)$. We may assume that I is not the zero ideal, for this ideal clearly arises from the full space X. We define F by

$$F = \{x : f(x) = 0 \text{ for every } f \in I\}.$$

It is easy to see that F is a proper closed subset of X; and since I is contained in some maximal ideal, it follows that F is non-empty. Our task is to prove that $I(F) = I$, and since it is obvious that $I \subseteq I(F)$, the real problem is to prove that $I(F) \subseteq I$. If f is any function which vanishes on F, we must show that f lies in I. We may evidently assume that $f \neq 0$, so that $\{x : f(x) = 0\}$ is a proper subset of X.

In the first part of our proof, we assume that f vanishes on some open set G which contains F. Since $f \neq 0$, G' is non-empty and is thus a compact subspace of X. For each point x in G', there exists a function g in I such that $g(x) \neq 0$. The technique used in the proof of Theorem A can now be applied again, to yield a finite number of functions g_1, g_2, $\ldots$, g_n in I with the property that at least one is non-zero at every point of G'. We next define a function g_0 by $g_0 = \Sigma_{i=1}^n g_i \overline{g_i} = \Sigma_{i=1}^n |g_i|^2$, and we observe that g_0 is in I and that $g_0(x) > 0$ for every x in G'. By the Tietze extension theorem, the function whose values on G' are given by $1/g_0(x)$ can be extended to a function h in $\mathcal{C}(X)$. It is easily seen that $g_0 h$ is in I, that it equals 1 on G', and that $f = f g_0 h$, so f is in I.

We now turn to the general case. For each $\epsilon > 0$, the sets K and L defined by $K = \{x : |f(x)| \leq \epsilon/2\}$ and $L = \{x : |f(x)| \geq \epsilon\}$ are disjoint closed subsets of X. K is clearly non-empty, and since $f \neq 0$, L is also non-empty for every sufficiently small ϵ. We assume that ϵ has been chosen at least this small, so that K and L constitute a disjoint pair of closed subspaces of X. By Urysohn's lemma, there exists a function g in $\mathcal{C}(X)$ such that $g(K) = 0$, $g(L) = 1$, and $0 \leq g(x) \leq 1$ for every x. We now define a function h in $\mathcal{C}(X)$ by $h = fg$, and we note that

$$\|f - h\| = \|f(1 - g)\| \leq \epsilon.$$

It is evident that h vanishes on the set $G = \{x : |f(x)| < \epsilon/3\}$; and since G is an open set which contains F, it follows from the preceding paragraph that h is in I.   This shows that for every sufficiently small positive number ϵ there exists a function h in I such that $\|f - h\| \leq \epsilon$, and since I is closed, we conclude that f is in I.

We give the following formal statement of our result.

Theorem B.   *Let X be a compact Hausdorff space.   Then to each non-empty closed set F in X there corresponds a proper closed ideal $I(F)$ in $\mathcal{C}(X)$, defined by $I(F) = \{f : f \in \mathcal{C}(X)$ and $f(F) = 0\}$; and further, $F \to I(F)$ is a one-to-one mapping of the class of all non-empty closed subsets of X onto the set of all proper closed ideals in $\mathcal{C}(X)$.*

As an easy consequence of this, we have

Theorem C.   *If X is a compact Hausdorff space, then every closed ideal in $\mathcal{C}(X)$ is the intersection of the maximal ideals which contain it.*

PROOF.   Since the intersection of the empty set of maximal ideals is $\mathcal{C}(X)$ itself, we may confine our attention to a proper closed ideal I. By Theorem B, $I = I(F)$ for some non-empty closed set F.   It is clear that the maximal ideals which contain I are precisely those associated with the points of F.   It therefore suffices to observe that a function in $\mathcal{C}(X)$ vanishes on $F \Leftrightarrow$ it vanishes at each point of F.

We have seen in Theorem A that the points and the topology of a compact Hausdorff space X can be recovered from the maximal ideals in $\mathcal{C}(X)$.   Since the maximal ideals in $\mathcal{C}(X)$ are objects of a purely algebraic nature, it follows that the compact Hausdorff space X is fully determined, both as a set and as a topological space, by the algebraic structure of $\mathcal{C}(X)$.   These observations lead us directly to

Theorem D (the Banach-Stone Theorem).   *Two compact Hausdorff spaces X and Y are homeomorphic $\Leftrightarrow$ their corresponding function algebras $\mathcal{C}(X)$ and $\mathcal{C}(Y)$ are isomorphic.*

75. THE STONE-ČECH COMPACTIFICATION (continued)

It is natural to wonder what can be said along the lines of Theorem 74-A in the case of a topological space X which is not necessarily compact Hausdorff.   Regardless of the properties of X, we know from our previous work that $\mathcal{C}(X)$ is a commutative B^*-algebra, that its maximal ideal space $\mathfrak{M}$ is a compact Hausdorff space, and that $x \to M_x$ is a mapping of X into $\mathfrak{M}$.   Our difficulty is that without restrictions of some kind on X, we know practically nothing about the properties of the mapping $x \to M_x$.   If it happens that this mapping is one-to-one and is also a homeomorphism of X onto a subspace of $\mathfrak{M}$, then we observe that X

is necessarily completely regular. It is therefore reasonable to assume at the outset that X is completely regular, and we shall see that several interesting conclusions follow from this hypothesis.

Theorem A. *Let X be a completely regular space and $\mathfrak{M}$ the space of maximal ideals in the commutative B^*-algebra $\mathcal{C}(X)$. Then the mapping $x \to M_x$ is a homeomorphism of X onto a subspace of $\mathfrak{M}$. Furthermore, if this mapping is used to identify X with its image in $\mathfrak{M}$, then (1) X is a dense subspace of $\mathfrak{M}$; (2) each function in $\mathcal{C}(X)$ has a unique extension to a function in $\mathcal{C}(\mathfrak{M})$; and (3) if Y is a compact Hausdorff space with the properties of $\mathfrak{M}$ stated in (1) and (2), then there exists a homeomorphism of $\mathfrak{M}$ onto Y which leaves the points of X fixed.*

PROOF. The fact that $x \to M_x$ is one-to-one is immediate from the complete regularity of X, so we may identify X as a set with its image in $\mathfrak{M}$. The subset X of $\mathfrak{M}$ has two topologies: its own, and its relative topology as a subspace of $\mathfrak{M}$. The following arguments show that these topologies are equal. We know that the Gelfand mapping $f \to \hat{f}$ is an isomorphism of $\mathcal{C}(X)$ onto $\mathcal{C}(\mathfrak{M})$. Also, just as in the proof of Theorem 74-A, we have $\hat{f}(x) = f(x)$ for each f in $\mathcal{C}(X)$ and each x in X. These observations imply that $\mathcal{C}(X)$ is precisely the set of all restrictions to X of functions in $\mathcal{C}(\mathfrak{M})$; and since both topologies are completely regular, it follows from Problems 19-1c and 27-4 that each is the weak topology generated by $\mathcal{C}(X)$, so they are equal and X can be regarded as a subspace of $\mathfrak{M}$. These observations also show that X is dense in $\mathfrak{M}$—for if $\hat{f}$ vanishes on X, then $f = 0$, $\hat{f} = 0$, and $\hat{f}$ also vanishes on $\mathfrak{M}$—and that each function f in $\mathcal{C}(X)$ has a unique extension $\hat{f}$ in $\mathcal{C}(\mathfrak{M})$. All that remains is to prove (3). We know that $\hat{f} \to f$ is an isomorphism of $\mathcal{C}(\mathfrak{M})$ onto $\mathcal{C}(X)$; and by the assumptions about Y, the mapping $f \to f'$, which assigns to each f in $\mathcal{C}(X)$ its extension f' in $\mathcal{C}(Y)$, is an isomorphism of $\mathcal{C}(X)$ onto $\mathcal{C}(Y)$. Thus $\hat{f} \to f \to f'$ is an isomorphism of $\mathcal{C}(\mathfrak{M})$ onto $\mathcal{C}(Y)$. If x is a point of X, then this isomorphism clearly carries the maximal ideal in $\mathcal{C}(\mathfrak{M})$ corresponding to x over to the maximal ideal in $\mathcal{C}(Y)$ corresponding to x; so by the Banach-Stone theorem, it induces a homeomorphism of $\mathfrak{M}$ onto Y which leaves the points of X fixed.

On comparing this result with Theorem 30-A, we see that $\mathfrak{M}$ is homeomorphic, in the manner described, to the Stone-Čech compactification $\beta(X)$. In this sense, therefore, $\mathfrak{M}$ and $\beta(X)$ can be considered equal to one another, and also to any other compact Hausdorff space which contains X as a dense subspace and has the required extension property. In effect, we have shown that the Stone-Čech compactification of a completely regular space X is unique and can equally well be regarded as the maximal ideal space of $\mathcal{C}(X)$.

76. COMMUTATIVE C^*-ALGEBRAS

In this final section, we apply the results of the preceding two chapters to the theory of operators on a non-trivial Hilbert space H. We know that $\mathcal{B}(H)$ and all its self-adjoint Banach subalgebras (that is, all C^*-algebras of operators on H) are B^*-algebras. As a special case of the Gelfand-Neumark theorem, we therefore have

Theorem A. *Let A be a commutative C^*-algebra of operators on H, and $\mathfrak{M}$ its space of maximal ideals. Then the Gelfand mapping $T \to \hat{T}$ is an isometric *-isomorphism of A onto $\mathcal{C}(\mathfrak{M})$.*

If $\{T_i\}$ is a non-empty set of operators on H, then the smallest Banach subalgebra of $\mathcal{B}(H)$ which contains every T_i is called the Banach subalgebra of $\mathcal{B}(H)$ *generated by* the T_i's. It is easy to see that this Banach subalgebra of $\mathcal{B}(H)$ is the closure of the set of all polynomials in the T_i's. If N is a normal operator on H, then the Banach subalgebra of $\mathcal{B}(H)$ generated by N and N^* is clearly a commutative C^*-algebra, and is called *the commutative C^*-algebra generated by N*. We now specialize Theorem A to

Theorem B. *Let N be a normal operator on H, and A the commutative C^*-algebra generated by N. If $\mathfrak{M}$ is the space of maximal ideals in A, then the Gelfand mapping $T \to \hat{T}$ is an isometric *-isomorphism of A onto $\mathcal{C}(\mathfrak{M})$.*

As it stands, this result is only a beginning. In order to exploit it effectively, our first task is to show that the spectrum of an operator in A—which is understood to be its spectrum as an element of $\mathcal{B}(H)$—equals its spectrum as an element of A. In proving this, we shall need the following preliminary fact.

Lemma. *Let X be a compact Hausdorff space and A a Banach subalgebra of $\mathcal{C}(X)$. If f is a real function in A which is regular in $\mathcal{C}(X)$, then it is also regular in A.*

PROOF. The range of f is clearly a compact subspace of the real line which does not contain 0. If $\epsilon > 0$ is given, then by the Weierstrass approximation theorem (see Problem 35-3) there exists a polynomial p such that $|p(t) - 1/t| < \epsilon$ for every t in $f(X)$. It follows from this that $|p(f(x)) - 1/f(x)| < \epsilon$ for every x, so $\|p(f) - 1/f\| < \epsilon$. Since $p(f)$ is in A and A is closed, we conclude that $1/f$ is in A.

Theorem C. *Let A be a commutative C^*-algebra of operators on H. If an operator T in A is regular in $\mathcal{B}(H)$, then it is also regular in A, and therefore the spectrum of T as an operator on H equals its spectrum as an element of A.*

PROOF. We begin by considering the special case in which T is assumed to be self-adjoint. Let B be the Banach subalgebra of $\mathfrak{B}(H)$ generated by T and T^{-1}. Since T and T^{-1} are self-adjoint and commute with one another, it is evident that B is a commutative C^*-algebra; and if $\mathfrak{M}$ is its space of maximal ideals, then B is isometrically *-isomorphic to $\mathcal{C}(\mathfrak{M})$ and T is represented by a real function in $\mathcal{C}(\mathfrak{M})$. $C = A \cap B$ is a Banach subalgebra of B and is therefore isomorphic to a Banach subalgebra of $\mathcal{C}(\mathfrak{M})$. Since T is in C and is regular in B, our lemma shows that T^{-1} is also in C and therefore lies in A.

We now turn to the general case, in which T is not assumed to be self-adjoint. It is clear that $U = TT^*$ is a self-adjoint operator in A, and since it has an inverse $U^{-1} = (TT^*)^{-1} = (T^*)^{-1}T^{-1} = (T^{-1})^*T^{-1}$ in $\mathfrak{B}(H)$, we know from the preceding paragraph that U^{-1} is in A. We now make use of the commutativity of A to write the relation $UU^{-1} = I$ in the form $T(T^*U^{-1}) = (T^*U^{-1})T = I$. This shows that $T^{-1} = T^*U^{-1}$, so T^{-1} lies in A and the proof is complete.

This result tells us, in particular, that if N is a normal operator on H, then its spectrum $\sigma(N)$ equals its spectrum as an element of the commutative C^*-algebra generated by N. Our next step is to provide a concrete representation for the space of maximal ideals in this algebra.

Theorem D. *Let N be a normal operator on H, A the commutative C^*-algebra generated by N, and $\mathfrak{M}$ the space of maximal ideals in A. Then the function $\hat{N}$ in $\mathcal{C}(\mathfrak{M})$ which corresponds to N under the Gelfand mapping is a homeomorphism of $\mathfrak{M}$ onto $\sigma(N)$.*

PROOF. It follows from Theorem C and part (4) of Theorem 70-B that $\sigma(N)$ is precisely the range of the continuous function $\hat{N}$ defined on $\mathfrak{M}$. Since both $\mathfrak{M}$ and $\sigma(N)$ are compact Hausdorff spaces, it suffices by Theorem 26-E to show that $\hat{N}$ is one-to-one. Let M_1 and M_2 be points of $\mathfrak{M}$ such that $\hat{N}(M_1) = \hat{N}(M_2)$. Then we also have

$$\widehat{N^*}(M_1) = \overline{\hat{N}(M_1)} = \overline{\hat{N}(M_2)} = \widehat{N^*}(M_2),$$

so each of the functions $\hat{N}$ and $\widehat{N^*}$ takes equal values at M_1 and M_2. Since A is the closure of the set of all polynomials in N and N^*, every function in $\mathcal{C}(\mathfrak{M})$ is a uniform limit of polynomials in $\hat{N}$ and $\widehat{N^*}$, and therefore every function in $\mathcal{C}(\mathfrak{M})$ takes equal values at M_1 and M_2. We conclude the proof by observing that since $\mathcal{C}(\mathfrak{M})$ separates the points of $\mathfrak{M}$, it follows that $M_1 = M_2$.

In accordance with this result, we may identify $\mathfrak{M}$ with the compact subspace $\sigma(N)$ of the complex plane; and when this identification is carried out, it is easy to see that $\hat{N}(z) = z$ for every z in $\mathfrak{M}$. We summarize our conclusions in

Theorem E. *Let N be a normal operator on H with spectrum $\sigma(N)$, and let A be the commutative C^*-algebra generated by N. Then the space $\mathfrak{M}$ of maximal ideals in A equals $\sigma(N)$, and the Gelfand mapping $T \to \hat{T}$ of A onto $\mathcal{C}(\mathfrak{M})$ is an isometric $*$-isomorphism which carries N into the function whose values are given by $\hat{N}(z) = z$ for every z in $\mathfrak{M}$.*

This theorem has a number of simple consequences, of which the following are only a few: (1) $N = 0 \Leftrightarrow \sigma(N) = \{0\}$; (2) N is singular $\Leftrightarrow \sigma(N)$ contains 0; (3) $\sigma(N^*) = \overline{\sigma(N)}$; (4) N is self-adjoint $\Leftrightarrow \sigma(N)$ lies on the real line; (5) N is unitary $\Leftrightarrow \sigma(N) \subseteq \{z : |z| = 1\}$; (6) N is a projection $\Leftrightarrow \sigma(N) \subseteq \{0,1\}$. If it happens that H is finite-dimensional, so that $\sigma(N)$ consists of a finite number of distinct complex numbers $\lambda_1, \lambda_2, \ldots, \lambda_m$, then we can write

$$\hat{N} = \sum_{i=1}^{m} \lambda_i \widehat{P_i},$$

where $\widehat{P_i}$ is the function in $\mathcal{C}(\mathfrak{M})$ defined by $\widehat{P_i}(\lambda_j) = \delta_{ij}$. It is evident from this that

$$N = \sum_{i=1}^{m} \lambda_i P_i,$$

where the P_i's are non-zero pairwise orthogonal projections in A such that $\sum_{i=1}^{m} P_i = I$. This is precisely the spectral resolution of N treated in Chap. 11, so Theorem E actually contains the finite-dimensional spectral theorem. We therefore have solid grounds for regarding Theorem E as the generalized form of the spectral theorem discussed in the last paragraph of Sec. 63, and all our promises are fulfilled.

Appendices

$\mathfrak{F}\text{ixed Point Theorems}$

and Some Applications to Analysis

Let f be a continuous mapping of the closed interval $[-1,1]$ into itself. Figure 39 suggests that the graph of f must touch or cross the indicated diagonal, or more precisely, that there must exist a point x_0 in

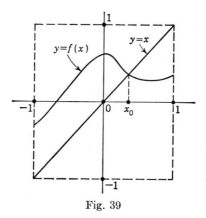

Fig. 39

$[-1,1]$ with the property that $f(x_0) = x_0$. The proof is easy. We consider the continuous function F defined on $[-1,1]$ by $F(x) = f(x) - x$, and we observe that $F(-1) \geq 0$ and that $F(1) \leq 0$. It now follows from the Weierstrass intermediate value theorem (see Theorem 31-C and the introduction to Chap. 6) that there exists a point x_0 in $[-1,1]$ such that $F(x_0) = 0$ or $f(x_0) = x_0$.

It is convenient to describe this phenomenon by means of the following terminology. A topological space X is called a *fixed point space* if every continuous mapping f of X into itself has a *fixed point*, in the sense that $f(x_0) = x_0$ for some x_0 in X. The remarks in the above paragraph show that $[-1,1]$ is a fixed point space. Furthermore, the closed disc $\{(x,y) : x^2 + y^2 \leq 1\}$ in the Euclidean plane R^2 is also a fixed point space (for a lucid elementary proof of this, see Courant and Robbins [6, pp. 251–255]). Both of these facts are special cases of

Brouwer's Fixed Point Theorem.   *The closed unit sphere $S = \{x : \|x\| \leq 1\}$ in R^n is a fixed point space.*

There are several proofs of this classic result, but since they all depend on the methods of algebraic topology, we refer the reader to Bers [3, p. 86]. Brouwer's theorem itself is a special case of

Schauder's Fixed Point Theorem.   *Every convex compact subspace of a Banach space is a fixed point space.*

For a proof, together with a discussion of other related results, see Bers [3, pp. 93–97]. Schauder's theorem was foreshadowed by the work of Birkhoff and Kellogg [5] on existence theorems in analysis. We illustrate the relevance of these ideas to such problems by giving a full treatment of Picard's theorem on the existence and uniqueness of solutions of first order differential equations.

We begin by considering an arbitrary metric space X with metric d. A mapping T of X into itself is called a *contraction* if there exists a positive real number $r < 1$ with the property that $d(Tx, Ty) \leq r\, d(x,y)$ for all points x and y in X. It is obvious that such a mapping is continuous. We shall need the following

Lemma.   *If T is a contraction defined on a complete metric space X, then T has a unique fixed point.*

PROOF.   Let x_0 be an arbitrary point in X, and write $x_1 = Tx_0$,

$$x_2 = T^2x_0 = Tx_1,$$

and, in general, $x_n = T^n x_0 = Tx_{n-1}$. If $m < n$, then

$$
\begin{aligned}
d(x_m, x_n) &= d(T^m x_0, T^n x_0) = d(T^m x_0, T^m T^{n-m} x_0) \\
&\leq r^m\, d(x_0, T^{n-m} x_0) = r^m\, d(x_0, x_{n-m}) \\
&\leq r^m\, [d(x_0, x_1) + d(x_1, x_2) + \cdots + d(x_{n-m-1}, x_{n-m})] \\
&\leq r^m\, d(x_0, x_1)[1 + r + \cdots + r^{n-m-1}] \\
&< \frac{r^m\, d(x_0, x_1)}{1 - r}.
\end{aligned}
$$

Since $r < 1$, it is evident from this that $\{x_n\}$ is a Cauchy sequence, and by the completeness of X, there exists a point x in X such that $x_n \to x$.

We now use the continuity of T to infer that x is a fixed point:

$$Tx = T(\lim x_n) = \lim Tx_n = \lim x_{n+1} = x.$$

We conclude the proof by showing that x is the only fixed point. If y is also a fixed point, that is, if $Ty = y$, then $d(x,y) = d(Tx,Ty) \leq r\, d(x,y)$; and since $r < 1$, this implies that $d(x,y) = 0$ or $y = x$.

This result is the key to

Picard's Theorem. *If $f(x,y)$ and $\partial f/\partial y$ are continuous in a closed rectangle $R = \{(x,y):a_1 \leq x \leq a_2 \text{ and } b_1 \leq y \leq b_2\}$, and if (x_0,y_0) is an interior point of R, then the differential equation*

$$\frac{dy}{dx} = f(x,y) \tag{1}$$

has a unique solution $y = g(x)$ which passes through (x_0,y_0).

PROOF. Since $f(x,y)$ and $\partial f/\partial y$ are continuous in R, they are bounded, and consequently there exist constants K and M such that

$$|f(x,y)| \leq K \tag{2}$$

and

$$\left| \frac{\partial}{\partial y} f(x,y) \right| \leq M \tag{3}$$

for all points (x,y) in R. We next observe that if (x,y_1) and (x,y_2) are in R, then the mean value theorem guarantees that

$$|f(x,y_1) - f(x,y_2)| = |y_1 - y_2| \left| \frac{\partial}{\partial y} f(x, y_1 + \theta(y_2 - y_1)) \right| \tag{4}$$

for some θ such that $0 < \theta < 1$. It now follows from (3) and (4) that

$$|f(x,y_1) - f(x,y_2)| \leq M|y_1 - y_2| \tag{5}$$

for all (x,y_1) and (x,y_2) in R.[1]

It is convenient at this stage to replace our problem by an equivalent problem relating to an *integral equation*. If $y = g(x)$ satisfies (1) and has the property that $g(x_0) = y_0$, then integrating (1) from x_0 to x yields

$$g(x) - g(x_0) = \int_{x_0}^{x} f(t,g(t))\, dt$$

or

$$g(x) = y_0 + \int_{x_0}^{x} f(t,g(t))\, dt. \tag{6}$$

Conversely, if $y = g(x)$ satisfies (6), then it is clear that $g(x_0) = y_0$, and on differentiating (6) we obtain (1). It therefore suffices to show that the integral equation (6) has a unique solution.

[1] The only use we make of the hypothesis that $\partial f/\partial y$ exists and is continuous in R is to derive the so-called *Lipschitz condition* (5).

To accomplish this, we choose a positive number a such that $Ma < 1$ and the closed rectangle R' determined by $|x - x_0| \leq a$ and $|y - y_0| \leq Ka$ is contained in R. We now let X be the set of all continuous real functions $y = g(x)$ defined on the closed interval $|x - x_0| \leq a$ such that $|g(x) - y_0| \leq Ka$. X is clearly a closed subspace of the complete metric space $\mathcal{C}[x_0 - a, \, x_0 + a]$ and is therefore itself a complete metric space. Our next step is to consider the mapping T of X into itself defined by $Tg = h$, where

$$h(x) = y_0 + \int_{x_0}^{x} f(t, g(t)) \, dt.$$

The fact that T maps X into itself is evident from (2), for

$$|h(x) - y_0| = \left| \int_{x_0}^{x} f(t, g(t)) \, dt \right| \leq Ka.$$

Furthermore, it follows from (5) that

$$|h_1(x) - h_2(x)| = \left| \int_{x_0}^{x} [f(t, g_1(t)) - f(t, g_2(t))] \, dt \right|$$
$$\leq Ma \sup |g_1(x) - g_2(x)|;$$

and since $Ma < 1$, this shows that T is a contraction on X. We now appeal to our lemma to conclude that the equation $Tg = g$ has a unique solution. Since this amounts to saying that the integral equation (6) has a unique solution, our proof is complete.

The ideas in this proof have a much wider scope than might be suspected, and can be applied to establish many other existence theorems in the theory of differential and integral equations.

Continuous Curves and the

Hahn-Mazurkiewicz Theorem

A continuous curve is usually thought of as "the path of a continuously moving point," and this rather vague notion is often felt to carry with it the even vaguer attribute of "thinness," or "one-dimensionality." For the case of plane curves, Jordan (in 1887) gave precise expression to this intuitive geometric concept by means of the following definition: if f is a continuous mapping of the closed unit interval $I = [0,1]$ into the

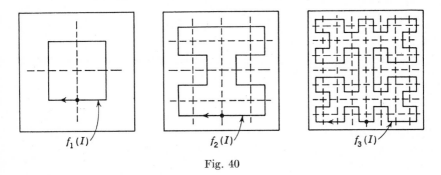

$f_1(I)$ $f_2(I)$ $f_3(I)$

Fig. 40

Euclidean plane R^2, then the subset $f(I)$ of R^2 is called a *continuous curve*. The fame of Jordan's definition rests mainly on Peano's discovery (in 1890) of a continuous curve which passes through every point of a closed square. Curves of this type have come to be called *space-filling curves*.

In Fig. 40, we show the first three stages in the construction of a particularly simple example known as *Hilbert's space-filling curve*. If the square under consideration is $S = \{(x,y) : 0 \leq x \leq 1 \text{ and } 0 \leq y \leq 1\}$, then the respective curves are the images of I under continuous mappings f_1, f_2, and f_3 of I into S. The process of constructing these curves can be continued in the same way, and it yields a sequence of continuous mappings f_n of I into S. By the manner in which each curve is constructed from its predecessor, it is clear that the sequence $\{f_n\}$ converges pointwise to a mapping f of I into S; and since this convergence is evidently uniform, f is continuous (see Problem 14-4) and $f(I)$ is a continuous curve in the sense of Jordan. Furthermore, each point of S lies in $f(I)$, so $f(I)$ is a space-filling curve.

Peano's discovery of space-filling curves was a shock to many mathematicians of the time, for it violated all their preconceived ideas of what a continuous curve ought to be. To a few of the others, however, it presented an opportunity. It suggested the very interesting problem of determining what a continuous curve actually is, or in other words, of finding intrinsic topological properties of a subset X of R^2 which are equivalent to the existence of a continuous mapping of I onto X.

Before describing the solution of this problem, we place it in a wider context by extending Jordan's definition. A topological space X is called a *continuous curve* if X is a Hausdorff space and there exists a continuous mapping of I onto X.[1] We know that I is compact and connected, so by Theorems 21-B and 31-B, any continuous curve is also compact and connected. In the lemmas below, we give two additional properties which every continuous curve must have.

It is convenient to begin by introducing the following concept. A mapping f of one topological space into another is said to be *closed* if it carries closed sets into closed sets, that is, if $f(F)$ is closed whenever F is closed. We shall use the fact that a continuous mapping of a compact space into a Hausdorff space is automatically closed (see the proof of Theorem 26-E).

Lemma. *Every continuous curve is second countable.*

PROOF. Let f be a continuous mapping of I onto a Hausdorff space X. We must show that X is second countable, that is, that it has a countable open base. I is a separable metric space, so it has a countable open base $\{B_i\}$, and it is easily seen that the class $\{G_i\}$ of all finite unions of the B_i's is also a countable open base for I. Since I is compact and X is Hausdorff, f is closed, and therefore each set $f(G_i')$ is closed. The class of all sets of the form $f(G_i')'$ is thus a countable class of open subsets of X, so it suffices to show that these sets constitute an open base for X.

Let x be a point of X with neighborhood G. The set $f^{-1}(\{x\})$ is closed and is therefore a compact subspace of I with neighborhood $f^{-1}(G)$. For each point y in $f^{-1}(\{x\})$, there exists a set in $\{G_i\}$ which contains y and

[1] A continuous curve in the sense of this definition is often called a *Peano space*.

is contained in $f^{-1}(G)$.   By the compactness of $f^{-1}(\{x\})$ and the fact that $\{G_i\}$ is closed under the formation of finite unions, there exists a G_i such that

$$f^{-1}(\{x\}) \subseteq G_i \subseteq f^{-1}(G).$$

On taking complements, we obtain

$$f^{-1}(\{x\})' \supseteq G_i' \supseteq f^{-1}(G)', \tag{1}$$

and since the complement of an inverse image equals the inverse image of the complement, we can write (1) in the form

$$f^{-1}(\{x\}') \supseteq G_i' \supseteq f^{-1}(G'). \tag{2}$$

If we now apply f to all members of (2), we get

$$ff^{-1}(\{x\}') \supseteq f(G_i') \supseteq ff^{-1}(G')$$

or

$$\{x\}' \supseteq f(G_i') \supseteq G',$$

so

$$\{x\} \subseteq f(G_i')' \subseteq G,$$

and the proof is complete.

Lemma.   *Every continuous curve is locally connected.*

PROOF.   Let f be a continuous mapping of I onto a Hausdorff space X.   We must show that X is locally connected.   By Problem 34-1, it suffices to show that if C is a component of an open subspace G of X, then C is open.

Let A be a component of $f^{-1}(G)$.   Then A is connected, and therefore $f(A)$ is connected in G; and since C is a component of G, we see that $f(A)$ is either disjoint from C or contained in C.   It follows from this that $f^{-1}(C)$ is a union of components of $f^{-1}(G)$.   Since $f^{-1}(G)$ is open and I is locally connected, Theorem 34-A tells us that the components of $f^{-1}(G)$ are open, so $f^{-1}(C)$ is open and $f^{-1}(C)' = f^{-1}(C')$ is closed.   We conclude the proof by observing that since the mapping f is closed, the set $ff^{-1}(C') = C'$ is closed, so C is open.

The above remarks and lemmas establish the easy half of the following famous characterization of continuous curves.

The Hahn-Mazurkiewicz Theorem.   *A topological space X is a continuous curve $\Leftrightarrow X$ is a compact Hausdorff space which is second countable, connected, and locally connected.*

For the remainder of the proof, we refer the reader to Wilder [43, p. 76].   Additional discussions of a descriptive and historical nature can be found in Wilder [44] and Hahn [15].

Boolean Algebras, Boolean Rings, and Stone's Theorem

We saw in Sec. 2 that a *Boolean algebra of sets* can be defined as a class of subsets of a non-empty set which is closed under the formation of finite unions, finite intersections, and complements. Our purpose in this appendix is threefold: to define abstract Boolean algebras by means of lattices; to show that the theory of these systems can be regarded as part of the general theory of rings; and to prove the famous theorem of Stone, which asserts that every Boolean algebra is isomorphic to a Boolean algebra of sets.

The reader will recall that a *lattice* is a partially ordered set in which each pair of elements x and y has a greatest lower bound $x \wedge y$ and a least upper bound $x \vee y$, and that these elements are uniquely determined by x and y. It is easy to show (see Problem 8-5) that the operations $\wedge$ and $\vee$ have the following properties:

$$x \wedge x = x \quad \text{and} \quad x \vee x = x; \tag{1}$$
$$x \wedge y = y \wedge x \quad \text{and} \quad x \vee y = y \vee x; \tag{2}$$
$$x \wedge (y \wedge z) = (x \wedge y) \wedge z \quad \text{and} \quad x \vee (y \vee z) = (x \vee y) \vee z; \tag{3}$$
$$(x \wedge y) \vee x = x \quad \text{and} \quad (x \vee y) \wedge x = x. \tag{4}$$

We shall see in the next paragraph that these properties are actually characteristic of lattices. Before proceeding further, however, we remark that

$$x \leq y \Leftrightarrow x \wedge y = x.$$

This fact serves to motivate the following discussion.

Let L be a non-empty set in which two operations $\wedge$ and $\vee$ are defined, and assume that these operations satisfy the above conditions. We shall prove that a partial order relation $\leq$ can be defined in L in such a way that L becomes a lattice in which $x \wedge y$ and $x \vee y$ are the greatest lower bound and least upper bound of x and y. Our first step is to notice that $x \wedge y = x$ and $x \vee y = y$ are equivalent; for if $x \wedge y = x$, then $x \vee y = (x \wedge y) \vee y = (y \wedge x) \vee y = y$, and similarly $x \vee y = y$ implies $x \wedge y = x$. We now define $x \leq y$ to mean that either $x \wedge y = x$ or $x \vee y = y$. Since $x \wedge x = x$, we have $x \leq x$ for every x. If $x \leq y$ and $y \leq x$, so that $x \wedge y = x$ and $y \wedge x = y$, then $x = x \wedge y = y \wedge x = y$. If $x \leq y$ and $y \leq z$, so that $x \wedge y = x$ and $y \wedge z = y$, then

$$x \wedge z = (x \wedge y) \wedge z = x \wedge (y \wedge z) = x \wedge y = x,$$

so $x \leq z$. This completes the proof that $\leq$ is a partial order relation. We now show that $x \wedge y$ is the greatest lower bound of x and y. Since $(x \wedge y) \vee x = x$ and $(x \wedge y) \vee y = (y \wedge x) \vee y = y$, we see that $x \wedge y \leq x$ and $x \wedge y \leq y$. If $z \leq x$ and $z \leq y$, so that $z \wedge x = z$ and $z \wedge y = z$, then $z \wedge (x \wedge y) = (z \wedge x) \wedge y = z \wedge y = z$, so $z \leq x \wedge y$. It is easy to prove, by similar arguments, that $x \vee y$ is the least upper bound of x and y.

This characterization of lattices brings the theory of these systems somewhat closer to ordinary abstract algebra, in which operations (instead of relations) are usually placed in the foreground.

A lattice is said to be *distributive* if it has the following properties:

$$x \wedge (y \vee z) = (x \wedge y) \vee (x \wedge z) \tag{5}$$

and
$$x \vee (y \wedge z) = (x \vee y) \wedge (x \vee z). \tag{6}$$

It is useful to know that (5) and (6) are equivalent to one another. For if (5) holds, then

$$\begin{aligned}
(x \vee y) \wedge (x \vee z) &= [(x \vee y) \wedge x] \vee [(x \vee y) \wedge z] \\
&= x \vee [(x \vee y) \wedge z] \\
&= x \vee [(x \wedge z) \vee (y \wedge z)] \\
&= [x \vee (x \wedge z)] \vee (y \wedge z) \\
&= x \vee (y \wedge z),
\end{aligned}$$

and a similar computation shows that (6) implies (5). We shall say that a lattice is *complemented* if it contains distinct elements 0 and 1 such that

$$0 \leq x \leq 1 \tag{7}$$

for every x (these elements are clearly unique when they exist), and if each element x has a *complement* x' with the property that

$$x \wedge x' = 0 \qquad \text{and} \qquad x \vee x' = 1. \tag{8}$$

We now define a *Boolean algebra* to be a complemented distributive lattice.

It is quite possible for an element of a complemented lattice to have many different complements. In a Boolean algebra, however, each element has only one complement.   To prove this, we suppose that x^* is also an element with the property that $x \wedge x^* = 0$ and $x \vee x^* = 1$. Then

$$x^* = x^* \wedge 1 = x^* \wedge (x \vee x') = (x^* \wedge x) \vee (x^* \wedge x')$$
$$= 0 \vee (x^* \wedge x') = x^* \wedge x',$$

so $x^* \leq x'$.   If we now reverse the roles of x' and x^*, we obtain $x' \leq x^*$, so $x^* = x'$.   In the light of this result, it is evident from (8) that x is the complement of x':

$$x'' = x. \tag{9}$$

Furthermore, it follows from (7) that $0 \wedge 1 = 0$ and $0 \vee 1 = 1$, so we have

$$0' = 1 \qquad \text{and} \qquad 1' = 0. \tag{10}$$

The identities

$$(x \wedge y)' = x' \vee y' \qquad \text{and} \qquad (x \vee y)' = x' \wedge y' \tag{11}$$

are also true in every Boolean algebra.   We shall prove the first part of (11).   Our principal tool will be the fact that

$$x \leq y \Leftrightarrow y' \leq x'. \tag{12}$$

To establish (12), it suffices to show that $x \leq y$ implies $y' \leq x'$, and the proof of this is easy: if $x \leq y$, then $x \wedge y' \leq y \wedge y' = 0$, so

$$y' = y' \wedge 1 = y' \wedge (x \vee x') = (y' \wedge x) \vee (y' \wedge x') = 0 \vee (y' \wedge x') = y' \wedge x',$$

and therefore $y' \leq x'$.   We now turn to the proof of $(x \wedge y)' = x' \vee y'$. Our first step is to observe that if $x' \leq z$ and $y' \leq z$, so that $z' \leq x$ and $z' \leq y$, then $z' \leq x \wedge y$ or $(x \wedge y)' \leq z$.   This shows that $(x \wedge y)'$ is less than or equal to any upper bound of x' and y', so $(x \wedge y)' \leq x' \vee y'$.   We conclude the proof by showing that $x' \vee y' \leq (x \wedge y)'$.   This follows at once from the relations $x' \leq (x \wedge y)'$ and $y' \leq (x \wedge y)'$, which, since they are equivalent to $x \wedge y \leq x$ and $x \wedge y \leq y$, are evidently true.   The second part of (11) can be proved in essentially the same way.

One of the basic facts about Boolean algebras is that these systems can be identified with a certain class of rings.   This enables us to study Boolean algebras by means of powerful techniques which are already available in the general theory of rings.

A *Boolean ring* is a ring with identity in which every element is idempotent (i.e., $x^2 = x$ for every x).   It is a surprising fact that multiplication in a Boolean ring is automatically commutative and that $x + x = 0$ (or equivalently, $x = -x$) for every x.   The proof of these statements rests on the relation

$$x + y = (x + y)^2 = (x + y)(x + y) = x^2 + xy + yx + y^2$$
$$= x + xy + yx + y,$$

which implies that $xy + yx = 0$, so $xy = -yx$. If we put $y = x$, this yields $x^2 = -x^2$, so $x = -x$; and from this we obtain $xy = -yx = yx$.

In order to make a Boolean algebra A into a Boolean ring R, we define addition and multiplication by

$$x + y = (x \wedge y') \vee (x' \wedge y) \quad \text{and} \quad xy = x \wedge y. \tag{13}$$

(For the motivation behind these definitions, see Example 40-3.) To verify that R actually is a Boolean ring, we proceed as follows. It is clear that

$$\begin{aligned} x + y = (x \wedge y') \vee (x' \wedge y) &= (y' \wedge x) \vee (y \wedge x') \\ &= (y \wedge x') \vee (y' \wedge x) \\ &= y + x, \end{aligned}$$

that

$$\begin{aligned} x + 0 = (x \wedge 0') \vee (x' \wedge 0) &= (x \wedge 1) \vee 0 \\ &= x \wedge 1 = x, \end{aligned}$$

and that

$$x + x = (x \wedge x') \vee (x' \wedge x) = 0 \vee 0 = 0.$$

The proof that addition is associative is more complicated. It is convenient to begin with the observation that

$$\begin{aligned} (x + y)' = [(x \wedge y') \vee (x' \wedge y)]' &= (x' \vee y) \wedge (x \vee y') \\ &= [(x' \vee y) \wedge x] \vee [(x' \vee y) \wedge y'] \\ &= [(x' \wedge x) \vee (y \wedge x)] \vee [(x' \wedge y') \vee (y \wedge y')] \\ &= (x \wedge y) \vee (x' \wedge y'). \end{aligned}$$

Now, using this, we have

$$\begin{aligned} x + (y + z) &= [x \wedge (y + z)'] \vee [x' \wedge (y + z)] \\ &= [x \wedge ((y \wedge z) \vee (y' \wedge z'))] \vee [x' \wedge ((y \wedge z') \vee (y' \wedge z))] \\ &= (x \wedge y \wedge z) \vee (x \wedge y' \wedge z') \vee (x' \wedge y \wedge z') \vee (x' \wedge y' \wedge z). \end{aligned}$$

It is clear by inspection that the expression last written is unaltered by interchanging x and z, so $x + (y + z) = z + (y + x)$; and since, by commutativity, we have $z + (y + x) = (x + y) + z$, it follows that addition is associative. The relevant properties of multiplication are fairly easy to establish. It is immediate from the definition that

$$x(yz) = (xy)z,$$

that $x^2 = x$, and that 1 is an identity. In view of the fact that multiplication is obviously commutative, all that remains is to verify that $x(y + z) = xy + xz$, and this is a consequence of the following computations:

$$\begin{aligned} x(y + z) = x \wedge (y + z) &= x \wedge [(y \wedge z') \vee (y' \wedge z)] \\ &= (x \wedge y \wedge z') \vee (x \wedge y' \wedge z), \end{aligned}$$

and

$$xy + xz = (x \wedge y) + (x \wedge z)$$
$$= [(x \wedge y) \wedge (x \wedge z)'] \vee [(x \wedge y)' \wedge (x \wedge z)]$$
$$= [(x \wedge y) \wedge (x' \vee z')] \vee [(x' \vee y') \wedge (x \wedge z)]$$
$$= (x \wedge y \wedge x') \vee (x \wedge y \wedge z') \vee (x' \wedge x \wedge z) \vee (y' \wedge x \wedge z)$$
$$= (x \wedge y \wedge z') \vee (x \wedge y' \wedge z).$$

Thus R is a Boolean ring.

We now reverse this process; that is, we start with a Boolean ring R, and we show that the definitions

$$x \wedge y = xy \qquad \text{and} \qquad x \vee y = x + y + xy \qquad (14)$$

convert it into a Boolean algebra A. If we keep in mind the fact that multiplication in R is commutative and that for every x we have $x^2 = x$ and $x = -x$, then (1) and (2) are evident. Property (3) follows from the associativity of multiplication and the computations

$$x \vee (y \vee z) = x \vee (y + z + yz)$$
$$= x + y + z + yz + xy + xz + xyz$$

and
$$(x \vee y) \vee z = (x + y + xy) \vee z$$
$$= x + y + xy + z + xz + yz + xyz.$$

Property (4) is also true, for

$$(x \wedge y) \vee x = xy \vee x = xy + x + xyx = x + xy + xy = x$$

and

$$(x \vee y) \wedge x = (x + y + xy) \wedge x = x^2 + yx + xyx = x + xy + xy = x.$$

These remarks show that A is a lattice. Further, this lattice is distributive, for

$$x \wedge (y \vee z) = x(y + z + yz) = xy + xz + xyz$$
$$= xy + xz + xyxz$$
$$= xy \vee xz$$
$$= (x \wedge y) \vee (x \wedge z).$$

It is easy to see that the elements 0 and 1 have the property that $0 \leq x \leq 1$ for every x, and that $x' = 1 + x$ acts as a complement for x, so A is a Boolean algebra.

It is worth noting that the two processes we have described are inverses of one another. Suppose we start with a Boolean algebra A and use (13) to make it into a Boolean ring R:

$$x + y = (x \wedge y') \vee (x' \wedge y) \qquad \text{and} \qquad xy = x \wedge y.$$

Next, we use (14) to convert R back into a Boolean algebra $\bar{A}$:

$$x \bar{\wedge} y = xy \qquad \text{and} \qquad x \bar{\vee} y = x + y + xy.$$

It is apparent that $x \bar{\wedge} y = xy = x \wedge y$; and since

$$1 + x = (1 \wedge x') \vee (1' \wedge x) = (1 \wedge x') \vee (0 \wedge x)$$
$$= x' \vee 0 = x',$$

we also have

$$x \bar{\vee} y = x + y + xy = 1 + (1 + x)(1 + y)$$
$$= 1 + x'y'$$
$$= (x' \wedge y')'$$
$$= x \vee y.$$

This shows that the operations in $\bar{A}$ coincide with those in A. Conversely, if we start with a Boolean ring R, make it into a Boolean algebra A, and then convert A back into a Boolean ring $\bar{R}$, then the operations in $\bar{R}$ coincide with those in R. We leave the verification of this to the reader.

The ideas developed above show that Boolean algebras are essentially identical with Boolean rings. The practical effect of this is a considerable saving of labor, for it allows us to transpose our study of Boolean algebras into the more familiar context of the theory of rings, where many standard tools—ideals, homomorphisms, etc.—lie ready at hand. We illustrate this principle by proving the basic representation theorem for Boolean algebras in two steps: first, we prove the corresponding theorem for Boolean rings; and second, we translate this result back into the language of Boolean algebras.

Before entering into the details, we give a brief description of the type of representation we are aiming at. If X is a compact Hausdorff space, then each of the sets $\emptyset$ and X is both open and closed (or more briefly, *open-closed*), and the class **A** of all such sets is a Boolean algebra of subsets of X. If X is disconnected, then **A** contains at least three sets; and if X is totally disconnected, then **A** may contain a great many sets, for, by Theorem 33-C, it is an open base for the topology of X. Furthermore, we know that **A** becomes a Boolean ring of sets if addition and multiplication are defined by

$$A + B = (A \cap B') \cup (A' \cap B) \qquad \text{and} \qquad AB = A \cap B.$$

Our basic representation theorem states that every Boolean algebra (Boolean ring) is isomorphic to the Boolean algebra (Boolean ring) of all open-closed subsets of some totally disconnected compact Hausdorff space.

Now for the details. The simplest of all Boolean rings is the ring $\{0,1\}$ of integers mod 2, and this ring is evidently a field. Conversely, any Boolean ring which is a field necessarily equals $\{0,1\}$.   To see this, it suffices to observe that if x is a non-zero element in such a ring, then

$$1 = xx^{-1} = x^2x^{-1} = x(xx^{-1}) = x1 = x.$$

If I is a proper ideal in a Boolean ring R, then the quotient ring R/I is also a Boolean ring; for R/I clearly has an identity, and

$$(x + I)^2 = x^2 + I = x + I$$

for every x in R.   Thus, by Theorem 41-C, $R/I = \{0,1\} \Leftrightarrow I$ is maximal. Since every homomorphism of R arises from an ideal in R, this tells us that the homomorphisms of R onto $\{0,1\}$ are precisely those of the form $R \to R/M$, where M is a maximal ideal in R.   A standard application of Zorn's lemma shows that R has maximal ideals, so there do exist homomorphisms of R onto $\{0,1\}$.   We shall need the following stronger statement.

Lemma.   *If x is a non-zero element in a Boolean ring R, then there exists a homomorphism h of R onto $\{0,1\}$ such that $h(x) = 1$.*

PROOF.   By the above remarks, it suffices to show that there exists a maximal ideal in R which does not contain x.   Since $x \neq 0$, there clearly exists at least one ideal which does not contain x.   If Zorn's lemma is applied to the set of all ideals which do not contain x, we obtain an ideal M which is maximal with respect to the property of not containing x. We conclude the proof by showing that M actually is a maximal ideal. To prove this, it suffices to show that M contains $1 + x$ (for it will then follow that any strictly larger ideal contains both x and $1 + x$, and so contains 1).   We therefore assume that M does not contain $1 + x$, and we deduce a contradiction from this assumption.   It is clear that

$$I = \{m + r(1 + x) : m \; \varepsilon \; M \text{ and } r \; \varepsilon \; R\}$$

is the smallest ideal containing both M and $1 + x$, so I properly contains M.   However, I does not contain x; for if it did, we would have

$$m + r(1 + x) = x$$

for some m and r, and this implies that

$$\begin{aligned} x = x^2 &= [m + r(1 + x)]x \\ &= mx + r(x + x^2) \\ &= mx + r(x + x) \\ &= mx, \end{aligned}$$

contrary to the fact that x is not in M.   This contradicts the maximality property of M, and the proof is complete.

We are now in a position to prove our principal theorem.

The Stone Representation Theorem. *If R is a Boolean ring, then there exists a totally disconnected compact Hausdorff space H such that R is isomorphic to the Boolean ring of all open-closed subsets of H.*

PROOF. Let H^* be the set of all mappings of R into the Boolean ring $\{0,1\}$. If for each x in R we define H_x by $H_x = \{0,1\}$, then H^* is the product set $P_{x \epsilon R} H_x$. We now impose the discrete topology on each H_x, and thus convert it into a totally disconnected compact Hausdorff space. This permits us to regard H^* as a product space, and it is also a totally disconnected compact Hausdorff space. For use in the next paragraph, we note that if x is any given element of R, then each of the sets

$$\{f : f(x) = 0\}$$

and $\{f : f(x) = 1\}$ is open-closed. This follows at once from the fact that each is the inverse image of an open-closed set in H_x under the projection of H^* onto H_x.

We now pass to the subspace H of H^* which consists of all homomorphisms of R onto $\{0,1\}$. It is clear that H is a totally disconnected Hausdorff space. To prove that it is also compact, it suffices to show that it is closed in H^*, and this we do as follows. A homomorphism of R onto $\{0,1\}$ is of course a mapping f in H^* such that $f(x + y) = f(x) + f(y)$ and $f(xy) = f(x)f(y)$ for all x and y and such that $f(1) = 1$. It is evident from this that H is the intersection of the following three subsets of H^*:

$$\bigcap_{x,y \epsilon R} \{f : f(x + y) = f(x) + f(y)\}, \tag{15}$$
$$\bigcap_{x,y \epsilon R} \{f : f(xy) = f(x)f(y)\}, \tag{16}$$
and
$$\{f : f(1) = 1\}. \tag{17}$$

We know from our remark in the preceding paragraph that (17) is closed; and if we can show that the other two sets are also closed, then it will follow at once that H is closed. We inspect (15). If x and y are any given elements of R, then it is easy to see that

$$\{f : f(x + y) = f(x) + f(y)\}$$

is the union of the following four sets:

$$\{f : f(x) = 0, f(y) = 0, \text{ and } f(x + y) = 0\},$$
$$\{f : f(x) = 0, f(y) = 1, \text{ and } f(x + y) = 1\},$$
$$\{f : f(x) = 1, f(y) = 0, \text{ and } f(x + y) = 1\},$$
and
$$\{f : f(x) = 1, f(y) = 1, \text{ and } f(x + y) = 0\}.$$

Each of these sets, being itself the intersection of three closed sets, is closed, so $\{f : f(x + y) = f(x) + f(y)\}$ is closed, and consequently (15) is also closed. A similar argument shows that (16) is closed, so H is closed and therefore compact.

Our next step is to exhibit an isomorphism T of R into the Boolean ring **R** of all open-closed subsets of H.   We define T by

$$T(x) = \{f : f \, \varepsilon \, H \text{ and } f(x) = 1\}.$$

It is clear that T maps R into **R**. T is also a homomorphism, for

$$
\begin{aligned}
T(x + y) &= \{f : f(x + y) = 1\} \\
&= \{f : f(x) + f(y) = 1\} \\
&= \{f : f(x) = 1\} + \{f : f(y) = 1\} \\
&= T(x) + T(y)
\end{aligned}
$$

and
$$
\begin{aligned}
T(xy) &= \{f : f(xy) = 1\} \\
&= \{f : f(x)f(y) = 1\} \\
&= \{f : f(x) = 1\} \cap \{f : f(y) = 1\} \\
&= T(x)T(y).
\end{aligned}
$$

Our lemma tells us that $T(x)$ is non-empty whenever $x \neq 0$, so T is an isomorphism of R into **R**.   It will be useful in the next paragraph if we also note here that

$$T(1) = \{f : f(1) = 1\} = H,$$

for it follows from this that

$$T(1 + x) = T(1) + T(x) = H + T(x) = T(x)'$$

for every x in R.

Finally, we show that T maps R onto **R**.   We begin by observing that the topology of H is defined by means of basic open sets of the form

$$B = \{f : f(x_i) = \epsilon_i, \, i = 1, \, \ldots \, , n\},$$

where $\{x_1, \, \ldots \, , x_n\}$ is an arbitrary finite subset of R and each ϵ_i equals 0 or 1.   These sets are evidently closed as well as open.   Furthermore, every set of this kind is in the range of T; for since

$$\{f : f(x_i) = 0\} = \{f : f(1 + x_i) = 1\},$$

if we define y_i to be x_i or $1 + x_i$ according as ϵ_i equals 1 or 0, then

$$
\begin{aligned}
B &= \bigcap_{i=1}^{n} \{f : f(x_i) = \epsilon_i\} \\
&= \bigcap_{i=1}^{n} \{f : f(y_i) = 1\} \\
&= \bigcap_{i=1}^{n} T(y_i) \\
&= T(y_1 \cdots y_n).
\end{aligned}
$$

We now consider an arbitrary open-closed set S in **R**.   Since S is compact and the B's constitute an open base, S is the union of a finite number of B's, say $B_1, \, \ldots \, , B_m$; and by the above result, each B_j is expressible in the form $B_j = T(z_j)$ for some element z_j in R.   It now follows that

$$\begin{aligned}
S = \bigcup_{j=1}^{m} B_j &= \left(\bigcap_{j=1}^{m} B_j'\right)' = \left(\bigcap_{j=1}^{m} T(z_j)'\right)' \\
&= \left(\bigcap_{j=1}^{m} T[1 + z_j]\right)' \\
&= (T([1 + z_1] \cdots [1 + z_m]))' \\
&= T(1 + [1 + z_1] \cdots [1 + z_m]).
\end{aligned}$$

This shows that T is an isomorphism of R onto $\mathbf{R}$, so the proof is complete.

We now conclude our theory by translating Stone's theorem into the language of Boolean algebras.

Let A and A^* be Boolean algebras.   A mapping h of A into A^* is called an *isomorphism* (or a *Boolean algebra isomorphism*) if it is one-to-one and has the following three properties: $h(x \wedge y) = h(x) \wedge h(y)$,

$$h(x \vee y) = h(x) \vee h(y),$$

and $h(x') = h(x)'$. A is said to be *isomorphic* to A^* if there exists an isomorphism of A onto A^*.   If A and A^* are converted into Boolean rings R and R^*, then it is easy to show that every Boolean algebra isomorphism of A onto A^* is a Boolean ring isomorphism of R onto R^*, and conversely.   We leave the details to the reader.

These ideas make it possible for us to state the following equivalent form of Stone's theorem: *If A is a Boolean algebra, then there exists a totally disconnected compact Hausdorff space H such that A is isomorphic to the Boolean algebra of all open-closed subsets of H.*

Bibliography

1. Achieser, N. I.: "Vorlesungen über Approximationstheorie," Akademie, Berlin, 1953.
2. Alexandroff, P., and H. Hopf: "Topologie," Springer, Berlin, 1935.
3. Bers, L.: "Topology," lecture notes, New York University Institute of Mathematical Sciences, New York, 1957.
4. Birkhoff, G.: "Lattice Theory," American Mathematical Society Colloquium Publications, vol. 25, New York, 1948.
5. Birkhoff, G. D., and O. D. Kellogg: Invariant Points in Function Space, *Trans. Amer. Math. Soc.*, 23 (1922), pp. 96–115.
6. Courant, R., and H. Robbins: "What Is Mathematics?" Oxford, London and New York, 1941.
7. Dixmier, J.: "Les algèbres d'opérateurs dans l'espace hilbertien (Algèbres de von Neumann)," Gauthier-Villars, Paris, 1957.
8. Dunford, N., and J. T. Schwartz: "Linear Operators, Part I: General Theory," Interscience, New York, 1958.
9. Fraenkel, A. A.: "Abstract Set Theory," North-Holland, Amsterdam, 1953.
10. Fraenkel, A. A., and Y. Bar-Hillel: "Foundations of Set Theory," North-Holland, Amsterdam, 1958.
11. Gál, I. S.: On Sequences of Operations in Complete Vector Spaces, *Amer. Math. Monthly*, 60 (1953), pp. 527–538.
12. Gödel, K.: What Is Cantor's Continuum Problem? *Amer. Math. Monthly*, 54 (1947), pp. 515–525.
13. Goffman, C.: Preliminaries to Functional Analysis, in "Studies in Mathematics," vol. 1, Mathematical Association of America, 1962.
14. Goldberg, R. R.: "Fourier Transforms," Cambridge, New York, 1961.
15. Hahn, H.: The Crisis in Intuition, in "The World of Mathematics," Simon and Schuster, New York, 1956.
16. Halmos, P. R.: "Naive Set Theory," Van Nostrand, Princeton, N.J., 1960.
17. ———: "Finite-dimensional Vector Spaces," Van Nostrand, Princeton, N.J., 1958.
18. ———: "Measure Theory," Van Nostrand, Princeton, N.J., 1950.

19. Hewitt, E.: The Rôle of Compactness in Analysis, *Amer. Math. Monthly*, 67 (1960), pp. 499–516.
20. Hille, E., and R. S. Phillips: "Functional Analysis and Semi-groups," American Mathematical Society Colloquium Publications, vol. 31, Providence, R.I., 1957.
21. Hurewicz, W., and H. Wallman: "Dimension Theory," Princeton, Princeton, N.J., 1941.
22. Kadison, R. V.: Order Properties of Bounded Self-adjoint Operators, *Proc. Amer. Math. Soc.*, 2 (1951), pp. 505–510.
23. Kakutani, S., and G. W. Mackey: Ring and Lattice Characterizations of Complex Hilbert Space, *Bull. Amer. Math. Soc.*, 52 (1946), pp. 727–733.
24. Kamke, E.: "Theory of Sets," Dover, New York, 1950.
25. Kelley, J. L.: "General Topology," Van Nostrand, Princeton, N.J., 1955.
26. Kolmogorov, A. N., and S. V. Fomin: "Elements of the Theory of Functions and Functional Analysis," 2 vols., Graylock, Rochester and Albany, 1957 and 1961.
27. Loomis, L. H.: "An Introduction to Abstract Harmonic Analysis," Van Nostrand, Princeton, N.J., 1953.
28. Lorch, E. R.: The Spectral Theorem, in "Studies in Mathematics," vol. 1, Mathematical Association of America, 1962.
29. Lorentz, G. G.: "Bernstein Polynomials," University of Toronto Press, Toronto, 1953.
30. Mackey, G. W.: Functions on Locally Compact Groups, *Bull. Amer. Math. Soc.*, 56 (1950), pp. 385–412.
31. McCoy, N. H.: "Rings and Ideals," Carus Mathematical Monographs, no. 8, Mathematical Association of America, 1948.
32. Naimark, M. A.: "Normed Rings," Noordhoff, Groningen, Netherlands, 1959.
33. Niven, I.: "Irrational Numbers," Carus Mathematical Monographs, no. 11, Mathematical Association of America, 1956.
34. Rickart, C. E.: "General Theory of Banach Algebras," Van Nostrand, Princeton, N.J., 1960.
35. Riesz, F., and B. Sz.-Nagy: "Functional Analysis," Frederick Ungar, New York, 1955.
36. Russell, B.. "My Philosophical Development," Simon and Schuster, New York, 1959.
37. Sierpinski, W.: "Cardinal and Ordinal Numbers," Monografie Matematyczne, vol. 34, Warszawa, 1958.
38. Smirnov, Y. M.: A Necessary and Sufficient Condition for Metrizability of a Topological Space, *Dokl. Akad. Nauk SSSR*, 77 (1951), pp. 197–200.

39. Stone, M. H.: On the Compactification of Topological Spaces, *Ann. Soc. Pol. Math.*, 21 (1948), pp. 153–160.

40. ———: A Generalized Weierstrass Approximation Theorem, in "Studies in Mathematics," vol. 1, Mathematical Association of America, 1962.

41. Taylor, A. E.: "Introduction to Functional Analysis," Wiley, New York, 1958.

42. Wilder, R. L.: "Introduction to the Foundations of Mathematics," Wiley, New York, 1952.

43. ———: "Topology of Manifolds," American Mathematical Society Colloquium Publications, vol. 32, New York, 1949.

44. ———: The Origin and Growth of Mathematical Concepts, *Bull. Amer. Math. Soc.*, 59 (1953), pp. 423–448.

45. Zaanen, A. C.: "An Introduction to the Theory of Integration," North-Holland, Amsterdam, 1958.

46. Zygmund, A.: "Trigonometric Series," 2 vols., Cambridge, New York, 1959.

Index of Symbols

$A = B,\ A \neq B$	Equality and inequality for sets, 5
$A \subseteq B$	Set inclusion, 5
$A \subset B$	Proper set inclusion, 6
$A \cup B$	Union of two sets, 8
$A \cap B$	Intersection of two sets, 9
A'	Complement of a set, 10
$A - B$	Difference of two sets, 13
$A \bigtriangleup B$	Symmetric difference of two sets, 13
$a \equiv b \pmod{m}$	Congruence modulo m for integers, 30
$[a,b],\ (a,b)$, etc.	Intervals on the real line, 5, 57
$\aleph_0$	Cardinal number of a countably infinite set, 34
$\bar{A}$	Closure of a set, 68, 96
$A_1 \leq A_2$	Order relation for self-adjoint operators, 268
A_n	Total matrix algebra of degree n, 284
$\hat{A}$	Function algebra representing a commutative Banach algebra, 319
$\mathfrak{B}(N,N')$	Space of bounded (or continuous) linear transformations of N into N', 221
$\mathfrak{B}(N)$	Algebra of operators on N, 222
$\beta(X)$	Stone-Čech compactification of X, 139, 141
c	Cardinal number of the continuum, 39
C	Complex number system, 23, 52–54, 214
C^n	n-dimensional unitary space, 23–24, 89–90, 214
C^∞	Infinite-dimensional unitary space, 90
C_∞	Extended complex plane, 162–163
$\mathfrak{C}[0,1]$	Space of bounded continuous real functions on $[0,1]$, 56
$\mathfrak{C}[a,b]$	Space of bounded continuous real functions on $[a,b]$, 84
$\mathfrak{C}(X,R)$	Space of bounded continuous real functions on X, 82, 106

L_p	Banach space of measurable functions, 215
l_∞^n	Banach space of n-tuples, 216
l_∞, c, c_0	Banach spaces of sequences, 216
L/M	Quotient space of a linear space with respect to a subspace, 193–194
$L_1(G)$	Group algebra of a finite or discrete group, 303–305
$\mathfrak{M}$	Space of maximal ideals, 319
$\min A, \max A$	Minimum and maximum of a finite set of real numbers, 45
M_x	Maximal ideal associated with a point, 327
$M + N$	Sum of two subspaces of a linear space, 195
$M \oplus N$	Direct sum of two subspaces of a linear space, 195
$m < n, m \leq n$	Order relation for cardinal numbers, 35, 48
N^*	Conjugate space of a normed linear space, 224
N^{**}	Second conjugate space of a normed linear space, 231
N/M	Quotient space of a normed linear space with respect to a closed subspace, 213
P_iX_i	Product of a class of sets, 25
p_i	Projection of a product onto a coordinate set, 25
R	Real number system, 21, 52, 214
$R \times R$ (or R^2)	Coordinate plane, 22
R^n	n-dimensional Euclidean space, 23–24, 85–89, 214
R^∞	Infinite-dimensional Euclidean space, 90
$r(x)$	Spectral radius of x, 310
$\rho(x)$	Resolvent set of x, 309
R/I	Quotient ring of a ring with respect to an ideal, 187
S	Set of singular elements in a Banach algebra, 305
S^*	Closed unit sphere in a conjugate space, 233–234
$S^\perp$	Orthogonal complement, 249
$[S]$	Subspace spanned by S, 194
$S_r(x_0)$	Open sphere with radius r and center x_0, 59
$S_r[x_0]$	Closed sphere with radius r and center x_0, 66
$\sup A$	Supremum (or least upper bound) of a set of real numbers, 45, 56
$\sigma(T)$	Spectrum of an operator, 289, 296
$\sigma(x), \sigma_A(x)$	Spectrum of an element in a Banach algebra, 308

T^*	Conjugate of an operator, 241
T^*	Adjoint of an operator, 263
$\lVert T \rVert$	Norm of an operator, 220–221
$[T]$, $[T]_B$	Matrix of an operator, 281
U	Universal set, 5
$\bigcup_i A_i$, etc.	Union of a class of sets, 11
$x \to y$	Mapping notation, 17
$x_n \to x$	Convergent sequence, 50, 70–71, 132
X_∞	One-point compactification of X, 163
$X_1 \times X_2$	Product of two sets, 23
$[x]$	Equivalence set associated with x, 27
$\lVert x \rVert$	Norm of x, 54, 81, 212
$\lVert x + M \rVert$	Norm of coset $x + M$, 213
$x(M)$	Function on maximal ideals, 318–319
$\hat{x}$	Function on maximal ideals, 319
$x(\lambda)$	Resolvent of x, 309
$x \wedge y$, $x \vee y$	Meet and join of x and y, 46
(x,y)	Inner product of x and y, 245
$x \in A$, $x \notin A$	x is (is not) an element of A, 5
$x \perp y$	x is orthogonal to y, 249
$x \sim y$	x is equivalent to y, 27
$x \leq y$	Order relation for real numbers and partially ordered sets, 7, 43
$x \equiv y \pmod{I}$	Congruence modulo an ideal in a ring, 186
$x \equiv y \pmod{M}$	Congruence modulo a subspace in a linear space, 193
x^*	Adjoint of x, 324
Z	Set of topological divisors of zero in a Banach algebra, 307

Subject Index

Absolute value, on complex plane, 53
  of a function, 159
  on real line, 52
Achieser, N. I., 157
Adjoint, of element in Banach *-algebra,
  324
  of operator, 263
Adjoint operation (involution), on $\mathfrak{B}(H)$,
  265
  on Banach *-algebra, 324
Alexandroff, P., 130n.
Algebra, 106, 208
  B^*-, 324
  Banach, 302
  Banach *-, 324
  Boolean (see Boolean algebra)
  center, 210
  commutative, 106
  complex, 106
  C^*-, 303
  disc, 303
  division, 208
  group, 303–305
  homomorphism, 210
  ideal in, 209, 313
  with identity, 106
  isomorphism, 210
  quotient algebra of, 209
  radical, 314
  regular representation, 210
  semi-simple, 316
  subalgebra of, 106, 208
  total matrix, 284
  von Neumann, 303
  W^*-, 303
Antisymmetry, 43
Arzela's theorem, 128

Ascoli's theorem, 126, 128
Axiom of choice, 46

B^*-algebra, 324
  representation, 325–326
Baire's theorem, 74, 75n.
Banach algebra, 302
  Banach subalgebra of, 302
  representation, 305
Banach *-algebra, 324
  *-isomorphism, 324
Banach space, 82, 212
  closed unit sphere, 217, 232
  representation, 234
  uniformly convex, 248
Banach-Steinhaus theorem, 240
Banach-Stone theorem, 330
Bar-Hillel, Y., 7, 46n.
Base, closed, 112
  generated by subbase, 101, 112
  open, 99
Basis, 197
  orthonormal, 293
Bell, E. T., 37
Bernstein polynomials, 154
Bers, L., 338
Bessel's inequality, 252–253, 257
Birkhoff, G., 29, 46n., 47
Birkhoff, G. D., 338
Bolzano-Weierstrass property, 121
Bolzano-Weierstrass theorem, 121
Boolean algebra, 345
  as Boolean ring, 347–349
  isomorphism, 353
  representation, 353
  of sets, 12, 344